Introduction to Combustion Phenomena

Combustion Science and Technology Book Series

Editor
Irvin Glassman, *Guggenheim Laboratories, Princeton University*

VOLUME 1   Gasdynamic Theory of Detonation
               Heinz D. Gruschka and *Franz Wecken*
VOLUME 2   Introduction to Combustion Phenomena
               A. Murty Kanury

Introduction to Combustion Phenomena
(for Fire, Incineration, Pollution, and Energy Applications)

A. MURTY KANURY

Senior Mechanical Engineer
Fire Research Group
Stanford Research Institute
Menlo Park, California, 94025

GORDON AND BREACH SCIENCE PUBLISHERS
New York   Paris   London

Copyright © 1975 by
  Gordon and Breach, Science Publishers, Inc.
  One Park Avenue
  New York, N.Y. 10016

Editorial office for the United Kingdom
  Gordon and Breach Science Publishers, Ltd.
  42 William IV Street
  London W.C.2

Editorial office for France
  Gordon & Breach
  7–9 rue Emile Dubois
  Paris 75014

Library of Congress catalog card number 73–81393. ISBN 0 677 02690 0. All rights reserved. No part of this book may be reproduced or utilized in any form or by any means, electronic or mechanical including photocopying, recording, or by any information storage and retrieval system, without permission in writing from the publishers. Printed in Great Britain.

Dedicated to
my Mother—Punnamma

and

my Wife—Kathleen

for the life and love they have given me

"Everything that happens is due to the flow and transformation of energy. . . . . . . Control fire and you control everything. The discovery of fire . . . . . . lifted man from the level of the beast and gave him dominion over the Earth".

Morton Mott-Smith in his
Introduction to Heat and its Workings.
D. Appleton & Co. (1933)

Preface

This book is written with the belief that a need exists for it as a textbook in the academic community to introduce combustion to students in physical sciences and as a reference book in the research community to present a broad but coherent background in combustion to scientists involved in the study of destructive fires, of incineration and of efficient and harmless extraction of energy from conventional fuels.

As a textbook, this work finds its origin in the days of my teaching a senior level undergraduate combustion course at the University of Minnesota in the late Sixties. Aimed at students from Mechanical, Aero and Chemical engineering schools, this course has the objective, among others, of introducing to the student how the science of combustion is composed of complex interactions between several disciplines including thermodynamics, chemical kinetics, chemical mechanisms, fluid mechanics, heat and mass transfer and, to some extent, mathematics. In spite of the availability of a number of exemplary treatises both on these component pure disciplines and on combustion, I found it considerably difficult to prescribe a single textbook for my course. I was not willing to prescribe a chapter in one book and a chapter in another for the fear of fragmented styles, concerns, notions, notations and levels of treatment. Being a student myself, in not too distant a past, I was also concerned about the budgetary considerations of an average student.

And then, to pick an existing book on combustion I ran into somewhat related but different deliberations. If one looks around, one immediately finds out that most of the combustion books are written as reference books for a specialist rather than as textbooks for a student. Furthermore, some of them concentrate exclusively on experimental aspects of combustion, and others deal exclusively with theory. Some of them emphasize the chemical aspects of combustion while others stress the physical features. Some of them describe the mechanical "nuts and bolts" aspects of combustion equipment and others present the science.

Evolved out of these thoughts is the first generation version of the present manuscript. Its explicit goal is to elucidate the workings of different types of combustion phenomena and to balance in this elucidation, theory with experiment, chemistry with physics and engineering with science.

Not to divert attention from the principal topic of the book but to make available the prerequisite knowledge in a consistent framework, thermodynamics, thermochemistry and equilibria of reactive mixtures are presented in the first three appendices. With the conviction that it is a highly specialized aspect of combustion and that its exclusion does not hurt the well-roundedness of the present work, I deliberately limited here the scope of discussion of detonative combustion.

The Minnesota course covered much of the material presented here in a series of 30 hour-long lectures of roughly three hours for each of the ten topics: thermodynamics of mixtures, thermochemistry and equilibrium in combustion systems, chemistry and physics of combustion, ignition/extinction, burning of solids, liquids and jets and finally, premixed flames. Note that Appendix E presents some problems for use in pedogogy.

Revision, updating and finalizing my lecture notes into the present form was done with intentions of rendering combustion a scope broader than of interest merely to a specialist involved in research and development, mostly of power and propulsion equipment. These are precisely the intentions which, I hope, make this book a valuable reference to scientists and engineers involved in various combustion-related researches. Let me elaborate on this hope.

Although combustion is one of the oldest scientific disciplines, the greatest strides of progress occurred only in the past two decades, mainly due to the impetus given by aero- and space-propulsion. Having conquered space by means of combustion, man has recently turned his scientific attention to the earthly problems such as: (a) prevention and reduction of hostile fires which in the U.S. alone annually kill over twelve thousand people and leave many thousands more with life-long physical and psychological scars; (b) reduction of the harmful effluents which are slowly but surely snuffing out the very civilization that grew to depend on combustion; (c) design of efficient and safe incinerators to dispose wastes; (d) improvement of the efficiency with which our dwindling chemical energy resources are burned; and (e) innovation of methods to extract energy from sources which were, hitherbefore, considered uneconomical. All too often solution to one of these problems may offer solutions to others.

Especially in the energy and fire problem areas, almost orders of magnitude increase in the interest, concern and activity occurred just in the most recent months and years. The influx of scientists into these two areas from various, sometimes remotely, related fields is so phenomenal that technical progress is tortuous and slow. Vividly clear is the lack of a thorough and uniform background in the fundamental physical and chemical implications

Preface

of various combustion processes. Even though a scientist's contributions in his specialty are exceptionally superior in caliber, noticeable lack of the overall perspectives of combustion taints his vision. The need for an intelligible and coherent account of combustion phenomena is obvious.

To the student who is opening up his wings and to the researcher who is expanding his horizons, I hope the topical arrangement of this book is appealing and logical. The various combustion phenomena are sorted into three distinct but conceptually related groups—those in which physical mixing and flow exert control, those in which chemical kinetics and mechanisms exert control and the rest in which both physics and chemistry come into play with more or less equal importance. Upon becoming conversant with the thermodynamic concepts of the first three appendices, one can delve into the main text of the book. Chapters two and three deal respectively with the chemistry and physics of combustion. Chapter 4 deals with combustion phenomena in which chemistry dominates, whereas Chapters 5–7 deal with those in which physical processes of flow and mixing dominate. Chapter 8, finally, deals with premixed flames in which careful consideration of both the chemistry and physics is needed.

Visible throughout this book is the special spirit of my teacher Perry Blackshear who taught me much on combustion and much more on life itself. While I hold him not responsible for any disasters, I do hope that this book befits the high caliber and clarity which he always insisted upon. Special thanks are due to Professor Irvin Glassman of Princeton University for his constant encouragement during the preparation of this book and for his technical review. Professor A. M. Mellor of Purdue has read the manuscript and suggested many improvements. My colleagues at Stanford Research Institute, especially Stan Martin and Norm Alvares provided me with much encouragement when I needed it. And finally, my wife Kathleen whose patience, understanding and love far exceed her typing abilities did, in fact, type the final version of the manuscript. To these people, I am much indebted.

Palo Alto   Anjaneya Murty Kanury

Contents

Preface vii

Chapter 1   **Introduction** 1
     1.1   Scope of this Book 1
     1.2   Scope of this Chapter 2
     1.3   Combustion Equipment 2
     1.4   Some Hostile Fire Problems 3
          (a)   Dust Explosions 3
          (b)   Liquid Spill Fires 3
          (c)   Forest Fires 4
          (d)   Structural Fires 4
     1.5   Pollution Problems Arising from Combustion . . 4
          (a)   Particulate Pollutants 4
          (b)   Gaseous and Vapor Pollutants 5
     1.6   Concluding Remarks 7

Chapter 2   **Chemistry of Combustion** 8
     2.1   Factors Influencing the Reaction Rates . . . 9
     2.2   Molecularity and Order of a Reaction . . . 10
     2.3   Integration of Rate Equations for Simple Isothermal Reactions 12
     2.4   Reversible Reactions 15
     2.5   Chain Reactions 16
     2.6   Explosions—Branching Chain Reactions . . 21
     2.7   Effect of Pressure on Reaction Rates . . . 24
          (a)   Simple Thermal Reactions 24
          (b)   Chain Reactions 26
     2.8   Effect of Temperature on Rates of Simple Reactions 26
     2.9   Collision Theory of Reaction Rates . . . 30
          (a)   Maxwell-Boltzmann Velocity Distribution Law . 30

	(b) Collision Frequency	33
	(c) Velocity of Molecules as a Function of Temperature of the Gas	34
	(d) Synthesis of Collision Theory	36
	(e) Mean Free Path Length	37
	(f) Comments on the Collision Factor	37
	(g) Energy—Reaction Coordinate Diagrams for Endothermic and Exothermic Reactions	40
2.10	Concluding Remarks	40

Chapter 3   Physics of Combustion **43**

3.1	Fundamental Laws of Molecular Transfer	43
	(a) Newton's Law of Viscosity	43
	(b) Fourier's Law of Heat Conduction	44
	(c) Fick's Law of Species Diffusion	45
3.2	Kinetic Theory of Dilute Gases	48
	(a) Mean Free Path Length	48
	(b) Calculation of the Transport Coefficients of Monatomic Dilute Gases	49
	(c) Dimensional Properties—Pr, Sc and Le	54
3.3	Concept of Boundary Layer	55
3.4	Heat Transfer Across a Boundary Layer	57
3.5	Heat Transfer in Free Convection	59
3.6	Comments on Mass Transfer Problems	67
3.7	Turbulence	68
3.8	Heat and Mass Transfer with Non-negligible Interfacial Velocities	70
	(a) Heat Transfer	70
	(b) Mass Transfer	72
3.9	Conservation Equations	73
	(a) Conservation of Mass (Continuity Equation)	73
	(b) Conservation of Momentum (Equation of Motion)	75
	(c) Conservation of Energy	77
	(d) Conservation of Species	78
	(e) Comments	79
3.10	Conclusions	79

Chapter 4   Kinetically Controlled Combustion Phenomena . . **81**

4.1	Categorization of Combustion Phenomena	81
	(a) Kinetically Controlled Regime	84

	(b) Diffusionally Controlled Regime	85
4.2	Progress of a Combustion Reaction	87
4.3	Ignition	90
	(a) Thermal Ignition	90
	(b) Chemical Chain Ignition	91
	(c) Scope of the Present Treatment	91
	(d) Two Types of Ignition	91
4.4	Spontaneous Ignition Delay	92
	(a) The Criterion	92
	(b) Ignition Delay	93
4.5	Semenov Theory of Spontaneous Ignition	95
4.6	Application of Semenov Theory to Predict Ignition Range	99
4.7	Spontaneous Ignition: Frank-Kamenetski's Steady State Analysis	104
4.8	Comments on Spontaneous Ignition	110
4.9	Forced Ignition	112
	(a) Some Preliminary Concepts	112
	(b) Ignition by Heated Spheres and Rods	117
	(c) Ignition by Flames	118
	(d) Ignition by Sparks	121
4.10	Range of Ignition (or Flammability Limits)	126
4.11	Auto-catalytic Ignition	132
4.12	Chemical Chain Ignition and Explosions	132
4.13	Longwell's Well-stirred Reactor (Space Heating Rate)	135
4.14	Conclusions	137

Chapter 5	**Diffusion Flames in Liquid Fuel Combustion**	**142**
5.1	Preliminary Remarks on Solid and Liquid Fuels	143
5.2	Flash Point and Fire (or Ignition) Point Temperatures	143
5.3	Scope of this Chapter	145
5.4	Atomization of Liquids	146
5.5	Surface Conditions for Evaporation with or without Combustion	148
	(a) Energy Balance at the Wall	150
	(b) Species F Balance at the Wall	151
5.6	Simple Steady State Vaporization of a Droplet without Combustion	152
	(a) Equations	152

		(b) Solution	154
		(c) Discussion of the Vaporization Rate Equation	156
		(d) Heat Transfer Factor	157
		(e) Thermodynamic Factor: B	158
	5.7	Droplet Vaporization Time	164
	5.8	Evaporation Followed by Combustion	167
		(a) Calculation of the Burning Rate	171
		(b) Location of the Flame	172
		(c) Temperature Profile	174
		(d) Fuel and Oxygen Mass Fraction Profiles	177
	5.9	Droplet Burning Time	178
	5.10	Spray Combustion	179
	5.11	Some Examples	181
	5.12	Concluding Remarks	191
		(a) Transient Effects	191
		(b) Composition of the Burning Liquid	192
		(c) Suspension of Inert Solids in a Pure Fuel	192
		(d) Chemical Considerations	192

Chapter 6   Combustion of Solids **195**

	6.1	Pyrolyzing Solids	197
		(a) Nusselt's Shrinking Drop Theory	197
		(b) Comments on Nusselt's Theory	198
	6.2	Description of Carbon Sphere Combustion	200
	6.3	Diffusional Theory of Carbon Combustion	205
	6.4	Combustion of Carbon with CO Burning in Gas Phase	211
	6.5	Combustion of Pulverized Coal	213
	6.6	Concluding Remarks	215

Chapter 7   Combustion of Gaseous Fuel Jets **217**

	7.1	Plane Free Nonburning Jet	218
	7.2	Invariance of the Jet Momentum, Enthalpy and Species Contents: Partial Integration of the Equations	222
	7.3	Solution of Laminar Plane Free Inert Jet	225
	7.4	Turbulent Plane Free Inert Jet	229
	7.5	Cylindrical Free Inert Jet	231
	7.6	Combustion of a Free Jet of Fuel Issuing into Quiescent Air	236

	(a) Flame Shape	240
	(b) Composition Profiles Across the Flame	244
	(c) Temperature Profile Across the Flame	245
7.7	Discussion of Experimental Flame Heights	247
7.8	Some General Comments	249
	(a) On the Technique of Solution	250
	(b) Real Jet Flames	250
7.9	Confined Flames	256
7.10	Longitudinal Confined Jet Flames	257
7.11	Comparative Importance of Droplet Vaporization and Jet Mixing in Combustors	265
7.12	Concluding Remarks	266

Chapter 8 Flames in Premixed Gases   270

8.1	Scope of This Chapter	270
8.2	Detonations and Deflagrations	272
8.3	Deflagrations—Some Basic Characteristics	276
8.4	Some Experimental Details	283
8.5	Experimental Results	285
	(a) Influence of Fuel/oxidant Ratio	285
	(b) Influence of Fuel Structure	286
	(c) Influence of Pressure	287
	(d) Influence of Initial Mixture Temperature	289
	(e) Influence of Flame Temperature	289
	(f) Influence of Inert Additives	293
	(g) Influence of Reactive Additives	294
8.6	Flame Propagation Theory	296
	(a) Zeldovich—Frank-Kamenetski Thermal Theory	297
	(b) Diffusion Theory of Tanford and Pease	300
8.7	Propagation of Turbulent Premixed Flames	301
	(a) Small Scale Turbulence	302
	(b) Large Scale Turbulence	304
8.8	Flame Stabilization	305
	(a) Stability of a Bunsen Flame	305
	(b) Stabilization of Flames	306
8.9	Acoustical Instabilities	312
8.10	Concluding Remarks	313

Postscript   316

Appendix A — Review of Thermodynamics of Gases — 321

- A.1 Equation of State (Gas Law) — 322
- A.2 First Law of Thermodynamics — 323
- A.3 Second Law of Thermodynamics — 329
- A.4 Thermodynamics of Nonreacting Gaseous Mixtures — 333
  - (a) Gibbs-Dalton Law — 333
  - (b) Implications of Gibbs-Dalton Law — 335
- A.5 Concluding Remarks — 338

Appendix B — Thermochemistry (First Law of Thermodynamics Applied to Chemically Reacting Systems) — 340

- B.1 Enthalpy of Formation of Compounds — 341
- B.2 Enthalpy of Reaction — 343
  - (a) Enthalpy—Temperature Diagram — 343
  - (b) Reactions at Constant Volume or Constant Pressure — 345
- B.3 Enthalpy of Combustion — 346
- B.4 Calculation of Enthalpy of Reaction from Bond Energies — 347
- B.5 Thermochemical Laws — 349
  - (a) Lavoisier-Laplace Law — 349
  - (b) Hess' Law of Summation — 350
- B.6 The Effect of Physical State on Enthalpy of Reaction — 351
- B.7 Temperature Dependency of Heat of Reaction — 352
- B.8 Concluding Remarks — 356

Appendix C — Equilibrium (Application of the Second Law of Thermodynamics to Chemically Reacting Systems) — 357

- C.1 Concept of Minimum Energy for Equilibrium — 358
- C.2 Free Energy — 360
  - (a) Variation of Free Energy with Pressure — 360
  - (b) Variation of Free Energy with Temperature — 360
- C.3 Chemical Equilibrium — 361
- C.4 Law of Mass Action — 362
- C.5 Free Energy and Chemical Equilibrium — 363
- C.6 Equilibrium Constant and Standard Free Energy of Reaction — 364
- C.7 Equilibrium Constants Defined on the Basis of Concentrations — 369

C.8	Influence of Temperature and Pressure on the Equilibrium Constant	371
	(a) Temperature Dependence	371
	(b) Pressure Dependence	373
C.9	Adiabatic Flame Temperature	373
	(a) Method of Calculation	373
	(b) Comments	376
C.10	Dissociation of Gases	379
C.11	Concluding Remarks	380

Appendix D   **Transport Property Tables**   382

Appendix E   **Some Problems for the Student**   390

Subject Index   407

CHAPTER 1

Introduction

1.1 Scope of This Book

Flames and fires have played a very intimate role in man's life since ages unknown. Many a story may now be built around the circumstances surrounding man or man-like creature when he discovered fire. It could be the penetratingly warm and bright sun, the fierceful flow of lava, the dread of a forest fire or, equally, the glow of a "firefly" that induced the first thoughts of man on the subject of fire.

Our civilization is bred on combustion of fuel for useful heat. Power production by coal and oil burning power plants is yet a predominant feature of our society and it seems like we ought to learn efficient ways of this production. Gas and oil burning equipment will continue to be under use to heat and cook in many a home. Pyrometallurgy is the back-bone of our industrial development. Internal combustion engines constitute a large and integral fraction of our economy. Modern jet engines brought distant parts of this world closer. Finally, rockets for interplanetary travel evolved.

And then, there are many unwanted effects of combustion. The very civilization that was bred on combustion is also threatened by it. Fires in our forests destroy thousands of lives and many millions of dollars worth of property every year. New developments in materials of construction and methods of design constantly increase the fire hazard in our industrial and residential complexes. The probability of uncontrolled mass fires in our densely industrialized metropolitan areas is horrifying. More important and more immediate is the threat of air pollution rapidly degenerating this earth to the extent of being unfit for life. Also challenging are the problems related to safe and efficient ways of disposing of wastes by incineration. Note that most of these unwanted effects of combustion are strong, implicit functions of our standard of living in this highly technological era. These and other problems created by technology can surely be solved by technology itself, provided today's engineer is enough of historian to track back to the evolution of fire, enough of a scientist to determine its future course and enough of a conservationist to care about and devise proper techniques of more efficient and harmless combustion. He has the professional as well as social responsibility to handle the problems arising from combustion.

This book is intended for such an engineer to provide him with the basic understanding of various combustion phenomena. The scope of application of such an understanding is only limited by the engineer's imagination in a given situation. Combustion is a field which requires expertise in a variety of scientific disciplines. A typical problem warrants a thorough grasp of the basic principles of thermodynamics, fluid flow, transport of heat and mass, chemical kinetics, and, to a certain degree, mathematics. Keeping the use of advanced mathematics to a minimum, this book emphasizes upon the physical and chemical processes and their interactions in flames.

1.2 Scope of This Chapter

In this chapter we briefly enumerate a variety of power and propulsion machines. We also consider description of some typical hostile fire problems before closing the chapter with a brief discussion of the principal contributions of combustion to air pollution.

1.3 Combustion Equipment

These may be arbitrarily divided into stationary equipment, surface mobile equipment and propulsion equipment.

Heating and power production plants involve combustion of coal, oil or gas. In the early practice, combustion of coal is carried out by feeding crushed coal to a stoker bed. Efficiency is greatly increased by burning jets of finely pulverized coal in a manner somewhat similar to the burning of a jet of gaseous fuel or a spray of liquid fuel. Burning of isolated briquettes, a porous bed of coal on a stoker, a fine jet of pulverized coal, a liquid spray and a gaseous fuel jet are a few typical combustion problems associated with these equipments. Stationary diesel engines are often used in some small scale and emergency power production plants.

Automobiles, trucks, buses, railroad locomotives and marine propulsion engines belong to the surface class of mobile equipment. Conventionally, engines of these equipments belong to the reciprocating internal combustion family. They operate either on constant volume or on constant pressure heat addition principle. Constant volume heat addition (Otto) cycle involves striking a spark in an adiabatically compressed fuel vapor/air mixture. Upon ignition, the flame would propagate through the combustible mixture at

Introduction

constant volume. The mixture itself is formed in a carburetor which performs the dual function of evaporating the liquid fuel and mixing the vapor with air. Gasoline engines offer an example of the Otto type. Constant pressure heat addition (diesel) cycle, on the other hand, involves injection of a liquid fuel through an "atomizer" nozzle into hot compressed air. The conglomeration of droplets in the liquid spray evaporates in the hot air to spontaneously ignite and burn at constant pressure.

Radially reciprocating internal combustion engines were used in the early developments of aircraft power plant. Modern propulsion devices consist of duct type engines. Ram jets, pulse jets, turbo jets and rockets are the main members of the duct family engines. Gas, liquid and pulverized solid fuels are possible to be used in these engines.

The question to be asked in connection with all these different types of combustion machines is this: How can we best operate these engines at their utmost efficiency with the least of atmospheric contamination?

1.4 Some Hostile Fire Problems

(a) *Dust Explosions*

Finely dispersed dusts arise in many practical instances in metal, plastic, chemical, agricultural, food and mining industries. Pesticides, chemicals, drugs, dyes, starches, metallic and plastic powders, carbon, graphite, coal and coke dusts are some such examples. These dispersions are notoriously explosive. The explosiveness is a strong function of the oxygen and dust concentrations, inert content, volatile content, moisture content, average size and shape of the dust particles and finally, the temperature, pressure and heat transfer conditions in the confinement.

(b) *Liquid Spill Fires*

Spills and pools of highly volatile jet engine fuels as well as various grades of oil refinery products pose a strong fire hazard. Evaporative combustion of such fuel surfaces is a topic of importance in developing techniques to prevent and/or control the fires. Lately, it has become a common nuisance to contaminate our coastal areas with oil slicks from barges and under-sea wells. Controlled combustion, if feasible, might offer a possible technique to control or clean up such menacing slicks. What with the implications of the recent "energy shortage", liquified propane and natural gases are expected to be transported over long distances, thus raising the probability of some disastrous fires.

(c) *Forest Fires*

One of man's greatest natural resources is his forest land. Destructive conflagrations have taken and continue to take a large toll every year. An understanding of the effect of fuel distribution, of wind and atmosphere, of geography and of flying embers upon the rate of spread of a forest fire would enable design of effective fire breaks and other control and fighting techniques.

(d) *Structural Fires*

Fires in domestic and industrial buildings damage much property and life in this country and around the world every year. An understanding of the conditions conducive to the birth, survival and the growth of this sort of fire would greatly help in fire protection, and quite possibly, prevention. A sound basis to rate various construction materials as to their contribution to a given fire hazard is provided only by a thorough understanding of the fire/fuel/environment interactions.

1.5 Pollution Problems Arising from Combustion

As Lewis states in his "Outlook in Combustion Research", "In the area of pollution, where there is a strong social responsibility, it should be no special feat to burn materials completely. Success will often be a matter of funds and engineering design". Combustion of any gaseous, liquid or solid fuel is practically always accompanied by emission of smoke, ash, odors, noxious and benign gases. Considerable amount of the fuel is expelled unburnt in the exhaust of a combustor. Table 1.1 shows five types of emissions from five classes of sources in the United States. To reduce (or eliminate) these emissions is an important mission for all of us.

(a) *Particulate Pollutants*

Fly ash is a common solid particulate pollutant arising from the solid noncombustible (usually, inorganic) constituents of fuels. Unburnt carbonaceous solid and liquid particles and ash particles too small to settle in the combustion chamber are carried away into the atmosphere by the exhaust gas stream.

What is commonly known as smoke is an aggregation of submicron size solid and liquid particles formed in the combustion chamber due to poor mixing of the fuel with the air and due to delayed release of the carbon particles. When the air supply is limited or deficient, incompletely burned

Introduction

TABLE 1.1

Estimated Nationwide Emissions, 1968*

Source	Millions of tons emitted per year				
	Particulates	Carbon monoxide	Sulfur oxides	Hydro-carbons	Nitrogen oxides
Transportation	1.2	63.8	0.8	16.6	8.1
Stationary equipment	8.9	1.9	24.4	0.7	10.0
Industrial processes	7.5	9.7	7.3	4.6	0.2
Solid waste disposal	1.1	7.8	0.1	1.6	0.6
Miscellaneous†	9.6	16.9	0.6	8.5	1.7

* From National Air Pollution Control Agency, Inventory of Air Pollutant Emissions, 1970, as quoted in the First Annual Report of the Council on Environmental Quality, 1970.
† Mainly forest fires, agricultural burning and coal waste fires.

particles of fuel are emitted in the exhaust. Such an emission mostly consisting of hydrocarbons, is known as "cold smoke". If oxygen is completely consumed in certain pockets of the chamber, cold smoke may be expected to be formed. Smoke obscures light by scattering and absorbing it. When the particle size is too large, scattering is greatly reduced to make the smoke seem "clean" to the eye. Nevertheless, its undesirability is no less.

Scrubbing, filtering, centrifugal separation and electrostatic precipitation are among the methods currently used to capture particulate emissions. Deliberately burning with deficient air supply in a large well-mixed combustor reduces smoke formation but increases cold smoke and carbon monoxide emissions. One of the main reactions occurring under such deficient air conditions is suspected to be the oxidation of carbon particles by carbon dioxide to form carbon monoxide. An "after-burner" system may be then used to take care of the cold smoke and CO emissions.

If sufficient time were available to burn solid and liquid particulates, smoke may be completely eliminated. In diesel engines, gas turbines, incinerators and fires, the times available are of the order of 0.001, 0.01, 0.1 and 1 seconds respectively. Provision of a swirl injection system increases this residence time considerably.

(b) *Gaseous and Vapor Pollutants*

As evidenced in Table 1.1, considerable amounts of sulfur dioxide, trioxide, carbon monoxide, nitric oxide, nitrogen dioxide, aldehydes and other hydrocarbon residues in addition to the usually expected CO_2 and H_2O are produced in combustion. Most of these products are toxic and noxious.

Minimization of these emissions often requires complicated compromises invoking thermodynamic and chemical equilibrium considerations.

The oxides of sulfur are poisonous like CO. They are formed when fuels containing sulfur impurity are burned. Power generating plants burning coal or oil are the principal producers of SO_x. Industries processing sulfuric compounds are expected to produce considerable amounts of SO_x pollutants as well.

When combustion is incomplete due to poor mixing or due to deficient air supply, CO, a deadly compound, is produced in addition to various fractions of unburnt fuel (i.e., cold smoke). About two thirds of all CO emitted comes from transportation equipment—mainly from gasoline-powered internal combustion engines. Automobile engines may also be considered to be the main contributor of the hydrocarbon emissions which are in general neither toxic nor corrosive. These emissions not only mean a waste of precious fuel but also play a very important role with the sulfur and nitrogen oxides in the photochemical reactions leading to the formation of "smog".

When ordinary air is heated to about 1,500 to 2,000° C, N_2 and O_2 decompose to atoms which partly combine to form nitrogen oxides. If these oxides were slowly cooled they "redecompose" to atoms which then recombine to the original N_2 and O_2. However, operation of an engine usually involves an extremely rapid conversion of heat to work, allowing little time for the nitrogen oxides to reform to N_2 and O_2. The oxides, as a consequence, are expelled "frozen" in the exhaust gases up to a thousand parts per million contributing to the menacing smog problem.

If one can accomplish complete combustion of a hydrocarbon fuel in air, the resultant products are CO_2, H_2O and N_2. By accomplishing complete combustion the formation of toxic, noxious and corrosive products may be minimized or eliminated. However, so long as man depends upon combustion to extract energy from the fuels, production of CO_2 is inevitable. The earth's atmosphere presently contains some 300 ppm of carbon dioxide. Nearly ten billion tons of CO_2 is produced each year by the combustion equipment to raise the CO_2 content of the atmosphere at a rate of 1 ppm. It is known that the photosynthesis associated with vegetation is capable of consuming up to about ten billion tons of CO_2 per year. But this much of it is produced by human and animal respiration and by decaying vegetation. The question then is where and how do we accommodate 10^{10} tons of CO_2 per year without upsetting the long-term ecological balance of the earth's atmosphere?

In general, reduction of the noxious pollutants requires careful design and adjustment of the combustors. Usually, the conditions which reduce formation of certain emissions encourage the formation of other emissions. Naturally, a compromising design becomes necessary. Consider, for example, the role of fuel/air ratio. An excess of air supply is desirable to cut down the

Introduction

yield of CO, particulate and hydrocarbon emissions. Excess air, on the other hand, encourages the formation of nitrogen and sulfur oxides. In stationary equipment, nitrogen oxides may be reduced by careful adjustment of the combustor and exhaust gas temperatures. Control of these oxides in automobile emissions is more difficult since the conditions desired to do so contradict those desired to cut down other pollutants.

In summary, then, a properly mixed-combustion chamber with an efficient fuel system with optimum fuel/air ratio may be expected to yield the most satisfactory combustion.

1.6 Concluding Remarks

The role played by combustion in the modern civilization is briefly examined here to discover that it is now more important than ever before to design tools and techniques to burn fuels and wastes efficiently with a minimum production of unwanted products, and to prevent destructive fires.

References

Environmental Quality, The First Annual Report of the U.S. Council on Environmental Quality. Transmitted to the United States Congress (Aug., 1970).

Kanury, A. M., "The Science and Engineering of Hostile Fires," *Fire Research Abstracts and Reviews*, **15**, p. 188 (1974).

Lewis, B., "Outlook in Combustion Research," *Combustion and Flame*, **14**, p. 1 (1970).

Stern, A. C., (ed.), *Air Pollution*, Vols. I, II and III. Academic (1968).

Zukrow, M. J., *Principles of Jet Propulsion and Gas Turbines*, Chapman and Hall (1951).

CHAPTER 2

Chemistry of Combustion

A complete study of chemical reactions would consist of answers to three important questions: (a) known the initial state (composition of the reactant mixture, temperature and pressure), what is the final equilibrium state? what are the changes in various thermodynamic properties? (b) how fast is the reaction nearing the final equilibrium state? and (c) by what mechanism are the molecules interacting to cause the reaction to proceed?

Appendices A–C of this book try to answer the first question. The answer for the second question "how fast is the reaction proceeding?", consists of a satisfactory rate equation which makes it possible to predict the reaction rates from known conditions of concentrations, temperature and pressure. The third question involves rather detailed examination of the actual and exact sequence of steps in the reaction process and thus requires a knowledge of any intermediate substances which may exist between the reactants and the final products.

This chapter is intended to answer the second question (i.e., to obtain a simple rate equation for the over-all reaction). We will also introduce, but not go into the intricacies of, the concepts of chain reactions and reaction mechanisms in an attempt to elaborate on the third question.

All chemical reactions take place at a definite rate (fast or slow) depending upon many experimental parameters such as composition, pressure and temperature. Some reactions such as those of oxidation and combustion are so fast that they appear as though they occur instantaneously.

The *rate of a reaction* is the quantitative measure of the number of moles of the product produced (or reactant consumed) per unit time per unit volume.*

---

* The rates of disappearance of reactants and appearance of products in the reaction

$$aA + bB \cdots \longrightarrow mM + nN + \cdots$$

are interrelated by

$$-\frac{1}{a}\frac{dC_A}{dt} = -\frac{1}{b}\frac{dC_B}{dt} = \frac{1}{m}\frac{dC_M}{dt} = \frac{1}{n}\frac{dC_N}{dt}.$$

$$\text{Rate} = \frac{\text{change in moles of species}}{\text{time increment} \times \text{volume}}$$

Since the number of moles present in a unit volume is, merely, concentration, the rate of a reaction is the rate of change of concentration with time. By the law of mass action, because the concentrations of the reactants A and B are gradually decreasing, the rate of reaction would also gradually decrease. For a simple reaction

$$A + B \xrightarrow{k_f} C + D$$

the rate is given by

$$r = -\frac{dC_A}{dt} = -\frac{dC_B}{dt} = \frac{dC_C}{dt} = \frac{dC_D}{dt} = k_f C_A C_B$$

and since $C_j = n_j/V$ where n_j is the number of moles of j present in the volume V,

$$r = -\frac{dn_A/V}{dt} = -\frac{dn_B/V}{dt} = \frac{dn_C/V}{dt} = \frac{dn_D/V}{dt}. \tag{2.1}$$

If the reactions occur at constant volume (in a bomb), the volume in Eq. 2.1 can be taken out of the derivative sign. The result suggests a method to determine the reaction rate from a measurement of the number of moles present at various stages of the reaction in a bomb of known volume. If the reaction occurs at a constant temperature (i.e., isothermal), the measurement of the number of moles as a function of time would further be simplified to the measurement of total pressure with time because, then, by the ideal gas law, $n = P \cdot (V/RT)$.

By the principles of thermodynamics and thermochemistry, we know how to calculate the maximum extent to which the reaction can proceed. At this maximum extent, equilibrium is achieved and no further changes (physical as well as chemical) take place, and as a result, the net rate of the reaction is zero. We know from Appendix C how to determine the equilibrium composition and adiabatic flame temperature. Further, if heat is supplied to or taken away from a reacting system so that it can be kept at a given constant temperature, the extent of the reaction can be determined. The speed with which a system approaches chemical equilibrium is determined by the rate of reaction.

2.1 Factors Influencing the Reaction Rates

All substances are orderly entities of one or more atoms. As described in Appendix B, a chemical reaction is simply a process in which various atoms exchange their partners. Such an exchange may involve a decrease in energy that can be held in the bonds, the excess energy being liberated during the course of reaction. The rate of exchange of atoms would depend (a) on the abundance of the reactant substance in a given space (i.e., concentration); (b) on the frequency with which the molecules mutually meet in proper orientation (this is determined by the total population (i.e., pressure)); and (c) on how energetic the collisions are (this is determined by their temperature which is indicated by the mean arithmetic speed). In other words, the chemical reaction rate depends on such variables as the partial pressures, the total pressure, and the temperature of the system. This chapter is devoted to a discussion of these dependencies.

2.2 Molecularity and Order of a Reaction

The *molecularity of a reaction* is defined as the number of atoms or molecules taking part in each act leading to the chemical reaction. In the reaction of nitrogen pentoxide decomposition, $N_2O_5 \rightarrow N_2O_4 + \frac{1}{2}O_2$, only one molecule is required for the reaction to occur. Hence this reaction is unimolecular. On the other hand, the familiar reaction of hydrogen iodide decomposing into hydrogen and iodine, $2HI \rightarrow H_2 + I_2$, is bimolecular. Two hydrogen iodine molecules interact with one another to break apart into a hydrogen molecule and an iodine molecule. Writing this reaction as $HI \rightarrow \frac{1}{2}H_2 + \frac{1}{2}I_2$ gives the quantitative proportions of the masses of the reactant and products, but it does not give any information about how the reaction occurs. The well-known water–gas reaction, $CO_2 + H_2 \rightarrow CO + H_2O$, is another example of a bimolecular reaction. By the same token, if three molecules are involved in each chemical reaction, the process is called termolecular. Typifying such a class are the reactions: $2NO + O_2 \rightarrow 2NO_2$ and $2CO + O_2 \rightarrow 2CO_2$. Higher molecularity reactions can quickly be recognized in many instances.

From a quantitative viewpoint, it is often convenient to classify reactions by their order rather than molecularity. The *order of a reaction* is the number of atoms or molecules whose concentrations determine the rate of the reaction. The molecularity, while giving no information regarding the rate (quantitative) of the reaction, yields an insight into the mechanism (qualitative) of the process. The order, on the other hand, while failing to recognize the mechanism of the steps involved in a reaction, enables calculation of the net quantitative rate of the reaction.

When one measures the pressure (or concentration) of nitrogen pentoxide with time as it decomposes, and differentiates the result to get the rate, one

Chemistry of Combustion

finds that the rate is directly proportional to the instantaneous concentration.

$$-\frac{dC_{N_2O_5}}{dt} = k_1 C_{N_2O_5}$$

This behavior characterizes the reaction by the first order. By the same kind of empirical experience, the decomposition rate of hydrogen iodide is found to be proportional to the square of the HI molecule concentration; hence it is a second order reaction. The decomposition of nitrogen dioxide (which typifies the reactions in a flame), $2NO_2 \to 2NO + O_2$ is found experimentally to follow

$$-\frac{dC_{NO_2}}{dt} = k_2 C_{NO_2}^2$$

The reaction between methlamine and ethyl bromide in benzene solution, $(C_2H_5)_3N + C_2H_5Br \to (C_2H_5)_4NBr$ is found to follow the rate law

$$-\frac{dC_{C_2H_5Br}}{dt} = k_2 C_{C_2H_5Br} \cdot C_{(C_2H_5)_3N}$$

Such a reaction is also called as of second order. However, the reaction is said to be of first order with respect to either $(C_2H_5)_3N$ or C_2H_5Br; but the over-all reaction is of second order.

The reaction between nitric oxide and oxygen is of third order. In high temperature plasmas, the recombination of oxygen atoms in the presence of a third body M is very prominent. (The third body is any other particle in the system such as "O" atom, O_2 molecule, an inert gas molecule or the wall of the vessel where two oxygen atoms can meet.)

$$O + O + M \longrightarrow O_2 + M^*.$$

Its rate is given by

$$-\frac{dC_O}{dt} = k_3 C_O^2 C_M$$

(The asterisk on M in the stoichiometric equation is to indicate that M is energized in the reaction.) Higher orders and sometimes even fractional orders are possible for other reactions. For example, the decomposition of acetaldehyde to CH_4 and CO is of the order $\frac{3}{2}$.

$$-\frac{dC_{CH_3CHO}}{dt} = k_{3/2} C_{CH_3CHO}^{3/2}$$

Some reactions are of order zero, meaning that the rates of these reactions are independent of the quantity of unconsumed reactants. It is of interest to note here that for hydrocarbons burning in oxygen, the over-all order of the reaction is known to lie between 1.7 to 2.2; the order with respect to oxygen being near unity. Note further that only five homogeneous reactions in the entire field of chemistry are known to be of higher orders. These are the reactions of nitric oxide (NO) with hydrogen, deuterium, oxygen, chlorine, and bromine. In general, reactions of order three or higher are very rare. The reason for this is simply that the probability of three suitable molecules colliding at one time with adequate energy in proper orientation is very small.

In the foregoing discussion we saw that the order of a reaction is more often than not, same in numerical value as the molecularity. Even though all reactions of molecularity n are also of the order n, all reactions of order n are not necessarily of molecularity n. That is, knowing the precise sequence of steps involved in a reaction one can deduce the order of the over-all reaction. However, with the knowledge of the order of an over-all reaction one cannot necessarily deduce the exact mechanism followed by the reaction.

Note at this point that the units of the rate constant k differ for various orders. For a first, second and nth order reactions, these units respectively are 1/second, cm^3/mole/second and (moles/cm^3)$^{1-n}$/second.

Reactions can be classified broadly as simple or complex. A simple reaction just contains one step whereas a complex reaction contains several intermediate steps. Examples can be found to illustrate the point that chemical reactions, particularly those in combustion, are not always simple and amenable for a straight forward analysis. Complications due to reversible and consecutive reactions may occur. Later in this chapter, when we study explosions, we will deal with these complexities in some detail. But first consider the rates of simple reactions.

2.3 Integration of Rate Equations for Simple Isothermal Reactions

In general, for an nth order reaction,

$$-\frac{dC_A}{dt} = k'_n C_A^{v_A} C_B^{v_B} C_C^{v_C} \tag{2.2}$$

where v_i is the order of the reaction with respect to ith species and $v_A + v_B + v_C + \cdots = n$ is the over-all order. If the initial composition and stoichiometry are known, Eq. 2.2 can be reduced to a simpler form.

$$-\frac{dC_A}{dt} = k_n c_A^n \tag{2.3}$$

where k_n is a function of k'_n, the initial composition and the temperature. For a given mixture at a fixed temperature, k_n is a constant. k_n is called the *specific reaction rate constant*, because it is the rate when $C_A = 1$. Equation 2.3 may be integrated by separating the variables.

$$\frac{C_A^{-n+1}}{-n+1} = -k_n t + \mathscr{C} \tag{2.4}$$

The integration constant $\mathscr{C}$ is evaluated to be $C_{AO}^{-n+1}/(-n+1)$ by noting that at $t = 0$, $C_A = C_{AO}$, the initial concentration.

$$C_A^{1-n} - C_{AO}^{1-n} = k_n(n-1)t \tag{2.5}$$

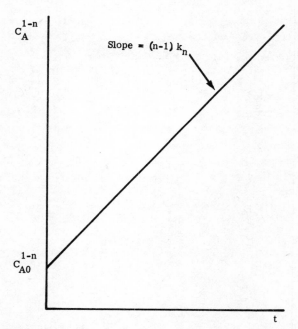

Figure 2.1 Variation of concentration with time in an nth order simple isothermal reaction

Equation 2.4 shows that if one plots C_A^{1-n} against t, a straight line is obtained as shown in Figure 2.1 with an intercept $= C_{AO}^{1-n}$ and slope $k_n(n-1)$.

It is often convenient to express the speed of a reaction towards completion in terms of a fractional lifetime. *Half-lifetime*, for example, is defined as the time required to consume half the initial reactants. Let us denote

TABLE 2.1

Order n	Rate	Integrated form	Half life	Graph
1	$k_1 C_A$	$\ln C_A = \ln C_{AO} - k_1 t$	$\dfrac{\ln 2}{k_1}$	$\ln C_A$ vs t, Slope $-k_1$
2	$k_2 C_A^2$	$\dfrac{1}{C_A} = \dfrac{1}{C_{AO}} + k_2 t$	$\dfrac{1}{C_{AO}} \cdot \dfrac{1}{k_2}$	$1/C_A$ vs t, Slope $+k_2$
3	$k_3 C_A^3$	$\dfrac{1}{C_A^2} = \dfrac{1}{C_{AO}^2} + 2k_3 t$	$\dfrac{3}{2} \cdot \dfrac{1}{C_{AO}^2} \cdot \dfrac{1}{k_3}$	$1/C_A^2$ vs t, Slope $+2k_3$
4	$k_4 C_A^4$	$\dfrac{1}{C_A^3} = \dfrac{1}{C_{AO}^3} + 3k_4 t$	$\dfrac{7}{3} \cdot \dfrac{1}{C_{AO}^3} \cdot \dfrac{1}{k_4}$	$1/C_A^3$ vs t, Slope $+3k_4$
n	$k_n C_A^n$	$\dfrac{1}{C_A^{n-1}} = \dfrac{1}{C_{AO}^{n-1}} + (n-1)k_n t$	$\left(\dfrac{2^{n-1}-1}{n-1}\right) \cdot \dfrac{1}{C_{AO}^{n-1}} \cdot \dfrac{1}{k_n}$	$1/C_A^{n-1}$ vs t, Slope $+(n-1)k_n$

Chemistry of Combustion

half-lifetime of an nth order reaction as $t_{1/2,n}$. From Eq. 2.5

$$t_{1/2,n} = \left(\frac{2^{n-1}-1}{n-1}\right)\frac{C_{AO}^{1-n}}{k_n} \qquad (2.6)^{*\dagger}$$

If $n = 1$, applying L'Hospital's rule to Eq. 2.5 we obtain

$$\ln C_A - \ln C_{AO} = -k_1 t; \quad C_A = C_{AO}\exp(-k_1 t) \qquad (2.7)$$

showing that for a first order reaction, the concentration of the reactant falls with time exponentially. Table 2.1 shows the above results and the following observations for various values of n.

(i) If C_A is measured as a function of time in an isothermal experiment, in order to find the order and specific reaction rate, plot graphs of $\ln C_A, 1/C_A, 1/C_A^2, \cdots 1/C_A^{n-1}$ etc. as functions of time and find which one yields a straight line; deduce n and k_n from this straight line.

(ii) Half-life period for a first order reaction is independent of the initial concentration. As the order becomes larger, the dependency of $t_{1/2}$ on C_{AO} becomes stronger.

(iii) The coefficient $(2^{n-1}-1)/(n-1)$ in the half-life expression is larger for larger orders. Consequently, reactions of higher orders proceed relatively slower if the initial concentrations are small.

2.4 Reversible Reactions

Many reactions involve three or more molecules in the chemical act, and usually proceed in several simultaneous or sequential stages. Several intermediate species may be released and consumed. In the integrations presented above, we have dealt with simple, unidirectional, isolated reactions. Complications often arise in experience with reactions which can proceed backwards as well. The rate constants k_f and k_b correspondingly are related to the

---

* In general for any other fractional life, Eq. 2.6 becomes $t_{1/z,n} = C_{AO}^{1-n}(z^{n-1}-1)/k_n(n-1)$ where $z = 2$ for half-life, 3 for one-third-life, 4 for one-quarter-life, etc.

† It becomes useful to remember here the L'Hospital's rule which yields the limits of indeterminate functions.

$$\lim_{x \to a}\frac{f_1(x)}{f_2(x)} = \lim_{x \to a}\frac{df_1(x)}{dx}\bigg/\lim_{x \to a}\frac{df_2(x)}{dx} = \cdots$$

Also recall the identity,

$$\frac{da^x}{dx} = a^x \ln a.$$

equilibrium constant K. (See Appendix C.) Consider a simple case represented by

$$A \underset{k_b}{\overset{k_f}{\rightleftarrows}} B$$

to derive an equation describing the temporal progress of this reaction towards equilibrium.

Suppose C_{AO} is the initial concentration of A and no B is present initially. After the reaction proceeds through a time t, let C_A and C_B be the concentrations. Then due to the particular stoichiometry chosen, $C_B = C_{AO} - C_A$. The forward reaction proceeds at a rate $k_f C_A$ whereas the backward reaction proceeds at a rate $k_b C_B = k_b(C_{AO} - C_A)$. The net rate of consumption of A is the difference between its consumption and generation rates.

$$-\frac{dC_A}{dt} = k_f C_A - k_b(C_{AO} - C_A) \qquad (2.8)$$

At equilibrium the net rate is zero, so that

$$\begin{aligned} k_f C_{A\,eq} &= k_b(C_{AO} - C_{A\,eq}) \\ k_b &= k_f C_{A\,eq}/(C_{AO} - C_{A\,eq}) \end{aligned} \qquad (2.9)$$

Substituting this value of k_b in Eq. 2.8,

$$-\frac{dC_A}{dt} = k_f C_{AO}\left(\frac{C_A - C_{A\,eq}}{C_{AO} - C_{A\,eq}}\right)$$

Integrating and applying the initial condition that $C_A = C_{AO}$ at $t = 0$,

$$\ln\left(\frac{C_A - C_{A\,eq}}{C_{AO} - C_{A\,eq}}\right) = -\left(\frac{k_f C_{AO}}{C_{AO} - C_{A\,eq}}\right)t \qquad (2.10)$$

Thus if we measure the concentration of A with time t till equilibrium is reached, k_f may be computed by Eq. 2.10 and thereby k_b from Eq. 2.9 and hence the equilibrium constant and standard free energy of reaction.

2.5 Chain Reactions

As mentioned earlier, very few combustion reactions being simple, the overall stoichiometric equation seldom represents the detailed mechanism of the reaction. In reality several reactions may proceed simultaneously or in a sequence. The intermediate species in such reactions are of utmost importance to derive the mechanism and the rate equation. Due to the presence of the intermediate species, the observed order of reaction often differs from the

Chemistry of Combustion

order calculated on the basis of the over-all (global) stoichiometric equation and law of mass action. It is frequently observed that many complex reactions show a nonintegral or fractional order.

A chain reaction occurs in a chain (or sequence) of steps. Oxidation of methane is a spectacular example to show how complex the seemingly simple reaction $CH_4 + 2O_2 \rightarrow CO_2 + 2H_2O$ can be. The following chain mechanism is found to be more or less satisfactory.

$$CH_4 + OH \longrightarrow CH_3 + O + H_2$$
$$CH_4 + O \longrightarrow CH_3 + OH$$
$$CH_3 + O_2 \longrightarrow H_2CO + OH$$
$$H_2CO + OH \longrightarrow HCO + H_2O$$
$$HCO + OH \longrightarrow CO + H_2O$$
$$CO + OH \longrightarrow CO_2 + H$$
$$H + O_2 \longrightarrow OH + O$$
$$H + H_2O \longrightarrow OH + H_2$$
$$O + O + M \longrightarrow O_2 + M^*$$
$$2OH + M \longrightarrow H_2O + O + M^*$$

Another example of chain reactions is the reaction of hydrogen and chlorine. If initially chlorine molecule is decomposed into atoms by exposure to light, $Cl_2 \xrightarrow{h\nu} 2Cl$, a chain results as below.

$$Cl + H_2 \longrightarrow HCl + H$$
$$+$$
$$Cl_2 \longrightarrow HCl + Cl$$
$$+$$
$$H_2 \longrightarrow HCl + H$$
$$+$$
$$Cl_2 \longrightarrow \cdots$$

If by chance the chain carriers (the atomic chlorine and hydrogen) are removed, the chain terminates. Such removal of chain carriers can occur by any of the following ways.

$$H + H \longrightarrow H_2$$
$$Cl + Cl \longrightarrow Cl_2$$
$$H + Cl \longrightarrow HCl$$

The chain carriers are usually free atoms or radicals which are very unstable. In hydrocarbon combustion such radicals as CH_3, C_2H_5, OH, etc. play the role of chain carriers. Whenever two or more of these transient species collide with a third body M (which could be any other atom or molecule in the system or the system walls) they are destroyed to form an inactive molecule. Thus if one desires to inhibit an unwanted reaction, he can accomplish this by any of the following three methods. First, he can increase the surface to volume ratio of the reaction vessel, thus giving more surface to act as the third body. A similar effect can also be achieved by adding a diluent to the reacting mixture. On the other hand, if the situation does not allow such alterations in the reaction system, one can operate the system at a higher pressure. At high pressure, the chances of two active species colliding simultaneously with a third body are greater—and thus chain breaking is more probable.* A third and slightly different method of terminating chain in a reaction is to introduce an inhibiting agent into the reacting system. The inhibiting agent reacts with the active species more readily and thus removes it from participating in the chain propagation.

Decomposition of propane and acetaldehyde show the existence of free radicals in chains depicted below.

Propane Decomposition:

$$C_3H_8 \longrightarrow CH_3 + C_2H_5$$
$$+$$
$$C_3H_8 \longrightarrow CH_4 + C_3H_7$$
$$\downarrow$$
$$CH_3 + C_2H_4$$
$$+$$
$$C_3H_8 \longrightarrow CH_4 + C_3H_7$$
$$\downarrow$$
$$CH_3 + C_2H_4$$
$$+$$
$$C_3H_8 \longrightarrow \cdots$$

Termination of the chain can occur if CH_3 radical accidentally meets either the vessel wall or the free radical C_3H_7.

$$CH_3 + C_3H_7 \longrightarrow C_4H_{10}$$

---

* For complex reactions involving active transient species, increased pressure thus reduces the over-all rate due to suppression of chain carriers. The effect of pressure on the rate of a simple one-step reaction, is contrary to this, as we have speculated in section 2.1 and proved in section 2.8.

Chemistry of Combustion

This mechanism offers an ingenious method of producing methane, ethylene and butane from propane.

Acetaldehyde Decomposition:

$$CH_3CHO \longrightarrow CH_3 + CHO$$
$$+$$
$$CH_3CHO \longrightarrow CH_4 + CH_3CO$$
$$\downarrow$$
$$CO + CH_3$$
$$+$$
$$CH_3CHO \longrightarrow \cdots$$

Combination of two CH_3 radicals, or two CH_3CO radicals or one of each of these two radicals would terminate the chain.

Consider the reaction between hydrogen and bromine. Knowing the simple second order behavior of hydrogen—iodine reaction, one may intuitively expect H_2—Br_2 reaction to behave in a similar fashion. To one's dismay, experimental measurements show that the rate of production of hydrogen bromide is reduced as its concentration is increased. Equation 2.11 gives this empirical behavior. If k and m are constants,

$$\frac{dC_{HBr}}{dt} = \frac{kC_{H_2}C_{Br_2}^{1/2}}{m + C_{HBr}/C_{Br_2}} \tag{2.11}$$

The following sequence of reactions is proposed to explain Eq. 2.11.

Chain initiation	(i)		Br_2	$\xrightarrow{k_i}$	$2Br$
Chain propagation	(ii)	$Br +$	H_2	$\xrightarrow{k_{ii}}$	$HBr + H$
Chain propagation	(iii)	$H +$	Br_2	$\xrightarrow{k_{iii}}$	$HBr + Br$
Chain inhibition	(iv)	$H +$	HBr	$\xrightarrow{k_{iv}}$	$H_2 + Br$
Chain inhibition	(v)	$Br +$	HBr	$\xrightarrow{k_v}$	$Br_2 + H$
Chain breaking	(vi)		$2Br$	$\xrightarrow{k_{vi}}$	Br_2
Chain breaking	(vii)		$2H$	$\xrightarrow{k_{vii}}$	H_2

Examination shows that steps (v) and (vii) are unimportant in view of the rare chances of their occurrence. Step (iv) is just a reverse of step (ii). Step (i) in this scheme begins the chain by generating two bromine atoms. The chain grows in steps (ii) and (iii) by the generation of new atomic species H and Br. However in step (iv) an active species (atomic hydrogen) is absorbed by the product HBr; thus the growth of the chain is hindered. The chain so strained by the inhibition step (iv) is broken by step (vi) in which two active bromine atoms combine to give a bromine molecule. Step (i) is called chain initiating; (ii) and (iii), chain propagating; (iv) chain inhibiting; and (vi), chain terminating. If one bromine molecule is broken by thermal decomposition into two atoms by step (i), steps (ii) and (iii) are automatically carried out leaving one bromine atom to begin the cycle of steps (ii) and (iii) again.

However, if a collision occurs between the hydrogen atom formed in step (ii), and a hydrogen bromide molecule, step (iii) is deprived of the hydrogen atom to continue the chain. Reaction (iv) thus inhibits the growth of the chain of events.

In general, the chain propagation represented by steps (ii) and (iii) can be written in illustrative form as below.

$$
\begin{aligned}
\text{Br} + \text{H}_2 &\longrightarrow \text{HBr} + \text{H} \\
&\quad + \\
\text{Br}_2 &\longrightarrow \text{HBr} + \text{Br} \\
&\quad + \\
\text{H}_2 &\longrightarrow \text{HBr} + \text{H} \\
&\quad + \\
\text{Br}_2 &\longrightarrow \cdots
\end{aligned}
$$

By considering the specific reaction rate constants for each step in the proposed mechanism, the over-all kinetics of the reaction can be determined. In spite of the fact that the mechanism seems to be complicated, such a kinetic analysis often takes a surprisingly simple form. Let us illustrate the point by examining the present hydrogen—bromine reaction.

If the transient active species are present in the system at a constant concentration (i.e., they are produced at the same rate as they are consumed),

$$\frac{dC_{\text{HBr}}}{dt} = k_{\text{ii}} C_{\text{Br}} C_{\text{H}_2} + k_{\text{iii}} C_{\text{Br}_2} C_{\text{H}} - k_{\text{iv}} C_{\text{HBr}} C_{\text{H}} \tag{2.12}$$

The steady state bromine atom and hydrogen atom concentrations are found as

$$\frac{dC_{\text{Br}}}{dt} = 0 = 2k_{\text{i}} C_{\text{Br}_2} - k_{\text{ii}} C_{\text{Br}} C_{\text{H}_2} + k_{\text{iii}} C_{\text{H}} C_{\text{Br}_2} + k_{\text{iv}} C_{\text{H}} C_{\text{HBr}} - 2k_{\text{vi}} C_{\text{Br}}^2$$

$$\frac{dC_{\text{H}}}{dt} = 0 = k_{\text{ii}} C_{\text{Br}} C_{\text{H}_2} - k_{\text{iii}} C_{\text{H}} C_{\text{Br}_2} - k_{\text{iv}} C_{\text{H}} C_{\text{HBr}}$$

Chemistry of Combustion

From these two equations, the two unknowns C_H and C_{Br} can be obtained in terms of the rate constants and the concentrations of H_2, Br_2, and HBr. Substituting these in Eq. 2.12 and rearranging, we get

$$\frac{dC_{HBr}}{dt} = \frac{\frac{2}{k_{iv}k_{vi}^{1/2}} \cdot k_{iii} k_{ii} k_i^{1/2} C_{H_2} C_{Br_2}^{1/2}}{\frac{k_{iii}}{k_{iv}} + \frac{C_{HBr}}{C_{Br_2}}} \qquad (2.13)$$

This result is the same as the empirically found rate equation given by Eq. 2.11.

Why is $H_2 + Br_2$ reaction so different from $H_2 + I_2$ reaction? We will try to answer this question later in this chapter.

2.6 Explosions—Branching Chain Reactions

Very rapid self-supporting chemical reactions whose rates perhaps become unmeasurably high, are called explosions. Chemical kinetic study of these important combustion reactions is of extreme concern in the design of combustion systems. A possible way to explain these explosions is by an extension of the concept of chain reactions. The reaction between hydrogen and oxygen to form water is a very convenient example to illustrate the phenomenon. If 2 moles of hydrogen and 1 mole of oxygen are contained in a vessel, the pressure and temperature of which can be varied and if the critical conditions to produce explosion are studied experimentally, a plot of explosion limits can be made as shown in Figure 2.2.

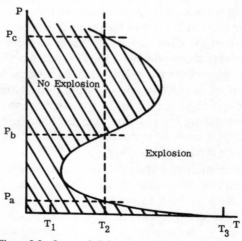

Figure 2.2 Inverted-S threshold of $H_2 + O_2$ reaction

If the vessel is kept at a temperature T_1 to observe if the reaction would occur or not, one finds that no matter how high the pressure is, measurable reaction is not possible. Instead, if one chooses a rather high temperature T_3, reaction is found to occur at all pressures no matter how high or low. An interesting point, however, is observed at moderate temperatures such as T_2.

At a temperature of T_2, if the pressure in the vessel is below P_a, no measurable reaction is found to occur. However, if pressure is increased slightly beyond P_a, the reaction rate becomes measurably high. (i.e., explosion occurs). As the pressure is further increased a rather surprising behavior is noticed. At pressures higher than P_b, no explosion occurs. P_b strongly depends on temperature. At very high pressures (i.e., pressures beyond $P = P_c$) explosions occur readily once again.

The occurrence of self-accelerating reactions can be explained by either of the following two limiting mechanisms. In reality, however, both the mechanisms come into play to result in an explosion.

(1) The heat produced in the slow initial reaction is accumulated due to inadequate heat loss to the environment. The accumulated heat, as a consequence would raise the temperature of the reactants. The higher the temperature of the reactants, the faster the reaction would proceed; thus liberating more heat. This heat in turn raises the temperature and the reaction accelerates to a very large rate. Explosions such as these are called "thermal explosions". A theory for this sort of explosion is first postulated by Semenov and is the subject matter for Chapter 4 of this book.

(2) Some reactions, once initiated, by some energetic process (i.e., by the initial supply of energy in the form of heat or light) proceed at a fast rate by chain processes. In the simple (or unbranching) chain reactions introduced in Section 2.5, one of the products in the chain propagating step is a mole of the chain carrier. If more than one chain carrier is produced in the chain, the so called "branching chain reaction" results. Branching chain explosions may be studied by following the concentration of the active species with time. Due to inadequate diffusional loss of these species to the environment, they accumulate in the system enhancing the reaction rate and therefore their own production rate. The process continues in a self-accelerating manner to culminate in an explosion.

Note the interesting similarity of the roles played by temperature and active species concentration in the determination of the rate of a simple thermal reaction and of a chain reaction, respectively.

Let $\mathcal{R}$, CC and $\mathcal{P}$ respectively stand for reactant, chain carrier and product. Then a dual branching chain reaction may be schematically illustrated as follows

Chemistry of Combustion

$$
\begin{array}{c}
CC + \mathscr{R} \longrightarrow \cdots \\
+ \\
CC + \mathscr{R} \longrightarrow \mathscr{P} \\
+ \\
CC + \mathscr{R} \longrightarrow \mathscr{P} \qquad CC + \mathscr{R} \longrightarrow \cdots \\
+ \qquad\qquad CC + \mathscr{R} \longrightarrow \cdots \\
+ \qquad CC + \mathscr{R} \longrightarrow \mathscr{P} \\
CC + \mathscr{R} \longrightarrow \mathscr{P} \qquad + \\
\qquad\qquad CC + \mathscr{R} \longrightarrow \cdots \\
+ \qquad CC + \mathscr{R} \longrightarrow \mathscr{P} \qquad CC + \mathscr{R} \longrightarrow \cdots \\
CC + \mathscr{R} \longrightarrow \mathscr{P} \qquad + \\
+ \qquad\qquad CC + \mathscr{R} \longrightarrow \cdots \\
CC + \mathscr{R} \longrightarrow \mathscr{P} \qquad CC + \mathscr{R} \longrightarrow \cdots \\
+ \\
CC + \mathscr{R} \longrightarrow \cdots
\end{array}
$$

For hydrogen/oxygen system for example,

$$
\begin{array}{c}
H_2 + O_2 \longrightarrow HO_2 + H \\
+ \qquad + \\
H_2 \qquad O_2 \longrightarrow \cdots \\
\downarrow \\
OH + H_2O \\
+ \\
H_2 \\
\downarrow \\
H + H_2O \\
+ \\
O_2 \\
\downarrow \\
OH + O \\
+ \qquad + \\
H_2 \qquad H_2 \longrightarrow OH + H \\
\downarrow \qquad\qquad + \qquad + \\
H + H_2O \qquad H_2 \qquad O_2 \longrightarrow \cdots \\
+ \qquad\qquad \downarrow \\
O_2 \qquad\qquad H + H_2O \\
\downarrow \qquad\qquad + \\
\cdots \qquad\qquad O_2 \\
\qquad\qquad \downarrow \\
\qquad\qquad \cdots
\end{array}
$$

chain branching reactions

For every H-atom consumed, two H_2O molecules and three H-atoms are generated.

In general for n-branching chain reaction, with R denoting a radical and n denoting their number,

$$A + B + C + \cdots \xrightarrow{k_i} R$$

$$R + A + B + C + \cdots \xrightarrow{k_{ii}} M + N + \cdots + nR. \quad \text{(branching)}$$

$$R \xrightarrow{k_t} A + B + C + \cdots \quad \text{(termination)}$$

For the concentration of R to be in steady state,

$$\frac{dC_R}{dt} = 0$$
$$= k_i C_A C_B C_C \cdots - k_{ii} C_R C_A C_B C_C \cdots + n k_{ii} C_R C_A C_B C_C \cdots - k_t C_R$$

Solving for C_R,

$$C_R = \frac{k_i C_A C_B C_C \cdots}{k_t - (n-1) k_{ii} C_A C_B C_C \cdots} \tag{2.14}$$

Obviously, when $(n-1) k_{ii} C_A C_B C_C \cdots = k_t$, the denominator goes to zero and hence C_R tends to infinity causing an explosion because the rate of reaction can be taken as roughly proportional to the radical concentration. As a matter of fact, when this occurs the steady state treatment becomes invalid except that we used it here to obtain an insight into the trend of the events. If $n = 1$, Eq. 2.14 gives the steady state radical concentration for a simple straight chain reaction to be $C_R = k_i C_A C_B C_C \cdots / k_t$.

2.7 Effect of Pressure on Reaction Rates

(a) *Simple Thermal Reactions*

Consider the simple reaction $A + B + C + \cdots \rightarrow$ Products. If n is the total number of moles in the volume, the ideal gas law gives $PV = nRT$ where P is total pressure. Let the temperature remain constant. If the partial pressure of species A is P_A and number of moles of A is n_A, $P_A = (n_A/V)RT = C_A RT$. Furthermore, $P_A = X_A P$ where X_A is the mole fraction of A. The concentration of A is therefore $C_A = X_A P / RT$. Similarly, concentrations of $B, C, \ldots$ can all be expressed as a product of the respective mole fraction and (P/RT).

Now for a first order equation,

$$-\frac{dC_A}{dt} = k_1 C_A = k_1 X_A P / RT.$$

Chemistry of Combustion

For a second order reaction,

$$-\frac{dC_A}{dt} = k_2 C_A^2 = k_2 X_A^2 (P/RT)^2.$$

For an mth order reaction,

$$-\frac{dC_A}{dt} = k_m C_A^m = k_m X_A^m (P/RT)^m.$$

Thus an mth order reaction rate is directly proportional to pressure raised to the power m. These relations reveal a new method of determining the order of a reaction. If the total pressure and rate are measured as the reaction proceeds at a constant temperature, a plot of the logarithm of rate versus logarithm of pressure would give a straight line with the slope equal to the order as shown in Figure 2.3.

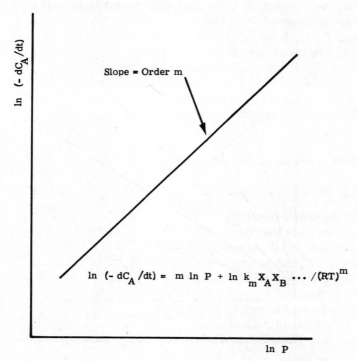

Figure 2.3 Influence of pressure on the rate of an mth order reaction

(b) Chain Reactions

The treatment given above is valid only for simple one-step reactions. When the reaction is complex with transient chain carrying species involved in the intermediate steps, the influence of pressure on the net rate will be contradictory to that given in Section 2.7(a); that is, while the rate of a simple reaction increases with increasing pressure, that of a chain reaction decreases with increasing pressure. The reason for this is that due to increased pressure, the collision frequency is increased. As a result, the transient species undergo an increased number of third-body collisions which suppress their concentration. The rate, hence is expected to be decreased.

2.8 Effect of Temperature on Rates of Simple Reactions

As described in Section 2.3, by conducting experiments at constant temperature and measuring concentration as a function of time, by methods of Table 2.1, the specific reaction rate constant at the particular temperature can be determined. Arrhenius performed a great many experiments in this

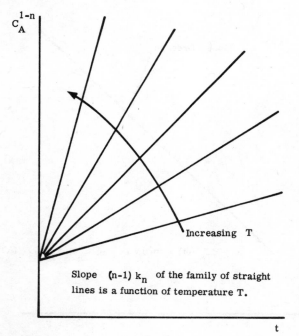

Figure 2.4 Influence of temperature on a chemical reaction

manner at various levels of temperature. His results can be qualitatively shown as in Figure 2.4.

The specific reaction rate constant indicated by the slope of the straight line is found to depend upon the temperature at which the reaction is allowed to occur. Plotting the logarithm of k_n against $1/T$ yields a result as shown in Figure 2.5.

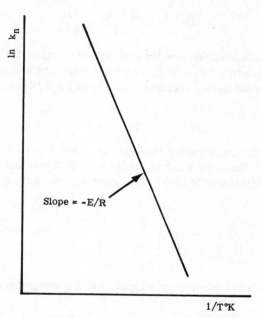

Figure 2.5 Specific reaction rate constant as a function of temperature yields the activation energy

Hence Arrhenius inferred that

$$-\frac{dC_A}{dt} = k_n C_A^n = \mathscr{A} e^{-E/RT} C_A^n \qquad (2.15)$$

$\mathscr{A}$ and E being empirical constants to be determined from Figure 2.5. The dependence of k_n on T,

$$k_n = \mathscr{A} e^{-E/RT} \qquad (2.16)$$

resembles Van't Hoff equation (Eq. C. 15) for equilibrium constant and Gibbs-Helmholtz equation (Eq. C.6) for free energy change. Noting this resemblance, Van't Hoff tried to examine the experimental observations of Arrhenius in the light of his own laws of chemical equilibrium.

The equilibrium constant K, ratio of forward and backward rate constants, is related to temperature and enthalpy of reaction as (cf. Eq. C. 15)

$$\frac{d \ln K}{dT} = \frac{\Delta H^\circ}{RT^2} \tag{2.17}$$

Hence,

$$\frac{d \ln k_f}{dT} - \frac{d \ln k_b}{dT} = \frac{\Delta H^\circ}{RT^2} \tag{2.18}$$

Van't Hoff at this point felt that the enthalpy of reaction ΔH° is the difference between two quantities E_f and E_b which are related to k_f and k_b respectively, in the same fashion as ΔH° is related to K. Thus Eq. 2.18 is written as

$$\frac{d \ln k_f}{dT} - \frac{d \ln k_b}{dT} = \frac{E_f}{RT^2} - \frac{E_b}{RT^2} \tag{2.19}$$

Furthermore, it is suspected that the dependency of k_f on E_f (and that of k_b on E_b) is of the same general nature as the dependency of K on ΔH°. Equation 2.19 thus is split into two independent equations.

$$\frac{d \ln k_f}{dT} = \frac{E_f}{RT^2}$$

$$\frac{d \ln k_b}{dT} = \frac{E_b}{RT^2}$$

If E_f and E_b are independent of temperature, an integration yields

$$\ln k_f = -\frac{E_f}{RT} + \text{Constant}$$

$$\ln k_b = -\frac{E_b}{RT} + \text{Constant}$$

so that

$$\begin{aligned} k_f &= \mathscr{A}_f e^{-E_f/RT} \\ k_b &= \mathscr{A}_b e^{-E_b/RT} \end{aligned} \tag{2.20}$$

Equation 2.20 is of the same form as the empirical correlation obtained by Arrhenius between the rate constant and temperature (Eq. 2.16). Dropping the subscripts f and b, in general,

$$k = \mathscr{A} e^{-E/RT} \tag{2.21}$$

This is called *Arrhenius' law of reaction rate* constant. For reasons which will be clear in later sections of this chapter, $\mathscr{A}$ is called the *frequency factor* (or pre-exponential factor) whereas E is called the *activation energy*. In view of equilibrium considerations, Arrhenius described the process of reaction as below. Of the many molecules present in the reacting system, he assumed that there exist some special molecules which have the ability to take part in a chemical reaction fruitfully, by virtue of their energy. He hypothetically considered a transformation of ordinary molecules into special molecules by a reaction.

$$\text{Common reactant molecules} \longrightarrow \text{special molecules.} \quad (2.22)$$

The enthalpy of formation of these special molecules is called activation energy. By the law of mass action, hence,

$$K = C^*/C$$

where K is the equilibrium constant for the transformation given by Eq. 2.22, C^* is the concentration of the special particles, and C is the concentration of inactive reactant particles. Then Van't Hoff relation for this reaction is

$$\frac{d \ln K}{dT} = \frac{E}{RT^2}$$

i.e., $K = \text{Constant} \times e^{-E/RT}$.

As, usually, the number of special molecules is much less than the inactive molecules, C can be considered to be a constant so that

$$\text{Constant} \times e^{-E/RT} = C^*/C$$

At infinitely high temperatures, all molecules will be special, i.e., $C^* \to C$ as $T \to \infty$.

Hence the constant is unity as $\exp(-E/RT) \to 1$ as $T \to \infty$. If C_t is the total number of molecules, $C_t = C^* + C$ so that $C_t \approx C$

$$C^* = KC \approx KC_t = C_t e^{-E/RT} \quad (2.23)$$

The quantity $\exp(-E/RT)$ indicates the fraction of molecules that are special (i.e., energetic) and capable of participating in a reaction. Or alternatively, $\exp(-E/RT)$ is the probability factor indicating the fraction of molecules whose energy is at least E. This is precisely the reason to call E, the activation energy. As shown in Figure 2.5 and Eq. 2.16, E/R is the slope of the plot $\ln k_n$ versus $1/T$. The factor $\mathscr{A}$ is a large number indicating the number of collisions between molecules in the reacting system. This constant hence is called pre-exponential collision factor.

Arrhenius-Van't Hoff analysis explains quantitatively the rate constant as a function of temperature even though it fails to reveal any intricacies of the reaction mechanism itself.

2.9 Collision Theory of Reaction Rates

In a vessel containing reactants, various molecules move at different speeds in a confusion, turmoil and chaos in various directions. The following prerequisite conditions for successful occurrence of a chemical reaction in this chaos are implicit in the Arrhenius-Van't Hoff arguments presented above.

(i) For a reaction to occur, suitable molecules must collide;
(ii) Only a fraction of these collisions will be in proper orientation; and
(iii) Only a fraction of the properly oriented collisions will be energetic enough for a chemical rupture.

The reaction rate then can be written as

$$\text{rate} \propto Z_{AB} \mathscr{S} e^{-E/RT}.$$

where Z_{AB} is collision frequency, (1/sec), $\mathscr{S}$ is the orientation factor and $\exp(-E/RT)$ is the energy probability factor. In what follows below, we will establish expressions for these factors by reviewing the essential kinetic theory of gases.

(a) Maxwell-Boltzmann Velocity Distribution Law

A series of collisions in the proper direction may give an abnormally high velocity to a molecule or a head-on collision may bring the molecule to a complete stop in a very short interval of time. The velocities that prevail among the molecules at any time are thus statistical and are given by Maxwell-Boltzmann distribution law. Books on kinetic theory of gases give details of derivation but here we simply state it as

$$\frac{dn'}{n'} = 4\pi \left(\frac{m}{2\pi RT}\right)^{3/2} \cdot e^{-mu^2/2RT} \cdot u^2 \, du \qquad (2.24)$$

where n' is total number of molecules present in the system, u is velocity, m is mass of a molecule, T is temperature and dn is the number of molecules having velocity between u and $u + du$. If one prescribes a velocity variation du, one can calculate for a given gas (i.e., given m) the velocity distribution at various temperatures by Eq. 2.24 as shown in Figure 2.6. Here, at each value of u, the number of particles which have velocity between u and $u + du$ are plotted. An example of the value for du is 0.01 meters. It can be seen that at 273 °K, the most common velocity is about 400 meters per sec and that particles at high velocities are almost none. At a higher temperature—say 1,273 °K, the most probable velocity is 1,000 m/sec and there do exist some particles at velocities as high as 2,500 m/sec. The most probable velocity increases as temperature increases.

Chemistry of Combustion

The activation energy for most combustion reactions is of the order of 40,000 calories per mole. Energy of the molecules is predominantly stored in them by virtue of translational movement, i.e., kinetic energy. If m is the mass of one molecule, and its velocity is u, the value of u corresponding to the energy of 40,000 calories is approximately 3,000 meters per second. For the reaction to proceed, molecules must possess energy higher than the

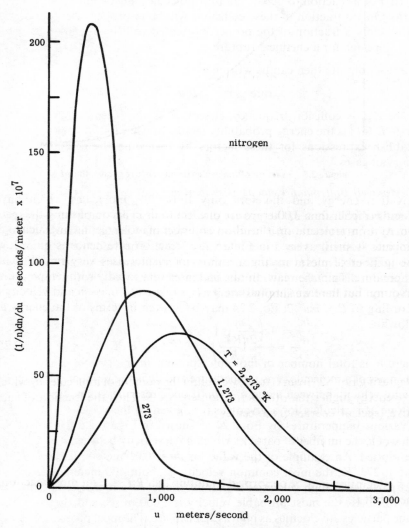

Figure 2.6 Probability of molecular velocities at different temperatures

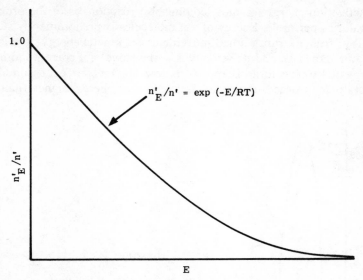

Figure 2.7 Fraction of molecules having energy greater than E

activation energy and therefore only those few molecules which have velocities higher than 3,000 mps are of effect to us in accomplishing the reaction. At room temperature, the relative number of molecules having such high velocities is small as seen in Figure 2.6. Reaction rate hence is quite low. The number of moles having a high velocity increases very rapidly with temperature. Thus the reaction rate increases very rapidly with temperature.

Noting that the translational energy E is related to mass m and velocity u according to $E = mu^2/2$, Eq. 2.24 may be written in terms of the energy as following.

$$\frac{dn'}{n'} = \frac{2}{\sqrt{\pi}} \left(\frac{1}{RT}\right)^{3/2} e^{-E/RT} \cdot \sqrt{E}\, dE \qquad (2.25)$$

Integrating Eq. 2.25 from E to ∞, we obtain the number of molecules n'_E which have energy higher than the activation energy E. Thus the fraction of such active, (special, energetic) molecules is

$$\int_E^\infty \frac{dn'}{n'} = \frac{2}{\sqrt{\pi}} \left(\frac{1}{RT}\right)^{3/2} \int_E^\infty e^{-E/RT} \cdot \sqrt{E}\, dE$$

This integration may be performed by noting that $RT \ll E$ for most reactions. The result is,

$$\frac{n'_E}{n'} = \frac{2}{\sqrt{\pi}} \cdot e^{-E/RT} \cdot \left(\frac{E}{RT}\right)^{1/2} \qquad (2.26)$$

Chemistry of Combustion

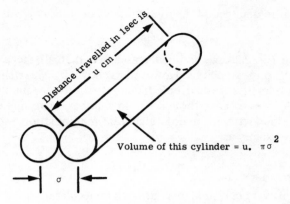

Figure 2.8 Collision diameter

If the calculations are made for molecules moving in a two dimensional plane, the following results replace Eq. 2.25 and 2.26 respectively.

$$\frac{dn'}{n'} = \left(\frac{1}{RT}\right)e^{-E/RT}\, dE \qquad (2.27)$$

$$\frac{n'_E}{n'} = e^{-E/RT} \qquad (2.28)$$

A plot of the thus obtained energetic fraction against E is shown in Figure 2.7. Also note the similarity between Eq. 2.28 and the empirical Arrhenius law (Eq. 2.16).

(b) Collision Frequency

Now that we obtained the fraction of molecules that have the required energy to participate fruitfully in a chemical reaction, next task we are faced with is the calculation of total collisions among molecules.

If the diameter of a molecule is σ, when two molecules of equal size touch one another, the distance between their centers is σ, as shown in Figure 2.8. Collision diameters for various gas molecules are tabulated in literature. A molecule will collide with any molecule traveling within a distance of σ from its center. If the velocity of the molecule is u, it travels u centimeters in one second, sweeping, as a result, a volume of $u\pi\sigma^2$ cubic centimeters.

If there are n' molecules in every cubic centimeter, the molecule A hits $n'u\pi\sigma^2$ molecules in one second as it sweeps through the cylindrical space. Each of the n' molecules undergoes, in a similar manner, collisions with others, so that in one second the total number of collisions will be

$$n'(n'u\pi\sigma^2) \tag{2.29}$$

In deriving Eq. 2.29, we implicitly assumed that all the molecules are traveling at the same velocity u. However, as shown by Maxwell–Boltzmann equation, the velocities are distributed; that is to say, that while some particles have low velocities, others, though few in number, have higher velocities. Considering this fact, with a little algebraic manipulation, we can derive Eq. 2.29 more exactly as

$$\tfrac{4}{3}\sqrt{3\pi}\, u\sigma^2 n'^2 \tag{2.30}$$

However, in every collision, there are two molecules involved. In Eqs. 2.29 and 2.30, we counted *all* the collisions suffered by each molecule. Thus the number of collisions occurring in one second are indeed only one half of those given by Eqs. 2.29 and 2.30.

$$Z'_{AA} = \tfrac{2}{3}\sqrt{3\pi}\, u\sigma^2 n'^2 \tag{2.31}$$

The subscript $_{AA}$ means that collisions are between molecules of the same kind A. If there are molecules of two kinds present, say A and B, Eq. 2.31 has to be modified as shown later.

(c) *Velocity of Molecules as a Function of Temperature of the Gas*

At this point, let us take a moment to relate the velocity of molecules to the pressure (and hence to the temperature) of the gas. Consider N of the molecules of A trapped in a cubical box of volume V and side S as shown in Figure 2.9. Suppose a molecule leaves the wall marked 1 in Figure 2.9 and travels to the opposite wall (marked 2) where it reflects to travel towards 1. It

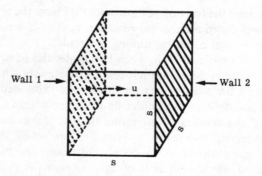

Figure 2.9 Motion of molecules in a box

travels back and forth if it does not collide with any other molecules. However, if it collides with other molecules, its path is deflected and can go in any of the six possible directions. One can say that on an average basis $N/6$ molecules at any given time are moving away from wall 1 normal to it and $N/6$ are moving towards it. Consider the collisions with the wall in this simplified model. One molecule hits the wall 1 once every time it travels a distance of $2S$. If its velocity is u, it takes an interval of time $2S/u$ to travel this distance. A particular molecule, therefore, hits a particular wall once every $2S/u$ seconds. What happens when it hits the wall?

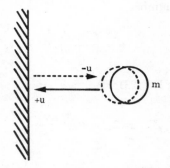

Figure 2.10 Momentum transfer when a molecule reflects off a wall

As shown in Figure 2.10, if the molecule and the wall are perfectly elastic, as a molecule reflects off a wall its velocity is just reversed. Thus if the mass of the molecule is m, the change in its momentum per collision is $mu - (-mu)$. Each collision takes $2S/u$ seconds, so that the force exerted by the molecule on the wall is the rate of momentum with time $= 2mu/(2S/u)$. There are $N/3$ molecules so colliding with the wall. The total force exerted on the wall thus is $(N/3)(2mu)/(2S/u)$. The pressure on the wall is defined as force per unit area of the wall.

$$P = \frac{N}{3} \cdot \frac{2mu}{2S/u} \bigg/ S^2 = \frac{Nmu^2}{3S^3} = \frac{Nmu^2}{3V} \qquad (2.32)$$

Where V = volume = S^3 of the box. What is m, the mass of one molecule of the gas? We know it by a familiar practical name—molecular weight M. Molecular weight is defined as the weight of one mole (which is the amount of gas that occupies 22.4 liters of volume at STP) of gas. Avogadro proved that one mole of any gas contains 6.023×10^{23} molecules. These considerations lead us to the relation between m and M as $M = m \cdot A_{va}$, where A_{va} is Avogadro's number. If N is equal to the Avogadro's number Eq. 2.32 becomes $PV = Mu^2/3$. But by the equation of state (for a perfect gas), $PV = RT$.

Combining these two equations we can relate the velocity u with the temperature T.

$$u = \left(\frac{3RT}{M}\right)^{1/2} \quad (2.33)$$

This equation shows that the mean velocity of the molecules of a gas is directly proportional to the square root of temperature and inversely proportional to the square root of the molecular weight of the gas. Remembering that the gas constant is simply the product of Boltzmann constant κ and the Avogadro's number,

$$u = \left(\frac{3\kappa T}{m}\right)^{1/2} \quad (2.34)$$

Substituting Eq. 2.34 in Eq. 2.31,

$$Z'_{AA} = 2\sqrt{\pi}\,\sigma^2 n'^2 \left(\frac{\kappa T}{m}\right)^{1/2} \quad (2.35)$$

(d) *Synthesis of Collision Theory*

Combining Eqs. 2.35 and 2.28, we get the rate constant

$$k' = \mathscr{S} 2\sqrt{\pi}\,\sigma^2 n'^2 \left(\frac{\kappa T}{m}\right)^{1/2} e^{-E/RT} \quad , K = k' N_0 \quad (2.36)$$

where $\mathscr{S}$ is a correction factor to account for the properly oriented fraction of the collisions. The rate equation for a chemical reaction of order n combining all the collision effects will hence become

$$-\frac{dC_A}{dt} = C_A^n \mathscr{S} 2 \left(\frac{\pi \kappa T}{m}\right)^{1/2} \sigma^2 n'^2 e^{-E/RT} \quad (2.37)$$

Writing in a simplified form by substituting $Z'_{AA}\mathscr{S} \equiv \mathscr{A}'n'^2$

$$-\frac{dC_A}{dt} = \mathscr{A} C_A^n e^{-E/RT} \quad , \mathscr{A} = \mathscr{A}' N_0 \quad (2.38)$$

In deriving Eq. 2.35, it is assumed that all the molecules present in the system are of the same kind A. However, in reacting mixtures there will be more than one kind of molecules. Let A and B be the two kinds of molecules present. For collisions between A and B, Eq. 2.35, is modified to

$$Z'_{AB} = \sigma^2_{AB} n'_A n'_B \left(\frac{8\pi\kappa T}{m_{AB}}\right)^{1/2} \qquad (2.39)*$$

where $\sigma_{AB} = (\sigma_A + \sigma_B)/2$, an effective collision diameter, n'_A and n'_B are number densities of A and B respectively, and m_{AB} is reduced or modified mass $= m_A m_B/(m_A + m_B)$. It could be noted that this expression does not contain the fraction $\frac{1}{2}$ because, for unlike molecules the collisions of the type $A \to B$ are distinguishable from those of the type $B \to A$.

(e) *Mean Free Path Length*

Parenthetically, if a molecule travels at a velocity u cm per sec, and collides with $\pi\sigma^2 un'$ molecules per second as in the derivation of Eq. 2.29, the average distance l between two collisions is given by

$$l = \frac{u}{\pi\sigma^2 un'} = \frac{1}{\pi\sigma^2 n'} \qquad (2.39a)$$

l is called the *mean free path length*. Equation 2.39a shows that the higher the total number of molecules in the box, the shorter the mean free path length and that the larger the molecular diameter, the more often the molecules collide. A more rigorous analysis yields

$$l = \frac{1}{\sqrt{2}\,\pi\sigma^2 n'} \qquad (2.39b)$$

The concept of mean free path length is of extreme use not only in chemical kinetics but also in the study of the transport phenomena.

(f) *Comments on the Collision Factor*

Collision factors calculated theoretically by Eqs. 2.35 and 2.39 differ greatly from the experimentally determined collision frequency. Sometimes, where polyatomic gases are involved, these equations yield values 10^6 times as large as those given by experiment. The reason for this was not clear for a long time. Not too long ago Eyring postulated the activated complex theory of reaction rates to clear these discrepancies. His treatment assumes that there exists an intermediate activated complex state of the reactants in all reactions.

---

* Arrhenius rate equation (Eq. 2.15) infers $\mathscr{A}$ to be independent of temperature, whereas Eq. 2.38 shows it to be proportional to $T^{1/2}$. Ignoring this square root dependence, one commits in $d\ln k/dT$ an error of about $RT/2E$. For most combustion reactions, this error seldom exceeds 2 to 3%.

In order to undergo a chemical reaction, molecules need not only to collide with one another, but also they should collide with proper energy in proper orientation so as to break the weakest of the bonds. The activated complex is a transient state of the reactants, which can spontaneously decompose into products at a constant rate. A frequent example given is that of hydrogen-iodide decomposition.

$$
\begin{array}{ccc}
\begin{array}{c} H \quad H \\ | + | \\ I \quad\; I \end{array} \longrightarrow &
\begin{array}{c} H----H \\ | \quad\quad | \\ I----I \end{array} \longrightarrow &
\begin{array}{c} H-H \\ + \\ I-I \end{array} \\
\text{Reactants} & \text{Activated Complex} & \text{Products}
\end{array}
$$

The concept of activated complex presents a valid physical meaning to the activation energy. It is the energy needed to raise the system from initial reactants to the activated complex. (In other words, taking the reactant state as datum, E is the energy of formation of the activated complex state.) On an energy—extent of reaction diagram, Figure 2.11 shows this graphically. If the reactants are provided with an energy of quantity E, they are converted

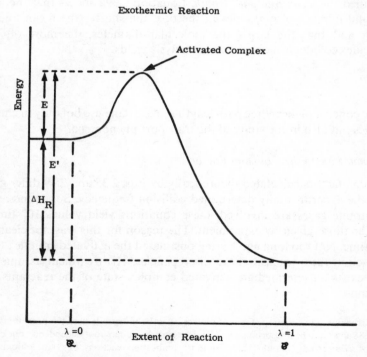

Figure 2.11 Activated complex for an exothermic reaction

Chemistry of Combustion

to the activated complex state. Activated complex can then either break up into products or return back to the initial reactant state, whereas the products cannot attain the state of the activated complex because the energy E' needed for such a reverse process is relatively much larger as shown in Figure 2.11. One can hence write

$$\text{Reactants} \rightleftharpoons \text{Complex} \longrightarrow \text{Products}$$

In terms of quantum theory, Eyring writes for the equilibrium between reactants and activated complex, the specific reaction rate constant

$$k = \frac{RT}{hA_{va}} K^\ddagger \tag{2.40}$$

where R is gas constant, T is absolute temperature, A_{va} is Avogadro's number and h is Planck's quantum constant. $K^\ddagger$ is the equilibrium constant given in terms of the free energy change as the reactants go to activated complex.

$$\Delta F^\ddagger = -RT \ln K^\ddagger = \Delta H^\ddagger - T \Delta S^\ddagger$$

Therefore,

$$K^\ddagger = e^{-\Delta F^\ddagger/RT} = e^{-\Delta H^\ddagger/RT} \cdot e^{\Delta S^\ddagger/R}$$

Substituting in Eq. 2.40, the rate constant is given by

$$k = \frac{RT}{hA_{va}} e^{\Delta S^\ddagger/R} e^{-\Delta H^\ddagger/RT} \tag{2.41}$$

In Eq. 2.41, $\Delta H^\ddagger$ is the enthalpy change between the states of activated complex and initial reactants. Hence it is equal, or approximately so, to the energy of activation.

$$\Delta H^\ddagger \approx E$$

Comparing Eqs. 2.36 and 2.41, the pre-exponential factor is given by

$$\mathscr{A} = \frac{RT}{hA_{va}} e^{\Delta S^\ddagger/R} \tag{2.42}$$

$\Delta S^\ddagger$ is the entropy of activation (i.e., change in entropy between the activated complex state and the reactant state). For simple monatomic or diatomic molecules, the frequency factor $\mathscr{A}$ computed by Eq. 2.42 is comparable to that obtained by Eq. 2.38 of collision theory. However, for complicated polyatomic molecules, the formation of activated complex is followed by a large decrease in entropy, making $\exp(\Delta S^\ddagger/R)$ relatively smaller. This accounts for the smaller values of $\mathscr{A}$ determined experimentally.

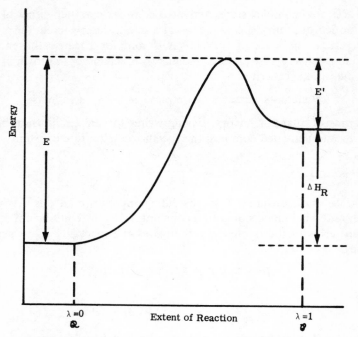

Figure 2.12 Activated complex for an endothermic reaction

(g) *Energy—Reaction Coordinate Diagrams for Endothermic and Exothermic Reactions*

Figures 2.11 and 2.12 give these diagrams respectively for an exothermic and endothermic reaction. If the reaction were a reversible one,

$$A \rightleftharpoons B$$

the activated complex for the forward and reverse will be the same. In such a case the enthalpy of reaction is the difference between the forward and backward activation energies. $E - E' = \Delta H_R$. For some simple reactions, the bond energies listed in Table B.3 can be of help to roughly estimate E by Pauling's Rule.

$$E = 0.28 \cdot (\text{Bond energy of the fuel molecule} + \text{bond energy of the oxidant molecule})$$

2.10 Concluding Remarks

Chemical kinetics is a very important branch of chemistry for use in predicting the behavior and rates of reactions in combustion systems. Design of

Chemistry of Combustion

combustion systems necessarily involves a need for the knowledge of how fast a combustion reaction takes place and by which specific mechanism. The concept of chain reactions elucidates the intricacies of explosions and detonations in combustible mixtures.

As shown in Figure 2.13 from A to B, the rate of a reaction initially increases due to rising temperature. Rising temperature simply provides more "active" molecules and thus makes $\exp(-E/RT)$ progressively larger in the rate equation. After a definite extent of consumption of the reactants, the concentration takes control over the reaction. Due to depletion of the reactant, C_A^n decreases rapidly, thus reducing the rate from B to C. If the reactant supply were abundant, the effect of rising temperature will make the rate grow indefinitely higher as shown in the figure, by the path ABD. Summarizing, in simple thermal reactions, initially the temperature controls the rate whereas in the later part, the depleting concentration of the reactant controls the rate. (Paranthetically, the period OA in which no vigorous reaction is noticed, is called the induction period or ignition delay, a topic to be considered in Chapter 4.)

The behavior of chain reactions may be explained in terms of the concentration of the free radicals playing a role similar to that played by temperature in simple thermal reactions.

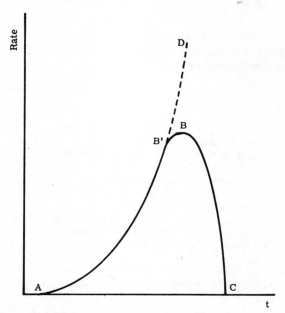

Figure 2.13 Rate of reaction initially increases exponentially due to the Arrhenius term and finally decreases due to the reactant depletion term

Here we may introduce the concept of the space heating rate (SHR). SHR is defined as the rate at which heat is released due to chemical reactions in a unit volume of the reaction space in a unit time. It is thus the product of the rate of reaction and enthalpy of combustion. The maximum space heating rate offers a convenient criterion to evaluate and grade combustion systems. The goal of the combustion engineer is to achieve as large a SHR as possible. For example, in the recent past, SHR in high performance propulsion engines is enhanced to tens of thousands of $BTU/ft^3/sec$. This figure, compared to 8 $BTU/ft^3/sec$ for coal-burning steam boilers, shows the spectacular progress achieved in the design of combustion chambers. Similarly, a modern structure afflicted by a destructive fire presents space heating rates of the order of thousands of $BTU/ft^3/sec$. Conventional construction materials, on the other hand, are known to lead to space heating rates of only hundreds of $BTU/ft^3/sec$. Many people hold the advent of synthetic materials to be responsible for these order of magnitude differences.

Also shown in this chapter is, that while being basically mechanistic in character, the simple collision theory predicts the reaction rates quite satisfactorily for reactions between simple molecules. However, the activated complex theory based on free energy of formation of the activated complex approximates the situation much closer to the actual reaction rates by considering the changes in entropy of formation of the activated complex from the reactant state. Steacie's book, "Atomic and Free Radical Reactions," provides a great variety of reaction kinetic data on various reactions of interest in combustion, whereas Rodiguin's book on Consecutive Chemical Reactions presents detailed methods of analysis.

References

Bird, Stewart and Lightfoot, *Transport Phenomena*.
Daniels, F. and Alberty, R. A., *Physical Chemistry*, Wiley (1955).
Glasstone, S., *Elements of Physical Chemistry*, Van Nostrand (1958).
Handbook of Physics and Chemistry, C.R.C., Cincinnati, Ohio.
Hirschfelder, et al., *Molecular Theory of Gases and Liquids*.
Moore, W. J., *Physical Chemistry*, Prentice Hall (1956).
Rodiguin, N. M. and Rodiguina, E. N., *Consecutive Chemical Reactions*, Van Nostrand (1964).
Semenov, N. N., *Chemical Kinetics and Chain Reactions*, Oxford (1935).
Semenov, N. N., *Some Problems of Chemical Kinetics and Reactivity*, Vols. 1 and 2, Pergamon (1959).
Smith, J. M., *Chemical Engineering Kinetics*, McGraw-Hill (1956).
Steacie, E. W. R., *Atomic and Free Radical Reactions*, ACS Monograph Series No. 125 and 102. Reinhold (1946) and (1954).

CHAPTER 3

Physics of Combustion

The discipline of fluid mechanics plays an important role in any study of flames. This chapter is intended to briefly present the fundamentals of fluid flow. We start out with the three basic laws of molecular transfer—Newton's law of viscosity for momentum transfer, Fourier law of conduction for heat transfer and Fick's law of diffusion for species transfer. These three laws give rise to the definitions of three physical properties of the fluid. viz., viscosity, thermal conductivity and species diffusivity. Using simple kinetic theory of gases, formulae to compute viscosity, conductivity and dilute gas diffusivity are derived.

The concept of boundary layer is then introduced. Equations for friction and heat transfer across a boundary layer are deduced from simple dimensional analysis. The effects of turbulence are briefly discussed.

The phenomenon of mass transfer is described and the differential equations of conservation of mass, momentum, heat and species are derived.

3.1 Fundamental Laws of Molecular Transfer

(a) *Newton's Law of Viscosity*

Consider two impervious flat plates of infinite width and length separated by a distance δ, the interspace being filled with a fluid, B. Let the system be isothermal. Now if the lower plate is fixed in position and the upper plate is moved, as shown in Figure 3.1, at a constant velocity u_∞ it is experimentally discovered that the velocity in the fluid varies from u_∞ at the upper plate to zero at the lower one.

By virtue of such a motion, it is found that the fluid exerts a shear force*

---

* Flow in which adjacent layers of fluid slide smoothly past one another is called "laminar." Such relative motion of the fluid layers involves shearing of the fluid due to an exchange of molecules among various layers, these exchanged molecules bearing momenta proportional to their chaotic random velocities.

44 *Introduction to Combustion Phenomena*

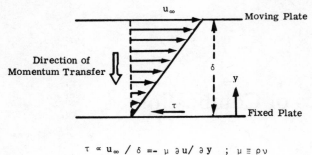

$$\tau \propto u_\infty / \delta = -\mu \,\partial u/\partial y \quad ; \quad \mu \equiv \rho \nu$$

Figure 3.1 Newton's law of viscosity

on the lower plate. The force per unit area is found to be directly proportional to the velocity u_∞ and inversely proportional to the separation distance δ.

$$\tau \equiv \frac{F}{A} \propto \frac{u_\infty}{\delta} \quad (\text{dynes/cm}^2 \propto \text{sec}^{-1})\dagger$$

In differential form, we obtain Newton's law of viscosity as

$$\tau = -\mu \frac{\partial u}{\partial y} \tag{3.1}$$

The constant of proportionality μ is called *dynamic viscosity* (units dyne sec/cm^2) which is sometimes expressed as the product of density ρ and kinematic viscosity ν (units cm^2/sec). τ is called *shear stress* and the velocity gradient $\partial u/\partial y$ is called *rate of shear*. The minus sign in Eq. 3.1 is to indicate that the shear stress is exerted in the direction of decreasing velocity as shown in Figure 3.1. The viscosity μ for gases does not depend upon pressure but varies linearly with square root of temperature. For liquids (which we well know, get "thin" if they are heated) viscosity decreases with increased temperature. Fluids obeying Eq. 3.1 are known as Newtonian fluids. All gases and most liquids obey Eq. 3.1. For some fluids, the viscosity depends upon the rate of shear. Such fluids are called *non-Newtonian*.

(b) *Fourier's Law of Heat Conduction*

Consider again the parallel plates. Let the distance of separation be δ_t as shown in Figure 3.2. Let both the plates be at rest. Let the temperature of the lower plate be T_W and that of the upper plate, T_∞. By virtue of the fact that the fluid temperature varies from T_∞ at the hot plate to T_W at the cold plate, heat is transferred across the fluid layers to the cold plate. This transfer of heat

---

† 1 dyne = 1 gm cm/sec^2.

Physics of Combustion

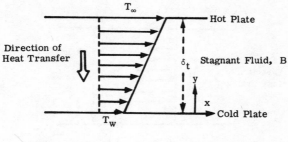

$$\dot{q}'' \propto (T_\infty - T_W)/\delta_t = -K \, \partial T/\partial y \, ; \quad K \equiv \rho C_p \alpha$$

Figure 3.2 Fourier's law of thermal conduction

across the fluid layers is once again due to an exchange of molecules, bearing energies proportional to the product of their mass, specific heat and temperature, the exchange being a consequence of the chaotic random movement of the molecules. The heat flow per unit area per unit time ($\dot{q}''$ cal/cm²/sec) is found to be proportional to the temperature difference $T_\infty - T_W$ and inversely proportional to the separation distance δ_t.

$$\dot{q}'' \equiv \frac{\dot{Q}}{A} \propto \frac{T_\infty - T_W}{\delta_t} \quad \text{(cal/cm}^2\text{/sec} \propto \text{°C/cm)}$$

In differential form, the heat flux across any fluid layer is given by Fourier's Law.

$$\dot{q}'' = -K \frac{\partial T}{\partial y} \tag{3.2}$$

The proportionality constant K is called *thermal conductivity* (units cal/cm/ °C/sec) which often is expressed as a product of density ρ, specific heat C_P (cal/gm/°C) and thermal diffusivity α (cm²/sec). Thermal conductivity is a physical property of the fluid just the same way as the viscosity is. The minus sign in Eq. 3.2 is to indicate that the heat flows in the direction of decreasing temperature as shown in Figure 3.2.

(c) *Fick's Law of Species Diffusion*

Once again, as shown in Figure 3.3, consider the parallel plate example. This time the system involves no movement nor any temperature differences. The separation distance is δ_D. Let the plates be porous. The space between the plates is filled with a fluid B. Let a different fluid A be transpired out of the lower plate and sucked into the upper plate at such a negligible rate that the fluid is not disturbed with motion. Let $C_{A\infty}$ be the concentration of the

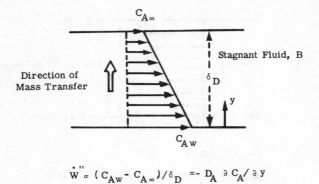

$$\dot{W}''_A \propto (C_{Aw} - C_{A\infty})/\delta_D = -D_A \, \partial C_A/\partial y$$

Figure 3.3 Fick's law of species diffusion

species A at the upper plate and let C_{Aw} be that at the lower plate. By virtue of the nonuniform concentration field, species A diffuses from the bottom plate to the top plate across the layers of fluid B. The mass flux of A received by the upper plate (or lost by the lower plate) is found to be proportional to the ratio of the concentration difference to the separation distance δ_D.

$$\dot{W}''_A \propto \frac{C_{Aw} - C_{A\infty}}{\delta_D} \qquad (\text{gm/cm}^2/\text{sec} \propto \text{gm/cm}^3/\text{cm})$$

In differential form, the mass of A diffusing across any layer of the fluid is related to the local concentration gradient by Fick's law as

$$\dot{W}''_A = -D_{AB} \frac{\partial C_A}{\partial y} \qquad (3.3)$$

D_{AB} is called *diffusivity* of species A through species B. It is commonly assumed as independent of the composition. Its units are (cm^2/sec). The minus sign in Eq. 3.3 is to indicate that species A transfers in the direction of decreasing concentration of A.

Equation 3.3 can be written in terms of the partial pressure gradient or mass fraction gradient as well. Let P, M, R, T and Y respectively be the pressure, molecular weight, universal gas constant, temperature and mass fraction. (Subscripts A and B refer to species A and B respectively. Unsubscripted quantities refer to the total mixture.) Assuming that the gases behave perfectly, Eq. 3.3 can be written as

$$\dot{W}''_A = -\frac{D_{AB} M_A}{RT} \frac{\partial P_A}{\partial y}$$

The mass fraction of A is related to the partial pressure by $Y_A = P_A M_A/PM$. Furthermore, $PM_A/RT = \rho M_A/M$ so that

$$\dot{W}_A'' = -\frac{\rho D_{AB}}{M}\frac{\partial Y_A M}{\partial y}$$

The mixture molecular weight itself is a function of the mass fractions and molecular weights of species A and B as

$$M = \left[\frac{Y_B}{M_B} + \frac{(1-Y_B)}{M_A}\right]^{-1}$$

Hence

$$\dot{W}_A'' = -\frac{\rho D_{AB}}{[1 - Y_A(M_B - M_A)/M_A]}\frac{\partial Y_A}{\partial y}$$

In general, in situations concerned with combustion, $Y_A(M_B - M_A)/M_A \ll 1$ so that

$$\dot{W}_A'' \approx -\rho D_{AB}\frac{\partial Y_A}{\partial y} \tag{3.4}$$

Flames are spatial zones, perhaps varying with time, in which nonuniformities of velocity, temperature and composition introduce momentum, heat and mass transfer. For this reason, any study of flames would involve the use of Eqs. 3.1 to 3.4. In general, these fundamental molecular transport equations can be written as

$$\mathscr{F} = -\mathscr{D}\frac{\partial \mathscr{P}}{\partial y} \tag{3.5}$$

where the flux $\mathscr{F}$, diffusivity $\mathscr{D}$ and the concentration $\mathscr{P}$ are as shown in Table 3.1.

TABLE 3.1

$$\mathscr{F} = -\mathscr{D}\frac{\partial \mathscr{P}}{\partial y}$$

Law	$\mathscr{F}$	$\mathscr{D}$	$\mathscr{P}$
Newton's Law of viscosity for the transfer of momentum	τ	ν	ρu (= momentum per unit volume)
Fourier's Law of conduction for heat transfer	$\dot{q}''$	α	$\rho C_p T$ (= enthalpy per unit volume)
Fick's Law of diffusion for species transfer	$\dot{W}_i''$	D_i	C_i (= amount of species i per unit volume)

As indicated by Eqs. 3.1 to 3.3 and Table 3.1 the larger the diffusivities v, α and D, the larger the rate of transfer. Frequently, momentum, heat and mass transfer occur simultaneously.

The magnitude of momentum transfer relative to that of heat transfer is a function of the ratio v/α. Recalling that the dimensions of v are same as those of α, (cm²/sec), v/α can be recognized as a nondimensional property of the fluid. It is called *Prandtl number* (denoted by Pr). Schmidt and Lewis numbers are defined similarly.

$$\text{Pr} \equiv v/\alpha$$
$$\text{Sc} \equiv v/D \tag{3.6}$$
$$\text{Le} \equiv \alpha/D \equiv \text{Sc}/\text{Pr}$$

The significance of these nondimensional properties will be obvious when we discuss the boundary layers in Section 3.3.

In order to calculate the diffusive flux of momentum, heat and species in flames, knowledge of accurate values of the transport properties μ, K and D is necessary. Whenever experimentally measured values are available for these coefficients they should be used. Many combustion phenomena, however, occur at high temperatures at which no experimental values are available. Under these conditions theoretical expressions, which agree satifactorily with the low temperature experimental data, are extrapolated.

Experimental data for solids, liquids, and relatively cool vapors (i.e. dense gases) are abundantly available in the literature. For dilute gases an excellent kinetic theory is available. Research is active on the prediction and measurement of transport properties of dissociated gases. In Section 3.2 we present a method of calculation of μ, K and D by a simple kinetic theory. Appendix D gives some data of interest in combustion.

3.2 Kinetic Theory of Dilute Gases

(a) *Mean Free Path Length*

In the random chaotic movement of molecules in a gas-body, the mean distance an average molecule would travel between two successive collisions is called mean free path length, L. We derived an equation in Chapter 2 to relate L to the size of the molecule and the number density.

$$L = \frac{1}{\sqrt{2\pi\sigma^2 n'}} \tag{2.39b}$$

Physics of Combustion

σ is the molecular diameter and n' is the number density. Equation 2.39b shows that the larger the molecular diameter or the pressure the shorter the mean free path length. This is consistent with intuition. If l is the characteristic physical dimension of our system, then the ratio L/l is known as Knudsen number. When $L/l \gg 1$ individual molecular motion has to be followed to understand the behavior of the system (Knudsen's Free molecule flow). If $L/l \ll 1$, the behavior of the system can be found by integrating the movement of an average pocket of molecules (continuum flow).

(b) Calculation of the Transport Coefficients of Monatomic Dilute Gases

Consider here continuum flow. Consider n' molecules trapped in a unit volume. Let the intermolecular forces be approximated by supposing the molecules to behave as rigid non-attracting spheres. Each particle is assumed to travel at an arithmetic speed of U. Then, on an average, one-sixth of the n' particles will be traveling in each of the six directions $\pm x, \pm y$ and $\pm z$. Consider, as shown in Figure 3.4, three planes located at $y = -L, 0$ and $+L$, L being the mean free path.

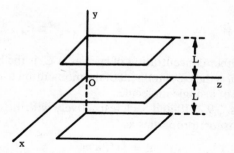

Figure 3.4 Derivation of transport coefficients (Redrawn with permission, from J. O. Hirchfelder, et. al. *The Molecular Theory of Gases and Liquids*. John Wiley and Sons, New York, (1954)).

Assume that the values of some property $\mathscr{P}$ (which could be mass, momentum or enthalpy as shown in Table 3.1), of the gas is everywhere constant in any of the three planes. Particles leaving the plane $y = +L$ in the $-y$ direction carry the value of $\mathscr{P}_+$ to $y = 0$ before undergoing a collision. Similarly particles having the property $\mathscr{P}_-$ at $y = -L$ carry this value to $y = 0$ by moving in the $+y$ direction. Recalling that only $\frac{1}{6}$ of the particles move in any one direction on an average

$$\text{Net flux} = v(\mathscr{P}_- - \mathscr{P}_+)/6 \qquad (3.7)$$

If $\mathscr{P}$ is a continuous function of y, by Taylor series, we can approximately write

$$\mathscr{P}_- \approx \mathscr{P}_0 - L\frac{\partial \mathscr{P}}{\partial y}\bigg|_0$$

$$\mathscr{P}_+ \approx \mathscr{P}_0 + L\frac{\partial \mathscr{P}}{\partial y}\bigg|_0$$

Substituting in Eq. 3.7, the net flux across the plane $y = 0$ is given by

$$\mathscr{F} = -\frac{UL}{3}\frac{\partial \mathscr{P}}{\partial y} \tag{3.8}$$

If we now associate the flux $\mathscr{F}$ and the property $\mathscr{P}$ with momentum, heat and mass transfer as shown by Eq. 3.5 and Table 3.1, Eq. 3.8 gives

$$\tau = -\frac{ULn'm}{3}\frac{\partial u}{\partial y}\ ;\ \mathscr{F} \equiv \tau;\ \mathscr{P} \equiv n'mu \tag{3.9}$$

$$\dot{q}'' = -\frac{ULn'\hat{C}_v}{3}\frac{\partial T}{\partial y}\ ;\ \mathscr{F} \equiv \dot{q}'';\ \mathscr{P} \equiv n'\hat{C}_v T \tag{3.10}$$

$$\dot{W}_A'' = -\frac{UL}{3}\frac{\partial n_A'}{\partial y};\ \mathscr{F} \equiv \dot{W}_A'';\ \mathscr{P} \equiv n_A' \tag{3.11}$$

Here n_A' is the number concentration of species A, $\hat{C}_v$ is the heat capacity of a single molecule, mu is the x-component of the momentum of a single molecule, and m is the mass of a single molecule.

Comparing Eqs. 3.9, 3.10 and 3.11 respectively with Eqs. 3.1, 3.2 and 3.4, we obtain the transport properties as

$$\mu = \tfrac{1}{3}ULn'm$$
$$K = \tfrac{1}{3}ULn'\hat{C}_v$$
$$D = \tfrac{1}{3}UL$$

Substituting Eq. 2.29b for L, and denoting the constant $\tfrac{1}{3}\sqrt{2\pi}$ by a,

$$\mu = aUm/\sigma^2 \tag{3.12}$$

$$K = aU\hat{C}_v/\sigma^2 \tag{3.13}$$

$$D = aU/\sigma^2 n' \tag{3.14}$$

These expressions show that the transport coefficients are directly proportional to the arithmetic mean speed and inversely proportional to the molecular diameter. μ and K are independent of n' and hence pressure. If now the relatively simple model of rigid sphere were retained but a Maxwellian velocity distribution were used instead of the arithmetic mean speed U, Eqs. 3.12 to 3.14 are replaced by the following equations which are in excellent agreement with experimental measurements.

$$\mu = 2.6693 \cdot 10^{-5}\sqrt{TM}/\sigma^2 \text{ gm/cm/sec.} \tag{3.15}$$

$$K = 1.9891 \cdot 10^{-4}\sqrt{T/M}/\sigma^2 \text{ cal/cm/°C/sec.} \tag{3.16}$$
$$= 15R\mu/4M$$

$$D = 2.6280 \cdot 10^{-3}\sqrt{T^3/M}/P\sigma^2 \text{ cm}^2/\text{sec.} \tag{3.17}$$

where M is the molecular weight, T is absolute temperature °K, P is pressure in atmospheres and σ is the molecular diameter in Angstrom units*. The diffusion coefficient D given in Eqs. 3.14 and 3.17 is for the diffusion of a species into itself. Equation 3.17 can be modified for binary diffusion as shown by Eq. 3.22.

In real life, molecules are not rigid (but soft and squishy), not spherical (but clusters of atoms) and there do exist mutual attraction and repulsion forces between molecules. Taking these factors into account (with what is known as Lennard-Jones (6-12) force potential), Chapman and Enskog found that Eqs. 3.15 to 3.17 need a slight correction as below:

$$\mu = 2.6693 \cdot 10^{-5}\sqrt{TM}/\sigma^2\Omega_\mu \tag{3.18}$$

$$K = 1.9891 \cdot 10^{-4}\sqrt{T/M}/\sigma^2\Omega_\mu \tag{3.19}$$
$$= 15R\mu/4M$$

$$D = 2.6280 \cdot 10^{-3}\sqrt{T^3/M}/P\sigma^2\Omega_D \tag{3.20}$$

where Ω_μ and Ω_D are correction factors (called *collision integrals*) which are weakly dependent upon temperature as shown in Table 3.2. The molecular constants σ and ε/κ are listed for some species in Table 3.3.

A few comments on Eqs. 3.18 to 3.20 are in order here.

Recall that these equations are derived for dilute monatomic nonpolar gases. Equations 3.18 and 3.20 can be used for nonpolar dilute polyatomic gases also. For polyatomic gases Eq. 3.19 has to be corrected with Euken formula

$$\frac{K_{\text{polyatomic}}}{K_{\text{Chapman-Enskog}}} = \left(\frac{4}{15}\frac{C_v}{R} + \frac{3}{5}\right) \tag{3.21}$$

where C_v is molar specific heat at constant volume. For polar gases of interest in combustion (primarily water vapor), it is safer to rely on the available experimental data.

---

* 1 Å = 10^{-8} cm and 1 micron = 10^{-4} cm.

TABLE 3.2

Collision Integrals for Calculation of Transport Properties*

$\kappa T/\varepsilon$	Ω_μ	Ω_D	$\kappa T/\varepsilon$	Ω_μ	Ω_D
0.30	2.785	2.662	3.60	0.9932	0.9058
0.40	2.492	2.318	3.80	0.9811	0.8942
0.50	2.257	2.066	4.00	0.9700	0.8836
0.60	2.065	1.877	4.20	0.9600	0.8740
0.70	1.908	1.729	4.40	0.9507	0.8652
0.80	1.780	1.612	4.60	0.9422	0.8568
0.90	1.675	1.517	4.80	0.9343	0.8492
1.00	1.587	1.439	5.0	0.9269	0.8422
1.10	1.514	1.375	6.0	0.8963	0.8124
1.20	1.452	1.320	7.0	0.8727	0.7896
1.30	1.399	1.273	8.0	0.8538	0.7712
1.40	1.353	1.233	9.0	0.8379	0.7556
1.50	1.314	1.198	10.0	0.8242	0.7424
1.60	1.279	1.167	20.0	0.7432	0.6640
1.70	1.248	1.140	30.0	0.7005	0.6232
1.80	1.221	1.116	40.0	0.6718	0.5960
1.90	1.197	1.094	50.0	0.6504	0.5756
2.00	1.175	1.075	60.0	0.6335	0.5596
2.20	1.138	1.0410	70.0	0.6194	0.5464
2.40	1.107	1.0120	80.0	0.6076	0.5352
2.60	1.081	0.9878	90.0	0.5973	0.5256
2.80	1.058	0.9672	100.0	0.5882	0.5130
3.00	1.039	0.9490	200.0	0.5320	0.4644
3.20	1.022	0.9328	300.0	0.5016	0.4360
3.40	1.007	0.9186	400.0	0.4811	0.4170

* Adapted from Hirschfelder, J. O., Curtis, C. F., and Bird, R. B., *Molecular Theory of Gases and Liquids*, Wiley, New York, 1954, with the permission of Wiley.

Equation 3.20 gives the coefficient of self-diffusion. For binary diffusion, Eq. 3.20 can be modified as

$$D_{AB} = D_{BA} = 2.6280 \cdot 10^{-3} \frac{\sqrt{\frac{T^3}{2}\left(\frac{1}{M_A} + \frac{1}{M_B}\right)}}{P\sigma_{AB}^2 \Omega_D} \tag{3.22}$$

M_A and M_B are molecular weights. σ_{AB} and ε_{AB} are given by

$$\sigma_{AB} = (\sigma_A + \sigma_B)/2$$
$$\varepsilon_{AB} = (\varepsilon_A \varepsilon_B)^{1/2} \tag{3.23}$$

TABLE 3.3

Parameters for use in the Lennard-Jones (6–12) Potential*

Molecule	σ (Å)	ε/κ (°K)
Ne	2.789	35.7
A	3.418	124.0
Kr	3.610	190.0
Xe	4.055	229.0
N_2	3.681	91.5
O_2	3.433	113.0
F_2	3.653	112.0
Cl_2	4.115	357.0
Br_2	4.268	520.0
I_2	4.982	550.0
NO	3.470	119.0
CO	3.590	110.0
CO_2	3.996	190.0
CH_4	3.882	137.0
CF_4	4.700	152.0
CCl_4	5.881	327.0
SF_6	5.510	200.9
SO_2	4.290	252.0
CH—CH	4.221	185.0
CH_2—CH_2	4.232	205.0
C_2H_6	4.418	230.0
C_3H_8	5.061	254.0
n-C_4H_{10}	4.997	410.0
i-C_4H_{10}	5.341	313.0
n-C_5H_{12}	5.769	345.0
n-C_6H_{14}	5.909	413.0
cyclo-C_6H_{12}	6.093	324.0
C_6H_6	5.270	440.0
Air	3.617	97.0

* Adapted with permission from: J. O. Hirschfelder, C. F. Curtis, and R. B. Bird, *Molecular Theory of Gases and Liquids*, Wiley, New York (1954).

For mixtures of gases the viscosity and conductivity can be calculated by the following formula

$$\mu_{\text{mixture}} = \sum_{i=1}^{n} \frac{X_i \mu_i}{\sum_{j=1}^{n} X_j \phi_{ij}} \; ; \; K_{\text{mixture}} = \sum_{i=1}^{n} \frac{X_i K_i}{\sum_{j=1}^{n} X_j \phi_{ij}} \qquad (3.24)$$

where

$$\phi_{ij} = \frac{1}{\sqrt{8}} \left(1 + \frac{M_i}{M_j}\right)^{-1/2} \left[1 + \left(\frac{\mu_i}{\mu_j}\right)^{1/2} \left(\frac{M_j}{M_i}\right)^{1/4}\right]^2$$

X_i is mole fraction of the species i, n is the total number of species in the mixture. The diffusivity of a species "A" through a mixture $B, C, D \ldots$ is a very complicated property. Two approximate methods are available to handle this situation. Firstly, lump all the mixture $B, C, D \ldots$ etc. together and treat the lump as a hypothetical species B' and calculate the binary diffusion coefficient of A through B'. Or alternatively,

$$D_{A\text{-mixture}} = \frac{1 - x_A}{\sum_{j \neq A} \frac{x_j}{D_{Aj}}} \qquad (3.25)$$

(c) *Dimensionless Properties*—Pr, Sc *and* Le

The derivation of the molecular transport properties has also indicated a good deal of similarity between the transfer of momentum, heat and mass transfer. This similarity extends itself even so far as to be distinctly seen in the final expressions for μ, K and D. Consider the results of simple kinetic theory (i.e. Eqs. 3.12 to 3.14). Noting that for a monatomic gas, $C_v = 3R/2$,

$$\frac{\mu C_v}{KM} = \frac{\mu C_v}{K} \equiv f = 1$$

The property f is related to the Prandtl number. It depends upon the nature of the gas under consideration. If f, μ and C_V are known, thermal conductivity and Prandtl number can easily be calculated. For monatomic gases f is 0.4. For diatomic gases f is somewhat lower than 0.5. For triatomic gases $f = 0.57$. Chapman-Enskog formulae yield f as

$$f = \tfrac{2}{5}$$

Noting that $C_p = C_v + R = 5R/2$, the Prandtl number is given by

$$\text{Pr} \equiv \frac{\nu}{\alpha} = \frac{\mu C_p}{K} = \frac{\mu C_v}{K} \frac{C_p}{C_v} = 0.666 \qquad (3.26)$$

This result shows that in a nonpolar monatomic dilute gas, heat transfer is easier than momentum transfer. The Schmidt number is given by Eqs. 3.18 and 3.20 as

$$Sc \equiv \nu/D = \mu/\rho D = 0.83 \tag{3.27}$$

Similarly, Lewis number is

$$Le \equiv Sc/Pr = \alpha/D = 1.25 \tag{3.28}$$

From the foregoing discussion, it is obvious that even though Pr, Sc and Le vary with composition, temperature and pressure, they are often of the order of unity. It is customary in combustion calculations to assume them as equal to unity.

3.3 Concept of Boundary Layer

As shown in Figure 3.5, consider a flat thin rigid stationary plate situated parallel to the flow in a stream of isothermal fluid flowing at a velocity u_∞. Let x be the coordinate along the plate in the direction of the flow and y be the coordinate normal to the surface of the plate, the origin being fixed at the leading edge. Let the width of the plate in z-direction (i.e. normal to the plane of the paper) be infinite.

If one measures, at any chosen x, the velocity u as a function of y, a profile of $u = u(y)$ is obtained as shown in Figure 3.5.

The velocity is uniform ($=u_\infty$) everywhere in the fluid except in a very thin layer of fluid adjacent to the wall. This thin layer is called "*boundary layer*". In the boundary layer, the velocity decreases from the free stream value u_∞ to the wall value i.e., zero. This reduction in velocity is due to the resistance offered by the solid wall to the flow which in turn imposes a force called *drag* on the wall. In reality the effect of the presence of the wall is felt to

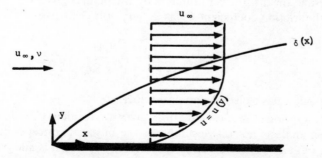

Figure 3.5 Boundary layer on a flat plate in an isothermal situation

quite large values of y into the fluid. That is, the thickness of the boundary layer is very large. For all practical purposes, however, the boundary layer thickness δ is defined as that distance from the wall beyond which the velocity is unmeasurably close to the free stream value.

The thickness of the boundary layer δ can be intuitively expected to increase with increasing x, with decreasing free stream velocity u_∞ and with increasing fluid viscosity. The magnitude of δ at any $x = l$ can be estimated by balancing the friction at the wall with the loss in the kinetic energy of the fluid. Consider a steady flow.

The mass flow through the boundary layer of thickness δ at an $x = l$ per unit depth in z-direction is approximately,

$$\dot{m} \approx \delta \cdot 1 \cdot \rho \cdot u_\infty/2$$

$(\delta \cdot 1)$ here is the area of flow and $u_\infty/2$ is the mean velocity of flow. This mass flow, before encountering the plate (i.e. for $x < 0$) has a velocity of u_∞ and hence a kinetic energy of $(\delta \cdot 1 \cdot \rho u_\infty/2)(u_\infty)$. As the fluid exits through the section at $x = l$, the kinetic energy is equal to $(\delta \cdot 1 \cdot \rho u_\infty/2)(u_\infty/2)$. The loss of kinetic energy thus is $(\delta \cdot 1 \cdot \rho u_\infty^2/4)$.

By Newton's law of viscosity, the force exerted on the wall in the length $x = 0$ to $x = l$ is given by $(l \cdot 1) \cdot (-\mu \, \partial u/\partial y)_w \approx l \cdot 1 \cdot \mu u_\infty/(\delta/2)$. Equating the loss of the kinetic energy of the fluid to the force on the wall,

$$\delta \rho u_\infty^2/4 \approx 2\mu l u_\infty/\delta \tag{3.29}$$

Resolving for δ,

$$\delta \approx \sqrt{8} \sqrt{\frac{\mu l}{\rho u_\infty}} \tag{3.30}$$

Equation 3.30 shows that δ will be larger for larger μ, larger l, smaller ρ and smaller u_∞.

The average shear force per unit area on the plate is given by $\tau \approx 2\mu u_\infty/\delta$. Defining the friction coefficient C_f as $2\tau/\rho u_\infty^2$, Eq. 3.30 gives,

$$C_f \approx \sqrt{2} \sqrt{\frac{\mu}{\rho u_\infty l}} = \frac{1.414}{\sqrt{\text{Re}_l}} \tag{3.31}$$

The group $\rho u_\infty l/\mu$ is called Reynolds number.

An exact solution of the flow equations gives this result as $C_f = 1.327/\sqrt{\text{Re}_l}$. Our simple analysis not only establishes the physical concepts involved in fluid flow, but also gives a solution the accuracy of which is adequate for an engineer's use.

3.4 Heat Transfer Across a Boundary Layer

Suppose now that the stream temperature is T_∞. Let the plate be at a temperature T_W ($< T_\infty$). Because of the nonuniformity of temperature in the fluid, heat flows from the fluid to the wall. Assume that the viscosity, thermal conductivity, specific heat and density of the fluid are all independent of temperature. If one measures the temperature as a function of y at any given x, one finds that it is everywhere close to T_∞ except in a thin layer adjacent to the cold wall as shown in Figure 3.6. This thin layer is called *thermal boundary layer*. The concept of thermal boundary layer is very similar to that of hydrodynamic boundary layer. The thermal boundary layer thickness δ_t is so defined that for $y > \delta_t$, the fluid temperature is unmeasurably close to T_∞. The magnitude of δ_t is a function of the flow boundary layer thickness δ and the Prandtl number. It can be estimated in the same way as δ is estimated, by equating the depletion of sensible enthalpy of the fluid with the heat gained by the wall. Consider a steady flow.

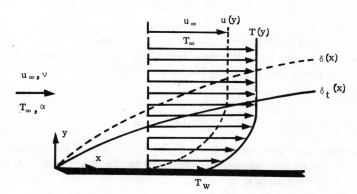

Figure 3.6 Flow and thermal boundary layers on a flat plate

The flow through the thermal boundary layer cross section at $x = l$ is $(\delta_t \cdot 1)(\rho u_\infty/2)(\delta_t/\delta)$. At $x = 0$, this mass flow brings in an enthalpy of

$$(\delta_t \cdot 1)(\rho u_\infty/2)(\delta_t/\delta) C_p T_\infty.$$

At $x = l$, the enthalpy carried out is

$$(\delta_t \cdot 1)(\rho u_\infty/2)(\delta_t/\delta) C_p (T_\infty + T_W)/2.$$

The difference between these two terms is the enthalpy loss of the fluid. The heat flux into the wall is given by Fourier's law of conduction. The total heat flow from $x = 0$ to $x = l$ is

$$(l \cdot 1)(-K\, \partial T/\partial y)_W \approx l \cdot 1 \cdot K(T_\infty - T_W)/(\delta_t/2)$$

Equating the enthalpy loss of fluid with the heat gain by the wall,

$$\delta_t \cdot 1 \cdot \frac{\rho u_\infty}{2} \cdot \frac{\delta_t}{\delta} \cdot \frac{C_p(T_\infty - T_W)}{2} \approx 2l \cdot 1 \cdot K \frac{(T_\infty - T_W)}{\delta_t}$$

Rearranging,

$$\frac{\delta_t^3}{\delta} \approx \frac{8lK}{\rho u_\infty C_p}$$

Recalling the definition of thermal diffusivity α as $K/\rho C_p$,

$$\left(\frac{\delta_t}{\delta}\right)^3 \approx \frac{8l\alpha}{u_\infty \delta^2}$$

Substituting Eq. 3.30 in the right hand side,

$$\frac{\delta_t}{\delta} \approx \frac{1}{\sqrt[3]{\text{Pr}}} \tag{3.32}$$

Equation 3.32 shows that $\delta_t = \delta$ if $\alpha = v$. The thermal boundary layer thickness will be greater than the flow boundary layer thickness if $\alpha > v$, i.e. Pr < 1. For gases, this is usually the case. Specifically,

$$\delta_t \approx \sqrt{\frac{8vl}{u_\infty}} \cdot \sqrt[3]{\frac{\alpha}{v}} \tag{3.33}$$

The heat flux into the wall is given by Fourier's law as the product of the fluid conductivity and temperature gradient in the fluid normal to the wall. This flux approximately is

$$\dot{q}'' = -K \left.\frac{\partial T}{\partial y}\right|_W \approx \frac{K(T_\infty - T_W)}{\delta_t} \approx \frac{K(T_\infty - T_W)}{\sqrt{\frac{8vl}{u_\infty}} \sqrt[3]{\frac{\alpha}{v}}} \tag{3.34}$$

K/δ_t is often called the *heat transfer coefficient*, denoted by h, so that

$$\dot{q}'' = h(T_\infty - T_W) \quad \text{cal/cm}^2/\text{sec} \tag{3.35}$$

$$h \equiv K/\delta_t \quad \text{cal/cm}^2/\text{sec}/^\circ\text{C} \tag{3.36}$$

Equation 3.35 is similar to the well-known Ohm's law in electricity. $\dot{q}''$ is analogous to the electric current. $(T_\infty - T_W)$ is the driving force analogous to the voltage potential and $1/h$ is analogous to the electrical resistance.

Physics of Combustion

Comparing Eqs. 3.34 and 3.35, the heat transfer coefficient is given by

$$h \approx \frac{K}{\sqrt{\dfrac{8vl}{u_\infty}} \sqrt[3]{\dfrac{\alpha}{v}}} \tag{3.37}$$

It is customary to express h as a fraction of K/l where l is the characteristic dimension of the flat plate. The fraction hl/K is dimensionless and is named as *Nusselt number*.

$$\text{Nu}_l \equiv \frac{hl}{K}$$

From Eq. 3.37,

$$\text{Nu}_l \approx \sqrt{\frac{u_\infty l}{8v}} \cdot \sqrt[3]{\frac{v}{\alpha}} \approx 0.354\, \text{Re}_l^{1/2}\, \text{Pr}^{1/3} \tag{3.38}$$

An exact solution of the problem gives the result as

$$\text{Nu}_l = 0.332\, \text{Re}_l^{1/2}\, \text{Pr}^{1/3} \tag{3.39}$$

Once again we can see that our approximate treatment gives acceptable accuracy in the final result. Note also that because $h \equiv K/\delta_t$, and $\text{Nu}_l \equiv hl/K$, the Nusselt number is simply the ratio l/δ_t of the characteristic length to the local thermal boundary layer thickness.

3.5 Heat Transfer in Free Convection

In Section 3.4 we considered the problem of heat transfer in forced flow. Assuming that viscosity and density are independent of temperature we first solved the problem of flow (Section 3.3) and the flow solution so obtained is used in heat transfer solution. Now consider as shown in Figure 3.7 a situation in which a different kind of flow occurs. In Figure 3.7, a thin flat plate whose temperature is T_W is shown standing vertically in an infinite stagnant ($u_\infty = 0$) medium of gas at a temperature $T_\infty (< T_W)$. If one measures the temperature of the fluid as a function of the distance from the plate surface, one discovers that drastic increase in temperature occurs (from T_∞ to T_W) in a relatively thin zone close to the wall. By perfect gas law, the density of the heated gas in the vicinity of the wall can be shown to be lower than the ambient density.

$$\rho = P/RT; \qquad \rho_\infty = P/RT_\infty$$

Due to the decreased density in the proximity of the wall, the local fluid rises upward, thus inducing an upward flow. Such an upward flow near a

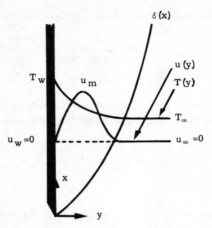

Figure 3.7 Free convective boundary layer about a vertical flat plate

heated solid object is called *free convective* flow. If one measures the upward velocity, one discovers that it is zero at $y = 0$, reaches a maximum somewhere in the thermal boundary layer and falls to the ambient value ($u_\infty = 0$) at distances far away from the wall.

It is logical to assume from the above description that the flow and thermal boundary layers are roughly equal in thickness. It is also clear that the flow field is strongly dependent upon the temperature field. Therefore, unlike in forced flow heat transfer, solution of the flow field has to be obtained simultaneously with that of heat transfer.

Let us approximate the profiles of temperature and velocity linearly as shown in Figure 3.8. Let u_m be the maximum velocity in the boundary layer. This velocity increases with height x in some yet undetermined fashion. To balance such an increase, fluid is entrained at the edge of the boundary layer from the ambient reservoir. The upward flow at $x = 0$ is zero because $u_m = 0$. At any section $x = l$, the upward flow is equal to $(u_m/2)(\delta \cdot 1)$, where $(u_m/2)$ is approximately the mean flow velocity through the section and $(\delta \cdot 1)$ is the area of flow. This flow is provided with entrainment which is equal to $v(l \cdot 1)$ where $(l \cdot 1)$ is the area of entrainment. The conservation of mass is satisfied by equating these two flows.

$$v \approx u_m \delta / 2l \tag{3.40}$$

Similarly for conservation of energy, heat convected upward by the flow is equal to heat lost by the wall by conduction.

$$\rho \frac{u_m}{2} \cdot (\delta \cdot 1) \cdot C_p \cdot (T_W - T_\infty) \approx (l \cdot 1) K \frac{(T_W - T_\infty)}{\delta/2} \tag{3.41}$$

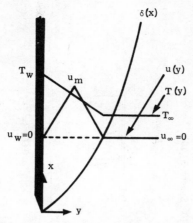

Figure 3.8 Approximate profiles of a free convective boundary layer about a vertical flat plate

In this equation, the left hand side is approximately equal to the difference between the enthalpy existing in the stream at $x = l$ and that entering into the stream at $x = 0$ whereas the right hand side is roughly equal to the heat conducted away from the wall. Resolving Eq. 3.41 for u_m

$$u_m \approx \frac{4\alpha l}{\delta^2} \qquad (3.42)$$

Considering the flow as dynamically in equilibrium, the buoyancy force is equal approximately to the viscous drag on the wall. The buoyancy force per unit volume of the boundary layer $\approx g(\rho_\infty - \rho)$; g being acceleration of gravity. Volume of the boundary layer from $x = 0$ to $x = l$ is approximately $(\delta \cdot 1 \cdot l/2)$. The total buoyancy hence $\approx g(\rho_\infty - \rho)(\delta \cdot 1 \cdot l/2)$. Taking $u_m/2$ to be roughly the mean flow velocity, the viscous drag on the wall of area $(l \cdot 1)$ is approximately $(l \cdot 1)\mu(u_m/2)/(\delta/2)$. Equating the buoyancy and viscous effects, the equation of motion gives

$$\delta^2 \approx \frac{2\nu u_m \rho}{g(\rho_\infty - \rho)} \qquad (3.43)$$

Substituting Eq. 3.42 for u_m in Eq. 3.43 and resolving for δ

$$\frac{\delta}{l} \approx 1.681 \left(\frac{\alpha}{\nu}\right)^{1/4} \left(\frac{\nu^2}{gl^3} \frac{\rho}{(\rho_\infty - \rho)}\right)^{1/4} \qquad (3.44)$$

It can be seen from Eq. 3.44 that δ increases linearly with $l^{1/4}$, $\alpha^{1/4}$, $\nu^{1/4}$ and $[\rho/(\rho_\infty - \rho)]^{1/4}$. The nondimensional group of variables $[gl^3(\rho_\infty - \rho)/(\rho\nu^2)]$ which appears in the right hand side of Eq. 3.44 is called *Grashof's number*.

TABLE 3.4

Some Recommended Heat Transfer Results for Fluid Flows

(Properties are recommended to be evaluated at some mean temperature)

$$\mathrm{Re} \equiv \frac{ux}{\nu}, \quad \mathrm{Nu}_x \equiv \frac{hx}{k}, \quad \overline{\mathrm{Nu}}_x \equiv \frac{\bar{h}x}{k}, \quad \mathrm{Gr} \equiv \frac{gx^3\beta\Delta T}{\nu^2}, \quad \mathrm{Pr} \equiv \frac{\nu}{\alpha}$$

Forced Convection

Critical Reynolds Numbers: Flat Plate: $3.2 \cdot 10^5$, Cylinder: $5 \cdot 10^5$, Sphere: $3 \cdot 10^5$.

Nature of Flow	Diagram	Result
1. *Laminar* flow parallel to a flat plate heated over $x \geq x_0$. $\mathrm{Re} = u_\infty x/\nu < 3 \cdot 10^5$	$U_\infty, T_\infty \rightarrow$; δ, δ_t, T_w, x_0	$\mathrm{Nu}_x = \dfrac{0.332\,\mathrm{Re}^{1/2}\mathrm{Pr}^{1/3}}{\sqrt[3]{1 - \left(\dfrac{x_0}{x}\right)^{3/4}}}$
2. Same as above but $x_0 = 0$	$U_\infty, T_\infty \rightarrow$; δ, δ_t, T_w	$\mathrm{Nu}_x = 0.332\,\mathrm{Re}^{1/2}\mathrm{Pr}^{1/3}$ $\overline{\mathrm{Nu}}_x = 0.664\,\mathrm{Re}^{1/2}\mathrm{Pr}^{1/3}$
3. Same as above but flow is *turbulent*.	$U_\infty, T_\infty \rightarrow$; δ, δ_t, T_w	$\overline{\mathrm{Nu}}_x = 0.037\,\mathrm{Re}^{4/5}\mathrm{Pr}^{1/3}$

TABLE 3.4 (Continued)

Nature of Flow	Diagram	Result
4. *Laminar Plane* stagnation point flow: $Re \equiv ux/\nu$	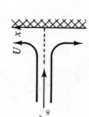	$Nu_x = 0.57\, Re^{1/2} Pr^{2/5}$
5. Axisymmetric stagnation point flow. $Re \equiv uR/\nu$. R = Radius of curvature.	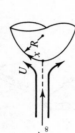	$Nu_x = 0.93\, Re^{1/2} Pr^{2/5}$
6. Flow around a sphere. $Re \equiv u_\infty d/\nu$ $17 < Re < 7 \cdot 10^4$		$\overline{Nu}_d = 0.37\, Re^{0.6}$
7. Same as above, General Equation		$\overline{Nu}_d = 2 + 0.6\, Re^{1/2} Pr^{1/3}$
8. Flow around a cylinder $Re \equiv u_\infty d/\nu$ $\quad 1 < Re < \quad\;\; 4$ $\quad 4 < Re < \quad\; 40$ $\quad 40 < Re < \quad 4{,}000$ $\;4{,}000 < Re < \;40{,}000$ $40{,}000 < Re < 250{,}000$	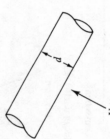	$\overline{Nu}_d = 0.891\, Re^{1/3}$ $\quad\quad = 0.821\, Re^{0.385}$ $\quad\quad = 0.615\, Re^{0.466}$ $\quad\quad = 0.174\, Re^{0.618}$ $\quad\quad = 0.024\, Re^{0.805}$
9. Same as above		$\overline{Nu}_d = 0.35 + 0.47\, Re^{1/2} Pr^{0.3}$

TABLE 3.4 (Continued)

Free Convection

Critical Value of $(GrPr) \approx 10^9$

Nature of Flow	Diagram	Result
10. *Laminar* Free convection on a vertical flat plate. $Gr \equiv gx^3\beta \Delta T/\nu^2$ $10^4 < GrPr < 10^9$		$\overline{Nu}_x = 0.59(GrPr)^{1/4}$
11. Same as above but *turbulent* flow.		$\overline{Nu}_x = 0.13(GrPr)^{1/3}$
12. Same as 10 and 11 but the plate is inclined.		Use above equations but $Gr \equiv \dfrac{gx^3\beta \Delta T}{\nu^2}\cos\alpha$
13. *Laminar* hot horizontal plate heated face up $Gr \equiv gl^3\beta \Delta T/\nu^2$ $10^5 < GrPr < 2 \cdot 10^7$		$\overline{Nu}_l = 0.54(GrPr)^{1/4}$

TABLE 3.4 (Continued)

Nature of Flow	Diagram	Result
14. Same as above but *turbulent* flow $2 \cdot 10^7 < \text{GrPr} < 3 \cdot 10^{10}$		$\overline{\text{Nu}}_l = 0.14(\text{GrPr})^{1/3}$
15. Horizontal plate heated face down. $\text{Gr} \equiv gl^3\beta\,\Delta T/\nu^2$ $3 \cdot 10^5 < \text{GrPr} < 3 \cdot 10^{10}$		$\overline{\text{Nu}}_l = 0.27(\text{GrPr})^{1/4}$
16. *Laminar* free convection around a heated horizontal cylinder $\text{Gr} \equiv gd^3\beta\,\Delta T/\nu^2$ $10^3 < \text{GrPr} < 10^9$	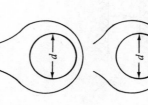	$\overline{\text{Nu}}_d = 0.525(\text{GrPr})^{1/4}$
17. Free convection around a sphere. $\text{Gr} \equiv gd^3\beta\,\Delta T/\nu^2$ $\text{Gr}^{1/4}\text{Pr}^{1/3} < 200$		$\overline{\text{Nu}}_d = 2 + 0.6\,\text{Gr}^{1/4}\text{Pr}^{1/3}$
18. Same as above but $3 \cdot 10^8 < \text{GrPr} < 5 \cdot 10^{11}$		$\overline{\text{Nu}}_d = 0.098(\text{GrPr})^{0.345}$

TABLE 3.4 (Continued)

Nature of Flow	Diagram	Result
19. Horizontal parallel plates. Lower plate hot. $Gr \equiv g\Delta^3 \beta \Delta T/\nu^2 < 10^3$		$Nu_\Delta = 1$
20. Same as above but $10^4 < Gr < 3.2 \cdot 10^5$		$Nu_\Delta = 0.21(GrPr)^{1/4}$
21. Same as above but $3.2 \cdot 10^5 < Gr < 10^7$		$Nu_\Delta = 0.075(GrPr)^{1/3}$
22. Vertical parallel plates. $Gr < 2 \cdot 10^3$		$Nu_\Delta = 1$
23. Same as above but $2 \cdot 10^3 < Gr < 2.1 \cdot 10^5$		$Nu_\Delta = 0.2(GrPr)^{1/4}$
24. Same as above but $2.1 \cdot 10^5 < Gr < 1.1 \cdot 10^7$		$Nu_\Delta = 0.071(GrPr)^{1/3}$

Physics of Combustion

It gives the magnitude of buoyant force relative to the viscous force. Hence

$$\frac{\delta}{l} \approx 1.681 \, \text{Pr}^{-1/4} \, \text{Gr}_l^{-1/4} \qquad (3.44a)$$

From Eqs. 3.40, 3.42 and 3.44, the variation of the maximum velocity u_m and the entrainment velocity v with respect to x can be obtained as

$$u_m \approx \frac{4\alpha l}{\delta^2} \approx 1.42 \frac{\alpha}{l} \, \text{Pr}^{1/2} \, \text{Gr}_l^{1/2} \qquad (3.42a)$$

$$v \approx \frac{u_m \delta}{2l} \approx \frac{2\alpha}{\delta} \approx 1.19 \frac{\alpha}{l} \, \text{Pr}^{1/4} \, \text{Gr}_l^{1/4} \qquad (3.40a)$$

u_m increases linearly with $l^{1/2}$ and v decreases linearly with $l^{1/4}$.

The heat transfer from the wall is given by a heat transfer coefficient $h \equiv K/\delta$. The Nusselt number is

$$\text{Nu}_l \equiv \frac{hl}{K} \equiv \frac{l}{\delta} \approx 0.595 \, \text{Pr}^{1/4} \, \text{Gr}_l^{1/4} \qquad (3.45)$$

The exact and more involved solution gives

$$\text{Nu}_l = 0.555 \, \text{Pr}^{1/4} \, \text{Gr}_l^{1/4} \qquad (3.46)$$

Table 3.4 gives a dictionary of heat transfer results for various flow situations.

3.6 Comments on Mass Transfer Problems

In the problems considered in Sections 3.3 to 3.5, the composition of the fluid is assumed to be spatially uniform. Suppose now that temperature is uniform in the entire space. Let spatial nonuniformities of composition be introduced into the flow field. Mass transfer occurs as a consequence of these nonuniformities. Define mass transfer coefficient h_D and mass transfer Nusselt number Nu_D as below.

$$h_D \equiv \frac{\rho D}{\delta_D} \qquad (\text{gm/cm}^2/\text{sec})$$

$$\text{Nu}_{D_l} \equiv \frac{h_D l}{\rho D} \equiv \frac{l}{\delta_D}$$

δ_D here is the thickness of the concentration boundary layer. The mass transfer results will be the same as Eqs. 3.39 and 3.46 except that the Prandtl number, ν/α, is replaced by the Schmidt number, ν/D.

For forced convection flat plate mass transfer problem,

$$\mathrm{Nu}_{D_i} = 0.332 \, \mathrm{Sc}^{1/3} \, \mathrm{Re}_l^{1/2} \qquad (3.47)$$

For free convection mass transfer from a vertical flat plate,

$$\mathrm{Nu}_{D_i} = 0.555 \, \mathrm{Sc}^{1/4} \, \mathrm{Gr}_l^{1/4} \qquad (3.48)$$

3.7 Turbulence

In the preceding sections we have introduced some important concepts of fluid flow. Flows in which adjacent layers of fluid slide past one another orderly and smoothly without extensive mixing is called *laminar* flow. The only mixing possible in laminar flow is due to molecular diffusion. The velocity, temperature and concentration profiles measured in laminar flow with a high sensitivity instrument will be smooth. In the preceding sections we considered laminar flow problems.

At very high Reynolds or Grashof's numbers (i.e., at high velocities of low viscosity fluids in wide passage ways or over large bodies) the flow becomes *turbulent*. In turbulent flow, pockets of the fluid (called *eddies*) move randomly back and forth and across the adjacent fluid layers. The flow no longer remains smooth and orderly. At any station it fluctuates with time randomly. If the velocity, temperature and composition profiles are measured with a high sensitivity instrument (such as a hot wire anemometer) the resultant profiles indicate much fluctuation as shown in Figure 3.9. If, on the other hand, the profiles are measured with an averaging instrument (such as a pitot static tube to measure velocity) the time-averaged profiles look somewhat like that shown in Figure 3.9.

From Figure 3.9 it can be seen that turbulence reduces the boundary layer thickness as a consequence of enhanced mixing due to eddies. The heat transfer and wall friction therefore increase.

In fully developed turbulent flow the molecular mixing (i.e. laminar "diffusion") may become negligibly small compared to the eddy mixing. In order to deal with turbulent flow problems (and most of the combustion problems are turbulent) we need to know the turbulent transport properties. Following the pattern of thought underlying Eq. 3.1, we can write for turbulent shear stress

$$\tau_T = -\rho\varepsilon\frac{\partial \bar{u}}{\partial y} \qquad (3.49)$$

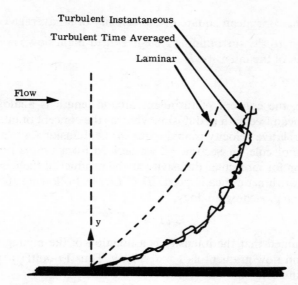

Figure 3.9 Laminar and turbulent velocity profiles on a flat plate

where ε is turbulent kinematic viscosity (units cm^2/sec) and $\bar{u}$ is time-averaged velocity. Similarly for heat and mass fluxes,

$$\dot{q}_T'' = -\rho C_p \varepsilon_t \frac{\partial \bar{T}}{\partial y} \tag{3.50}$$

$$\dot{W}_{A_T}'' = -\rho \varepsilon_D \frac{\partial \bar{Y}_A}{\partial y} \tag{3.51}$$

ε_t and ε_D are respectively called turbulent thermal diffusivity and species diffusivity. $\varepsilon/\varepsilon_t$ is called turbulent Prandtl number. Likewise, $\varepsilon/\varepsilon_D$ is termed as turbulent Schmidt number. It is customary to assume that $\Pr_T \approx \mathrm{Sc}_T \approx 1$. Unlike molecular diffusivities v, α and D, the turbulent diffusivities ε, ε_t and ε_D depend not only on the nature of the fluid but also on the flow situation. That is, they are not the properties of the fluid alone. Consider the following hypothesis to substantiate this concept.

Turbulence, classically, is described by two properties of an average eddy—the *scale* and the *intensity*. The scale l' of turbulence is defined as the distance an eddy would move through before "dissolving" ("dissipating" or "disappearing") to lose its identity. It is of the same order of magnitude as the size of the eddy itself. Scale thus is sometimes known as the "mixing length." The intensity of turbulence is a measure of the "violence" (or "vehemence") of the eddies, i.e., the time-mean fluctuation of the velocity in the fluid. The root mean square of the velocity of an average eddy pocket, $\sqrt{\overline{u'^2}}$, relative to the surrounding time-averaged main flow is often used as an indication of the intensity, I.

$$I \approx \sqrt{\overline{u'^2}}$$

In principle, the concept of turbulent mixing length is analogous to the molecular mean free path in diffusion whereas the concept of intensity as the mean eddy relative velocity is analogous to the diffusional arithmetic mean speed of a molecule. In Section 3.2 we derived from simple kinetic theory an expression for molecular diffusivity as the product of the mean free path length and arithmetic mean speed. ($D \approx LU/3$). Following the suggestion implicit in the preceding analogy,

$$\varepsilon \approx l'I \approx l'\sqrt{\overline{u'^2}}$$

Prandtl assumed that the intensity is a function of the mixing length itself and the mean flow gradient as $I \approx l' \cdot \partial \bar{u}/\partial y$. Thus Prandtl's expression for turbulent diffusivity is

$$\varepsilon \approx l'^2 \frac{\partial \bar{u}}{\partial y}$$

It is conceivable that the eddies will be suppressed near the solid walls of the flow system. ε, then will be zero at the walls. This fact could be taken into account by assuming that the mixing length is proportional to the distance from the wall.

$$\varepsilon \approx cy^2 \frac{\partial \bar{u}}{\partial y} \qquad (3.52)$$

This expression shows that the turbulent viscosity is a function of the fluid (c), of the geometry of flow apparatus (y^2) and the flow ($\partial \bar{u}/\partial y$) itself. (Recall that the molecular transport coefficients are functions of the fluid alone.) This fact complicates studies of turbulent flows. Many researchers attempted turbulent flow theories in which ε is formed of different combinations of the distance from the wall y and the derivatives of the mean flow $\partial^n \bar{u}/\partial y^n$. To date the methods used in turbulent combustion studies are predominantly of empirical nature.

3.8 Heat and Mass Transfer with Non-negligible Interfacial Velocities

(a) *Heat Transfer*

In Section 3.1, Fourier law of heat conduction is postulated by considering a temperature gradient between two infinite parallel plates the interspace δ_t

Physics of Combustion

between them being filled with a stagnant fluid, B. Consider now the following modification as shown in Figure 3.10. Let the plates be porous. Let a feeble flow of (ρv) gms/cm^2/sec of the fluid B be forced from the upper porous plate (temperature T_∞) normal to the plate surface. Let the fluid be sucked into the lower plate at the same rate so that a steady state flow is attained. The movement of the medium causes an enhancement of heat flow because heat now is carried not only by conduction but also by an extra mechanism—the gross movement of the fluid in a nonuniform field of temperature. As shown in Figure 3.10, the amount of fluid leaving a unit area of the upper surface in a unit time is (ρv) gms. If C_p is the specific heat of the fluid, the sensible enthalpy of this amount of fluid at the upper surface is equal to $\rho v C_p (T_\infty - T_a)$ cal. Here T_a is some arbitrary datum temperature.

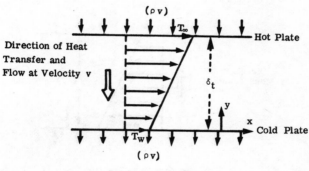

Figure 3.10 Heat transfer with non-negligible interfacial (normal) velocity

The same amount of fluid upon reaching the lower plate (whose temperature is T_W) has a sensible enthalpy of $\rho v C_p (T_W - T_a)$. The difference between these two sensible enthalpies is the enhancement of heat transfer. The net heat transfer hence is

$$\dot{q}'' \approx \frac{K(T_\infty - T_W)}{\delta_t} + \rho v C_p (T_\infty - T_W)$$

Recalling the definition of the heat transfer coefficient as $h \equiv K/\delta_t$,

$$\dot{q}'' = h(T_\infty - T_W)(1 + \text{Pe}) \tag{3.53}$$

where $\text{Pe} = (\rho v C_p \delta_t / K)$ is a nondimensional criterion known as Peclet number which is a measure of the convection heat flux in units of the conductive flux.

For most flow situations encountered in combustion, Pe ≪ 1. Hence

$$\dot{q}'' \approx h(T_\infty - T_W) \qquad (3.54)$$

(b) *Mass Transfer*

In an isothermal parallel plate situation Fick's law is postulated in Section 3.1 by considering a gradient of composition. Care was taken there in such a way that the flow of species A normal to the plate surface is so feeble as not to disturb the stagnant medium. However, in most combustion problems such a feeble flow is not possible. As a consequence, convective transfer of species A across the layer of fluid B augments diffusive transfer. An amount (ρv) gms of the mixture fluid at the upper wall in Figure 3.11 contains an amount of

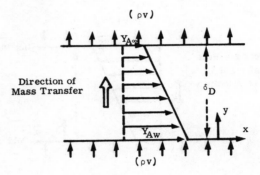

$$\dot{W}_A'' = \rho D_A (Y_{Aw} - Y_{A\infty})/\delta_D + (\rho v)(Y_{Aw} - Y_{A\infty})$$

Figure 3.11 Mass transfer with non-negligible interfacial velocity

species A equal to $(\rho v)Y_{A\infty}$ whereas an equal amount of the mixture at the lower wall contains $(\rho v)Y_{AW}$ gms of A. Thus the convected mass of A per unit area per unit time is equal to $(\rho v)(Y_{AW} - Y_{A\infty})$ so that the total rate of transfer is

$$\dot{W}_A'' \approx \rho D \frac{(Y_{AW} - Y_{A\infty})}{\delta_D} + (\rho v)(Y_{AW} - Y_{A\infty})$$

Recalling the definition of the mass transfer coefficient $h_D \equiv \rho D/\delta_D$, and defining mass transfer Peclet number as $\text{Pe}_D \equiv \rho v \, \delta_D/\rho D$,

$$\dot{W}_A'' = h_D(Y_{AW} - Y_{A\infty})(1 + \text{Pe}_D) \qquad (3.55)$$

Unlike in the case of heat transfer, Pe_D is not negligible.

Physics of Combustion

3.9 Conservation Equations

In deducing the results of heat, mass and momentum transfer in sections 3.3 to 3.6, implicitly used are the principles of conservation of mass, momentum, energy and species. Differential equations describing these conservation principles are customarily used to give a precise description of the problem. Because in the following chapters we will be dealing with several combustion problems described by boundary layer differential equations, we shall briefly derive these equations here.

(a) *Conservation of Mass (Continuity Equation)*

Consider an element of volume $\Delta x \cdot \Delta y \cdot \Delta z$ located at the origin as shown in Figure 3.12. The components of velocity in x, y, and z directions respectively

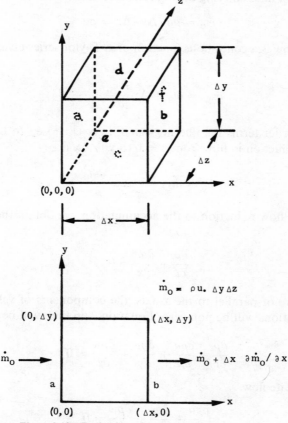

Figure 3.12 Derivation of the conservation equations

are u, v and w. Let the flow be in x-direction so that the components of velocity v and w in y and z directions respectively are zero. Let the control volume be so chosen that $\Delta y \cdot \Delta z$ is unity. The product $\Delta y \cdot \Delta z$ is the area of the two faces which are normal to the x-axis. (Face labeled "a" is at $x = 0$ and that labeled "b" is at $x = \Delta x$.) Let ρ be the density of the fluid.

Now, certain amount of the fluid enters the control volume through the face "a," and certain of this entered amount leaves through the face "b." The balance of the fluid accumulates in the control volume. The rate of accumulation is given by the time rate of increase in the density multiplied by the volume.

$$\text{Accumulation} = \frac{\partial \rho}{\partial t} \cdot \Delta x \cdot \Delta y \cdot \Delta z = \frac{\partial \rho}{\partial t} \Delta x$$

The amount of fluid entering through the face at $x = 0$ is

$$\dot{m}_0 = \rho u \cdot \Delta y \cdot \Delta z = \rho u$$

If the mass flow is a continuous function of x, Taylor series gives the exiting flow as

$$\dot{m}_{\Delta x} = \dot{m}_0 + \Delta x \frac{\partial \dot{m}_0}{\partial x} + \cdots$$

The higher order terms may be neglected if Δx is chosen to be extremely small. The reduction in flow from $x = 0$ to $x = \Delta x$ is

$$\dot{m}_0 - \dot{m}_{\Delta x} = -\Delta x \cdot \frac{\partial \dot{m}_0}{\partial x}$$

Equating the flow reduction to the accumulation, we obtain the continuity equation.

$$\frac{\partial \rho}{\partial t} + \frac{\partial \rho u}{\partial x} = 0 \tag{3.56}$$

If the flow is not parallel to the x-axis, the components of velocity in the y and z directions will be non-zero. Equation 3.56 then will be modified to

$$\frac{\partial \rho}{\partial t} + \frac{\partial \rho u}{\partial x} + \frac{\partial \rho v}{\partial y} + \frac{\partial \rho w}{\partial z} = 0 \tag{3.57}$$

For steady state flow,

$$\frac{\partial \rho u}{\partial x} + \frac{\partial \rho v}{\partial y} + \frac{\partial \rho w}{\partial z} = 0 \tag{3.58}$$

Physics of Combustion

For an incompressible steady flow,

$$\frac{\partial u}{dx} + \frac{\partial v}{\partial y} + \frac{\partial w}{\partial z} = 0 \qquad (3.59)$$

For the 1-dimensional situation

$$\frac{\partial u}{\partial x} = 0$$

This means for constant density one dimensional flow, the velocity u is independent of x.

(b) *Conservation of Momentum: (Equation of Motion)*

In fluid flow an element of the fluid is subjected to various kinds of forces. Forces can broadly be classified into two categories—surface forces and volumetric forces. Surface forces may be tangential (for example: viscous drag) or normal (example: hydrostatic pressure, surface tension, etc.). Volumetric (or body) forces most commonly encountered in fluid flow are gravitational or magnetic.

Most combustion problems involve geometries which enable us to neglect certain terms in the equations of conservation without introducing considerable errors. In the neighborhood of a wall, certain terms in the conservation equations are negligibly small whereas far from the wall certain other terms are negligible. In general, since the gradients in velocity, temperature and composition vanish far from the wall, the flow there can be considered as inviscid, non-conducting and non-diffusing. In the close proximity of the wall, however, the flow is viscous (conducting and diffusing). The notion of breaking complex flow problems into two regions, an inviscid region remote from the wall and a viscous region close to the wall, was first introduced by Ludwig Prandtl.

Prandtl's boundary layer equations are deduced from the following observations.

(i) In the boundary layer, velocities normal to the wall are small compared to those parallel to the wall.
(ii) Gradients in velocity, temperature and composition parallel to the wall are small compared to those normal to the wall.
(iii) Pressure gradients normal to the wall are vanishingly small.

Continuity equation for Prandtl's boundary layer becomes

$$\frac{\partial \rho u}{\partial x} + \frac{\partial \rho v}{\partial y} = 0$$

The conservation equations are commonly derived by first making the above boundary layer assumptions of Prandtl. Specifically the equation of motion is obtained by equating the sum of viscous, gravitational and pressure forces to the rate of change of momentum of the control volume. (Newton's Law of motion.) Consider a steady state flow. Momentum in x-direction entering through the face marked "a" in Figure 3.12 is

$$(\rho u) \cdot u \cdot (\Delta y \cdot \Delta z)$$

Momentum in x-direction exiting from the control volume through face marked "b" in Figure 3.12 is given by Taylor's series as

$$\rho u^2 \, \Delta y \, \Delta z + \Delta x \frac{\partial}{\partial x}(\rho u^2 \, \Delta y \, \Delta z)$$

Momentum in x-direction entering into the control volume through the face "c" is

$$(\rho u) \cdot v \cdot (\Delta x \, \Delta z)$$

The x-momentum exiting through face "d" by Taylor series is

$$\rho u v \, \Delta x \, \Delta z + \Delta y \frac{\partial}{\partial y}(\rho u v \, \Delta x \, \Delta z)$$

Viscous forces on face "c" are given by

$$-\mu \frac{\partial u}{\partial y} \cdot \Delta x \, \Delta z$$

On face "d", by Taylor series,

$$-\mu \frac{\partial u}{\partial y} \Delta x \, \Delta z + \Delta y \frac{\partial}{\partial y}\left(-\mu \frac{\partial u}{\partial y} \Delta x \, \Delta z\right)$$

The pressure force on face "a" is

$$P \, \Delta y \, \Delta z$$

On face "b", by Taylor series, it is

$$P \, \Delta y \, \Delta z + \Delta x \frac{\partial}{\partial x}(P \, \Delta y \, \Delta z)$$

Neglecting body forces, the equation of motion is given by equating the sum of all the above terms to zero.

$$\frac{\partial}{\partial x} \rho u^2 + \frac{\partial}{\partial y} \rho u v - \frac{\partial}{\partial y} \mu \frac{\partial u}{\partial y} + \frac{\partial P}{\partial x} = 0$$

Physics of Combustion

By differentiating the first two terms by chain rule and applying the continuity equation, we obtain

$$\rho u \frac{\partial u}{\partial x} + \rho v \frac{\partial u}{\partial y} = \frac{\partial}{\partial y}\left(\mu \frac{\partial u}{\partial y}\right) - \frac{\partial P}{\partial x} \qquad (3.60)$$

A similar derivation for conservation of momentum in y-direction gives

$$\frac{\partial P}{\partial y} = 0 \qquad (3.61)$$

(c) Conservation of Energy

Neglecting radiation, frictional dissipation (i.e. generation of heat due to friction among adjacent fluid layers) and work terms, we shall here derive the energy equation for a boundary layer. Consider a steady state flow. Consider the faces marked "a", "b", "c" and "d" in the control volume shown in Figure 3.12. Through these faces heat flows into or out of the control volume by convection and conduction. Heat may be accumulated in the control volume the consequence being a rise in its temperature as a function of time. If we assume a steady state ($\partial/\partial t = 0$) the accumulation is zero. Furthermore, heat may either be produced or absorbed due to chemical reactions in the control volume.

The amount of heat entering through face "a" is

$$(\rho u \cdot \Delta y \, \Delta z \cdot C_p T) + \left(-K \frac{\partial T}{\partial x}\right) \cdot \Delta y \, \Delta z$$

The amount of heat existing through face "b" is equal to the above quantity plus the change in the above quantity in a distance Δx.

$$\left[(\rho u \cdot \Delta y \, \Delta z \cdot C_p T) + \left(-K \frac{\partial T}{\partial y}\right) \cdot \Delta y \, \Delta z\right]$$
$$+ \Delta x \cdot \frac{\partial}{\partial x}\left[(\rho u \cdot \Delta y \, \Delta z \cdot C_p T) + \left(-K \frac{\partial T}{\partial y}\right) \cdot \Delta y \, \Delta z\right]$$

Similarly the heat entering through face "c" and that exiting through face "d" are respectively given by

$$\left[(\rho v \cdot \Delta x \, \Delta z \cdot C_p T) + \left(-K \frac{\partial T}{\partial y}\right) \Delta x \, \Delta z\right]$$

and

$$\left[(\rho v \Delta x \Delta z C_p T) + \left(-K\frac{\partial T}{\partial y}\right)\Delta x \Delta z\right]$$
$$+ \Delta y \cdot \frac{\partial}{\partial y}\left[(\rho v \Delta x \Delta z C_p T) + \left(-K\frac{\partial T}{\partial y}\right)\Delta x \Delta z\right]$$

The amount of heat produced in the control volume of a unit area is given by the product of the rate of reaction (moles of A/cm^3/sec), the volume (cm^3) and the heat of reaction (cal/mole). That is

$$\dot{W}_A''' \cdot \Delta H \cdot (\Delta x \, \Delta y \, \Delta z)$$

Summing up all the above quantities for a heat balance, and recalling that $\partial T/\partial x \ll \partial T/\partial y$ by Prandtl approximation,

$$(\rho u C_p T \Delta y \Delta z) + \left[\rho v C_p T \Delta x \Delta z - K\frac{\partial T}{\partial y}\Delta x \Delta z\right] + \dot{W}_A'''\Delta H \Delta x \Delta y \Delta z$$
$$= (\rho u C_p T \Delta y \Delta z) + \Delta x \frac{\partial}{\partial x}(\rho u C_p T \Delta y \Delta z)$$
$$+ \left[\rho v C_p T \Delta x \Delta z - K\frac{\partial T}{\partial y}\Delta x \Delta z\right]$$
$$+ \Delta y \frac{\partial}{\partial y}\left[\rho v C_p T \Delta x \Delta z - K\frac{\partial T}{\partial y}\Delta x \Delta z\right]$$

Simplifying,

$$\frac{\partial}{\partial x}\rho u C_p T + \frac{\partial}{\partial y}\rho v C_p T - \frac{\partial}{\partial y}K\frac{\partial T}{\partial y} = \dot{W}_A'''\Delta H$$

Differentiating the first two terms and applying continuity equation, we obtain

$$\rho u \frac{\partial C_p T}{\partial x} + \rho v \frac{\partial C_p T}{\partial y} = \frac{\partial}{\partial y}K\frac{\partial T}{\partial y} + \dot{W}_A'''\Delta H \qquad (3.62)$$

(d) *Conservation of Species*

Following the sequence of steps as in energy conservation, one can now derive an equation for conservation of any chemical species in the boundary layer.

$$\rho u \frac{\partial Y_A}{\partial x} + \rho v \frac{\partial Y_A}{\partial y} = \frac{\partial}{\partial y}\rho D_A \frac{\partial Y_A}{\partial y} + \dot{W}_A''' \qquad (3.63)$$

An equation of this form occurs for each relevant species.

(e) Comments

For an unsteady state situation, Eqs. 3.60 to 3.63 become

$$\rho \frac{\partial u}{\partial t} + \rho u \frac{\partial u}{\partial x} + \rho v \frac{\partial u}{\partial y} = \frac{\partial}{\partial y} \mu \frac{\partial u}{\partial y} - \frac{\partial P}{\partial x} \qquad (3.64)$$

$$\rho \frac{\partial v}{\partial t} = -\frac{\partial P}{\partial y} \qquad (3.65)$$

$$\rho \frac{\partial C_p T}{\partial t} + \rho u \frac{\partial C_p T}{\partial x} + \rho v \frac{\partial C_p T}{\partial y} = \frac{\partial}{\partial y} K \frac{\partial T}{\partial y} + \dot{W}_A''' \Delta H \qquad (3.66)$$

$$\rho \frac{\partial Y_A}{\partial t} + \rho u \frac{\partial Y_A}{\partial x} + \rho v \frac{\partial Y_A}{\partial y} = \frac{\partial}{\partial y} \rho D_A \frac{\partial Y_A}{\partial y} + \dot{W}_A''' \qquad (3.67)$$

In addition to Eqs. 3.59 to 3.63, the statement of the problem includes a kinetic rate equation. Given adequate boundary conditions, transport properties and specific heat, these six equations can be solved for the six variables $u, v, P, T, Y_A, \dot{W}_A'''$ as functions of the space coordinates x and y. If a stoichiometric equation is available, the rates of production or consumption of other species can be calculated from $\dot{W}_A'''$.

An important observation can be made here regarding the momentum, energy and species equations. If the pressure is uniform (the case of flat plate flow) and if the flow is non-reactive ($\dot{W}_A''' = 0$) Equations 3.60, 3.62 and 3.63 assume a form similar to one another. The equations then express a balance between the convective flux and diffusional flux. If $Pr = Sc = Le$ and if the boundary conditions can be written in an analogous manner, solution of any of the above three equations would also give the solution of the other two. This parallelism of the conservation equations is often exploited in solving many important combustion problems.

The turbulent flow problems can also be described by the above equations provided the molecular transport properties are replaced by the eddy properties. The solution, of course, becomes more complex because the turbulent transport properties are functions of the flow geometry as well as the fluid under consideration.

3.10 Conclusions

Fluid mechanics become an integral part of any combustion problem. Extensive literature, partly cited in the references of this chapter, is available for further information. Appendix D gives the property values of various gases and liquids pertinent to combustion.

References

Bird, R. B., Stewart, W. E., and Lightfoot, E. N., *Transport Phenomena*, Wiley (1962).
Chemical Rubber Company, *Handbook of Physics and Chemistry*.
Eckert, E. R. G. and Drake, R. M., *Heat and Mass Transfer*, McGraw-Hill (1959).
Fristrom, R. M. and Westenberg, A. A., Molecular Transport Properties for Flame Studies, *Fire Research Abstracts and Reviews*, Vol. 8, No. 3, p. 155 (1966).
Hilsenrath, J., et al., Tables of Thermodynamic Properties of Gases, U.S.N.B.S. Circular No. 564, Washington, D.C. (1955).
Hirschfelder, J. O., Curtiss, O. F., and Bird, R. B., *Molecular Theory of Gases and Liquids*, Wiley (1954).
Industrial and Engineering Chemistry. March issues every year since 1953.
Jeans, J. H., *Kinetic Theory of Gases*, Cambridge University Press (1967).
Jeans, J. H., *The Dynamical Theory of Gases*, 4th edition, Dover (1955).
Reid, R. C. and Sherwood, T. K., *The Properties of Gases and Liquids*, McGraw-Hill (1958).
Schlichting, H., *Boundary Layer Theory*, McGraw-Hill (1960).
Thermodynamic and Transport Properties of Gases, Liquids and Solids. Proceedings of the A.S.M.E. Symposium on Thermal Properties. Touloukian, Y. S., Chairman (1959).
Tsederberg, N. V., *Thermal Conductivity of Gases and Liquids*, translated from Russian, MIT Press (1965).
Westenberg, A. A., Present Status of Information on Transport Properties Applicable to Combustion Research, *Combustion and Flame*, Vol. 1, p. 346 (1957).

CHAPTER 4

Kinetically Controlled Combustion Phenomena

Most of the practical combustion phenomena belong to one of the following three categories:

(i) Phenomena which are primarily controlled by chemical kinetics.
(ii) Phenomena which are primarily controlled by diffusion, flow and other physical mixing processes, and
(iii) Phenomena in which the roles played by chemical kinetics and physical mixing are more or less of equal importance.

Ignition, explosion, extinction and quenching of flames serve as examples of category (i) phenomena. The burning of a gaseous fuel jet, of a liquid fuel spill, spray, or drop, of a carbon sphere and of a candle illustrate the diffusionally controlled combustion phenomena. Flames in a gasoline engine, a Bunsen burner and other situations in which the fuel and oxidant are premixed belong to the third category.

In this chapter we explain briefly the basis upon which a simple categorization such as the one above is possible. The main body of this chapter considers in detail the kinetically controlled combustion phenomena such as ignition, explosion, flammability and blow-off. A true understanding of these phenomena is of obvious importance in many combustion and fire problems.

4.1 Categorization of Combustion Phenomena

Let A, B and C denote respectively the fuel, oxidant and product of combustion which are all in gas phase and are *uniformly* distributed* in a combustion chamber. If the temperature of this uniform mixture is also uniform, the rate of reaction in the vessel at any given instance will be independent of

---

* Such a "uniform" distribution may be accomplished by increasing the molecular mixing or turbulent intensity, say, by mechanical stirring.

location as well. Such a well-mixed homogeneous combustion is said to be *kinetically controlled*. Explicitly, kinetically controlled phenomena are those in which the reaction rate is slow compared to the rates of heat and species diffusion so that the species and temperature have adequate time available to smooth out any spatial nonuniformities.

On the other hand, nature frequently presents phenomena in which the reaction rates are not slow. When reactions are very fast, the spatial non-uniformities of composition and temperature fail to be washed out in the short available time. As a result, gradients of species and temperature are established in space; such gradients cause conduction of heat and diffusion of species towards the regions of lower temperatures and concentrations respectively. That is, the reactants diffuse into the flame zone whereas the products and heat diffuse away from the flame zone. Such a poorly mixed combustion is said to be *diffusion controlled*. Explicitly, diffusion controlled phenomena are those in which the rate of reaction is much faster than the rate of diffusion.

Consider an alternative description. In a given combustion phenomenon, the fuel and oxidant have to be supplied (by flow, diffusion and mixing) to a station where they react chemically. Heat and products have then to be removed from this station physically. There are two characteristic rates involved in this problem—the rate of supply and the rate of consumption. The lowest of these two rates governs the overall speed of the process. In kinetically controlled combustion phenomena, the rate of consumption of the fuel and oxidant by chemical reaction is much smaller than the rate of supply by flow, diffusion or mixing. In diffusion controlled combustion phenomena, the rate of flow, diffusion and mixing is much smaller than the chemical reaction rate.

It is clear from the above description that in kinetically controlled phenomena, the flame occurs more or less uniformly in the entire reaction space whereas in diffusionally controlled phenomena, it is located at some distinct station in the space. This point of distinction is schematically represented in Figure 4.1.

In order to quantitatively estimate whether a given combustion phenomenon is controlled kinetically or diffusionally, let us recall the definition of the half-life time of a simple chemical reaction from Chapter 2. It is the time required by a packet of the reactant mixture to consume half the initial reactant concentration. Table 2.1 gives half-life time for a first order reaction as $t_{1/2} = \ln 2/k_1$; for a second order reaction, $t_{1/2} = 1/C_{AO}k_2$ and for an nth order reaction, $t_{1/2} = (2^{n-1} - 1)/[(n - 1)(C_{AO}^{n-1}k_n)]$. It is obvious that if the reaction is infinitely fast, its half-life time will be infinitesimally short and vice versa. The half-life time thus can be used to indicate the speed of a given simple reaction.

Kinetically Controlled Combustion Phenomena

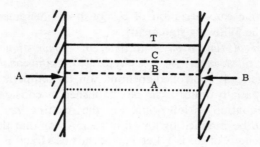

Kinetically Controlled

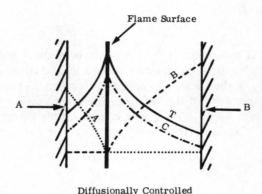

Diffusionally Controlled

Figure 4.1 Kinetically controlled and diffusionally controlled combustion profiles

Consider now a second characteristic of a reacting system. For a nonflow (or completely confined) system, the speed of diffusion can be described in terms of l^2/D where l is the characteristic linear dimension of the system and D is the coefficient of diffusion. The units of l^2/D are those of time. l^2/D is hence the diffusion (or physical) time. In a narrow reaction vessel the contents of which have a high mean mutual diffusivity the diffusion time is extremely short.

If our system involves flow of the reacting gases (examples: a tubular flow reactor, a turbojet combustion chamber, flow in the flame on a burning solid or liquid fuel wall, etc.), the physical time can be defined on the basis of the flow length (length of the tube, height of a wall, etc.) and the flow velocity. l/u, for instance, is called residence time which may be taken as equivalent to the physical time. A high velocity flow through a short combustion duct gives a very short physical time.

Our qualitative categorization of combustion phenomena can now be quantified by the following statement.

If the chemical time $\gg$ physical time, the combustion phenomenon is kinetically controlled; otherwise, it is physically (or diffusionally) controlled.

That in a kinetically controlled phenomenon, the temperature and composition remain spatially uniform can be established by considering the species conservation equation as following. Let the reacting gas mixture of the species A and B be confined by a wall in such a way that the characteristic thickness of the body of gas is δ. Let Y_{AW} be the mass fraction of A at the wall and Y_A be some characteristic mean mass fraction of A in the reacting body of gas. Then the amount of species A transferred from the wall to the gas phase can be written in terms of a mass transfer coefficient h_D (which is proportional to ρD) as

$$\dot{W}_A''' = h_D(Y_{AW} - Y_A) \quad \text{gm/cm}^2/\text{sec.*}$$

The amount of A consumed in the gas phase reaction is given by a simple rate law which we assume as of order unity and of Arrhenius type. $\dot{W}_A''' = k_1' C_A e^{-E/RT}$. ($C_A$ is the mean concentration (gm/cm^3) of species A and T is the mean temperature of the reacting gases.) Expressing C_A in terms of mass fraction, $\dot{W}_A''' = k_1 Y_A e^{-E/RT}$, where k_1 is simply k_1' multiplied by the mixture density. Equating the supply and reaction rates in a volume of $(\delta \cdot 1)/\text{cm}^3$

$$h_D(Y_{AW} - Y_A) = k_1 Y_A e^{-E/RT} \delta \tag{4.1}$$

Solving for Y_A,

$$Y_A = Y_{AW} \bigg/ \left[1 + \frac{k_1 \delta e^{-E/RT}}{h_D}\right] \tag{4.2}$$

The ratio in the denominator is known as *Damkohler number*, Da. Combining Eqs. 4.1 and 4.2,

$$\dot{W}_A''' = h_D Y_{AW} \left[1 - \frac{1}{1 + \frac{k_1 \delta e^{-E/RT}}{h_D}}\right] = \left[\frac{k_1 \delta e^{-E/RT} Y_{AW}}{1 + \frac{k_1 \delta e^{-E/RT}}{h_D}}\right] \tag{4.3}$$

Case (a): Kinetically (or Rate or Reaction or Chemically) Controlled Regime:

The physical rate of supply and mixing is much larger than the chemical rate of reaction if Da $\to$ 0.

$$h_D \gg k_1 \delta e^{-E/RT}$$

---

* The dot-prime notation is used throughout this book. A dot represents per unit time and a prime represents per unit length. Thus $\dot{W}''$ is in gm/cm^2/sec; $\dot{W}'''$ is in gm/cm^3/sec.

Kinetically Controlled Combustion Phenomena

If this is true, Eq. 4.2 indicates that $Y_A \approx Y_{AW}$; that is, the composition remains nearly uniform throughout the reaction space. The rate of depletion of the fuel, then, is given by the reaction rate (Eq. 4.3) as

$$\dot{W}_A''' \approx k_1 \delta Y_{AW} e^{-E/RT} \tag{4.4}$$

The conditions for such a kinetically controlled regime arise when the mixing is high, the diffusion coefficient is high, the gas body thickness is small, the pre-exponential collision factor is small, the activation energy is large and the gas body temperature is low. Combustion in well-mixed reactors, ignition and extinction processes and explosions are some examples of kinetically controlled phenomena.

Case (b): Diffusionally (or Diffusion or Flow or Physically) Controlled Regime:

If $Da \to \infty$, $h_D \ll k_1 \delta e^{-E/RT}$, the physical rate (of supply flow, mixing and diffusion) is much smaller than the chemical rate (of the reaction). Equation 4.2 then gives

$$Y_A \approx \frac{h_D Y_{AW}}{k_1 \delta e^{-E/RT}} \ll Y_{AW}$$

i.e. the gas body mass fraction of species A is negligible compared with the mass fraction at the wall. The rate of fuel depletion is given by the mass transfer rate (Eq. 4.3) as

$$\dot{W}_A''' \approx h_D Y_{AW} \tag{4.5}$$

When the mixing is poor, the flow and diffusion are slow, the gas body thickness is large and the chemical reaction is fast, then diffusion controlled combustion phenomena arise.

Diffusion flames can be converted into kinetic flames by either increasing the chemical time or decreasing the physical time. For example, if we blow out a match or a candle, or send a burning droplet into a fast moving air stream, the flame is extinguished due to the inability of the chemical reaction rate to keep in pace with the increased rate of diffusion of the fuel and oxygen into the reaction zone.* Similarly, if we turn the pilot light up too far on a gas stove (or for that matter, increase the flow velocity of fuel in any diffusion flame), a point is reached where the flame lifts away from the tube through which the fuel issues. Figure 4.2 shows how the temperature and composition gradients are decreased as we go from a relatively slow diffusion rate (Figure 4.2a) to a fast one (Figure 4.2b) and to a critical one (Figure 4.2c) which, if exceeded, causes the extinguishment of the flame.

---

* In fact, it is fair to say that all extinguishment processes are kinetically controlled.

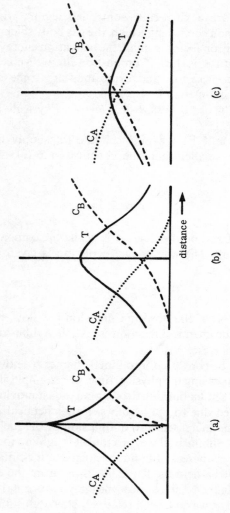

Figure 4.2 Profiles as the diffusion rates are increased from very low (a), to medium (b), to very fast (c)

4.2 Progress of a Combustion Reaction

The rate of a combustion reaction*, as discussed in Chapter 2, depends mainly on the temperature of the reacting gas mixture and the concentrations of the initial reactants and the intermediate and final products. The temperature dependency is often expressed by the Arrhenius' exponential $e^{-E/RT}$ or sometimes by a simple power law T^m. Those reactions whose rates are influenced only by the concentration of the initially present reactants, are called *simple* or *thermal*. Reactions whose rates are effected by the concentrations of the intermediate and final products are called *complex* or *auto-catalytic*. Chain reactions are a special type of auto-catalytic reactions. The most appropriate method of studying chain reactions is by keeping the temperature of the system constant during the reaction and thus eliminating the thermal effects on the development of the reaction. The rate equation is then formulated on the basis of a hypothetical "mechanism" of interaction of various initial, intermediate and final participant species.

In order to keep a reacting system isothermal, the heat generated due to the reaction is extracted out by a cooling provision. However, to achieve perfectly isothermal conditions is rather difficult when the system involves rapid oxidation reactions. Nonisothermal reactions (i.e. most combustion processes) are unsuitable for the study of chain mechanisms.

If the reactants are contained in a vessel of volume V and surface area S, the overall energy conservation equation for an isothermal reaction expresses the fact that all the heat released by the reaction is removed from the system.

$$V\dot{q}''' - hS(T - T_W) = 0 \qquad (4.6)$$

where $\dot{q}'''$ is the heat released due to reaction in a unit time in a unit volume (equal to the reaction rate multiplied by the heat of combustion) and h is the heat transfer coefficient associated with heat transfer from the reacting mixture to the vessel walls. T and T_W are respectively the average temperature of the reacting gases and the vessel wall.

Exothermic reactions, in which heat is released so rapidly that it is extremely difficult to maintain isothermal conditions, are studied in an alternative way. If *all* losses are eliminated, the chemical heat release raises the temperature of the reacting mixture. Reaction in a perfectly insulated vessel, thus, is called an adiabatic or isolated reaction. The overall energy equation for an adiabatic reaction is written as

$$V\dot{q}''' - \rho CV\frac{dT}{dt} = 0 \qquad (4.7)$$

---

* Needless to mention is that all combustion reactions are exothermic.

where ρ and C respectively are the mean mixture density and specific heat. Adiabatic reactions are particularly suitable for understanding thermal effects.

Real combustion chambers are very seldom isothermal or adiabatic. In them, the heat released by chemical reactions is partly lost to the vessel walls and the surroundings and partly utilized to heat up the mixture. The general overall energy equation for a real combustion process thus includes both the loss and dT/dt terms.

$$V\dot{q}''' - \left[\rho C V \frac{dT}{dt} + hS(T - T_W)\right] = 0 \qquad (4.8)$$

Since the reactant is continuously depleted with time, the rate of a *simple* isothermal reaction is a monotonically decreasing function of time as shown in Figure 4.3. This statement, as discussed in Chapter 2 and Sections 4.11 and 4.12, is not true for chain type reactions.

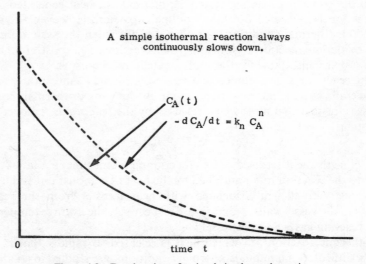

Figure 4.3 Deceleration of a simple isothermal reaction

Consider now an adiabatic reaction vessel. Let C_{AO} be the initial concentration of the species A in the mixture of gases enclosed in such a vessel. Let T_0 be the initial temperature of the mixture. Then recalling that $\dot{q}''' \equiv -\Delta H \cdot \dot{W}_A'''$ and that $\dot{W}_A''' \equiv dC_A/dt$, Eq. 4.7 can be rewritten as below.

$$\frac{dT}{dC_A} = -\frac{\Delta H}{\rho C} \qquad (4.9)$$

Kinetically Controlled Combustion Phenomena

With the initial condition that when $t = 0, T = T_0$ and $C_A = C_{AO}$, assuming $\Delta H/\rho C$ is a constant, integration of Eq. 4.9 yields

$$\frac{T - T_0}{C_{AO} - C_A} = \frac{\Delta H}{\rho C} \qquad (4.10)$$

Equation 4.10 indicates that the increase in temperature is linearly proportional to the decrease in the reactant concentration. When the reaction is complete $C_A = 0$ and the temperature is equal to the adiabatic flame temperature, T_f, so that $\Delta H/\rho C \equiv (T_f - T_0)/C_{AO}$. Substituting this result in Eq. 4.10,

$$\frac{T - T_0}{T_f - T_0} = \frac{C_{AO} - C_A}{C_{AO}} \qquad (4.11)$$

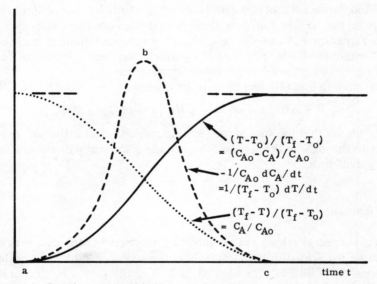

Figure 4.4 Complementary relation between the temperature rise and reactant depletion

Figure 4.4 shows qualitatively this complementary relation between the temperature rise and concentration depletion. Differentiating Eq. 4.11 with respect to time, the reaction rate can be expressed in terms of the rate of increase of temperature.

$$-\frac{dC_A}{dt} = \frac{C_{AO}}{T_f - T_0} \frac{dT}{dt} \qquad (4.12)$$

This rate is also qualitatively illustrated in Figure 4.4. Using Eqs. 4.11 and 4.12 to replace C_A and $-dC_A/dt$ the simple Arrhenius equation can be written to give the rate of increase of temperature.

$$-\frac{dC_A}{dt} = k_n C_A^n e^{-E/RT}$$

$$\frac{dT}{dt} = k_n \left(C_{AO} \frac{(T_f - T)}{(T_f - T_0)} \right)^n e^{-E/RT} \quad (4.13)$$

The characteristic steep increase (*a* to *b* in Figure 4.4) of the rate due to the exponential influence of temperature and the subsequent deceleration (*b* to *c* in Figure 4.4) due to the reactant depletion ($C_A \to 0$ and $(T_f - T) \to 0$) need no further elaboration here.

The complementary relation between T and C_A (given by Eqs. 4.10 and 4.11) is an important tool in the study of flames—because, then, solution of the energy or the species equation alone provides a complete behavior of the flame. Equation 4.11 can also be derived for a non-adiabatic flame in terms of the maximum flame temperature T_∞ which, because of the loss of energy, will be lower than the adiabatic flame temperature.

Rearranging Eq. 4.10, one obtains the relation

$$\rho C T + \Delta H C_A = \rho C T_0 + \Delta H C_{AO} = \rho C T_\infty + \Delta H C_{A\infty}$$

which implies that the sum of the sensible and chemical enthalpies remains invariant throughout the process. This is quite consistent with the first law of thermodynamics.

4.3 Ignition

Most of the energy released in a combustion reaction is in thermal form while a fraction is released in the form of light. Emission of light is either due to incandescent solid particles such as carbon in the flames or due to some unstable (excited) intermediate species. Of the heat generated, part is lost from the reacting mixture and part is retained by it.

(a) Thermal Ignition

Under certain conditions of heating brought about by an external source of energy such as a spark, hot vessel walls, compression, etc., there is always some temperature of the reacting mixture at which the rate of heat generation exceeds the loss rate. The excess heat increases the mixture temperature which in turn leads to higher reaction rate. The mixture temperature rises con-

Kinetically Controlled Combustion Phenomena

tinuously and acceleratively until a high heat evolution rate is attained. *Ignition* is then said to have occurred.

In reality, the accelerative rise of temperature is quite abrupt; the previously invisible slow reaction suddenly becomes visible and measurable.

An uncontrollably fast reaction is known as an *explosion*. Closed vessel explosions are very common in practice. At ignition, any combustion reaction seems as though it were an explosion. For this reason, superficially, the terms "explosion" and "ignition" are used synonymously in the combustion literature.

Under strictly adiabatic conditions the development of any exothermic reaction always culminates in a thermal ignition, the only factor opposing the progressive self-heating being the reactant depletion.

(b) Chemical Chain Ignition

If the combustion reaction involves intermediate chain carriers, ignition is possible even under isothermal conditions. If the rate of chain carrier generation exceeds the rate of their termination, the reaction becomes progressively fast and subsequently leads to ignition. As explained later in Sections 4.11 and 4.12, the chain initiation itself may require an external source of thermal or photon energy. Once the chain is initiated, the external source may be removed and ignition may be expected if the above criterion of positive chain carrier balance is fulfilled.

Determination of the conditions under which a given combustible mixture ignites, is an important topic in the design of combustion engines as well as in fire prevention.

(c) Scope of the Present Treatment

A major part of the rest of this chapter deals with ignition and extinction from a thermal viewpoint. The concepts of ignition delay, flammability limits, quenching distance and minimum ignition energy are presented through this thermal theory. Practical utility of these concepts is discussed wherever possible. The discussion naturally leads to the phenomena of ignition and explosions associated with chain reactions such as $H_2 + O_2$ and $CO + O_2$ combustion.

(d) Two Types of Ignition

Experience shows that there are two general modes of ignition—spontaneous and forced. *Spontaneous ignition* is sometimes called as auto-ignition or self-ignition. Spontaneous ignition occurs as a result of raising the temperature of a considerable volume of a combustible gas mixture by containing it in hot boundaries or by subjecting it to adiabatic compression. Because the heat generation rate is a strong exponential function of temperature whereas the heat loss rate is a simple linear function, even a slight increase in the temperature of the reacting mixture would greatly increase the rate of its temperature rise. As a consequence, once the generation rate exceeds the loss rate, ignition occurs in the whole volume almost instantaneously. The reaction then proceeds by itself without any further external heating.

Forced ignition, on the other hand, occurs as a result of *local* energy addition from an external source such as an electrically heated wire, an electric spark, an incandescent particle, a pilot flame, etc. A flame is initiated locally near the ignition source and it propagates into the rest of the mixture. Forced ignition is defined as the local initiation of a flame and its subsequent propagation.

There are many instances in which a fuel and an oxidant are rapidly mixed at a high temperature which can result in a spontaneous ignition. For example, a spray of diesel fuel into the hot compressed air is in part vaporized and mixed with the air in a very short time. Following a definite delay, the reaction would proceed rapidly enough to be considered a flame. In ramjets and in turbojet after-burners ignition of a fuel spray takes place spontaneously. And then there are those technically important instances of spontaneous ignition where it is *not wanted*, such as the legendary pile of oily rags, hay stacks, soybean drying bins and the cases of oil splashed on hot surfaces and the knock in a gasoline engine.

Examples of forced ignition are too many in number to enumerate here. The most studied forced ignition phenomenon deals with spark ignition of a gasoline-air mixture in automotive and aircraft engines.

The two kinds of ignition may seem to be different at a first look but at the cost of redundancy, it is worthwhile to repeat here that an initial, perhaps brief, supply of energy from an external source is involved in both of them.

4.4 Spontaneous Ignition Delay

(a) *The Criterion*

Consider a vessel of volume V and surface S containing a combustible mixture. Let T_0 be the initial temperature of the mixture. Assume that the temperature at any later time in the mixture is spatially uniform. Let the vessel walls be kept at T_0 for all times. The overall energy conservation is then given by Eq. 4.8.

$$V\dot{q}''' - \rho C V \frac{dT}{dt} - hS(T - T_0) = 0 \qquad (4.8)$$

Kinetically Controlled Combustion Phenomena

If the heat transfer coefficient is constant, at a very low pressure the reaction rate will be small because $\dot{q}''' \propto \dot{W}_A''' \propto C_A^n \propto P^n$. The system then remains at $T = T_0$. At a very high pressure, the heat generation overwhelms the loss term (i.e. the system approaches adiabatic conditions). The temperature and the reaction rate, then, enhance one another until spontaneous ignition occurs. Therefore, it is reasonable to expect that there exists a critical pressure below which the reaction behaves more like an isothermal (nonexplosive) process and above which it behaves like a spontaneously exploding adiabatic process. Similar arguments can be made by keeping the pressure and heat transfer fixed while varying the wall temperature T_0 or by keeping the pressure and T_0 fixed while varying the heat transfer coefficient. In any case, the result is the criterion of positive heat balance required for spontaneous ignition.

$$\Delta H V \dot{W}_A''' \geq h S (T - T_0) \tag{4.14}$$

(b) *Ignition Delay*

When the limiting case of adiabaticity is approached, Eq. 4.7 can be used to deduce the concept of *ignition delay* (sometimes called ignition lag, induction time, or ignition time) as following.

The strong influence of temperature on a simple thermal reaction rate $\dot{W}_A'''$ is often expressed by the conventional Arrhenius exponential or by a simple power law.

$$\dot{W}_A''' = k_n' C_A^n T^m \tag{4.15}$$

or

$$\dot{W}_A''' = k_n C_A^n e^{-E/RT} \tag{4.15a}$$

The power m is of the order 20 to 30 for most combustible mixtures whereas the activation energy E is of the order 20 to 60 kilocalories per mole. Incorporating Eq. 4.15 into Eq. 4.7 and integrating*, the time history of temperature is obtained as

$$\left(\frac{T}{T_0}\right)^{m-1} = \left[1 - \frac{t}{\left(\frac{\rho C T_0}{(m-1)\Delta H k_n' C_{AO}^n T_0^m}\right)}\right]^{-1} \tag{4.16}$$

Equation 4.16 indicates that as the time

$$t \to \left(\frac{\rho C T_0}{(m-1)\Delta H k_n' C_{AO}^n T_0^m}\right)$$

---

* It is assumed that the preignition reactant consumption is negligible.

the temperature T of the reacting mixture rises very steeply. The critical time

$$t_i \equiv \frac{\rho C T_0}{(m-1)\Delta H k'_n C^n_{AO} T^m_0} \tag{4.17a}$$

is called ignition delay. It can be seen that the ignition delay is short if the mixture has a low volumetric heat capacity, high temperature dependence of the rate, high heat of combustion and high initial reaction rate.

If Eq. 4.15(a) were substituted in Eq. 4.7 and integrated with the assumption of negligible reactant consumption during the ignition delay, the following results evolve.

$$\left(\frac{T}{T_0}\right)^2 \exp[E/RT_0(T_0/T - 1)] \approx 1 - \frac{t}{t_i} \tag{4.16a}$$

where

$$t_i \equiv \rho C \frac{RT_0^2}{E} \frac{e^{E/RT_0}}{\Delta H k_n C^n_{AO}} \tag{4.17a}$$

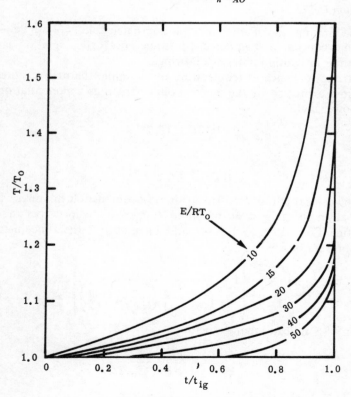

Figure 4.5 Ignition delay

Kinetically Controlled Combustion Phenomena

Equation 4.17 shows the influence of the initial heat evolution rate and of the kinetic parameters on t_i. Figure 4.5 illustrates T/T_0 as a function of t/t_i (i.e., Eq. 4.17(a)) for various values of E/RT_0.

4.5 Semenov Theory of Spontaneous Ignition

In this section the spontaneous ignition criterion developed in Section 4.4 is examined quantitatively to obtain a relation between the threshold (or critical) temperature, pressure and composition.

Denoting the heat generation rate $\Delta H V k_n C_A^n e^{-E/RT}$ by $\dot{q}_g$ and the heat loss rate $hS(T - T_0)$ by $\dot{q}_l$, Eq. 4.8 can be rewritten in the following simple form.

$$\rho C V \frac{dT}{dt} = \dot{q}_g - \dot{q}_l \tag{4.8}$$

Keeping the pressure fixed, $\dot{q}_g(T)$ will be a steeply rising function as shown in Figure 4.6(a). $\dot{q}_l(T)$ is a linear function of T with a slope of hS cal. per sec. per °K. Keeping hS fixed, three different $\dot{q}_l(T)$ functions are also shown in Figure 4.6(a) corresponding to three values of the wall temperature T_0. The right hand side of Eq. 4.8 as a function of T is shown in Figure 4.6(b) for the three values of T_0 corresponding to Figure 4.6(a).

Wall temperature relatively low: When $T_0 = T_{03}$, the $\dot{q}_g$ curve and $\dot{q}_l$ line intersect at two points, a and b; at a and b thus dT/dt is zero. If the starting (i.e., T at time $t = 0$) reacting mixture temperature T is less than T_a, $\dot{q}_g - \dot{q}_l$ (and by Eq. 4.8, dT/dt) is positive and $d(\dot{q}_g - \dot{q}_l)/dt$ is negative. That is, a mixture which is initially cooler than T_a slowly heats up until $T = T_a$, at a rate which continuously decreases with time. Curves of such heating for four different values of the starting temperature are shown in Figure 4.7(a).

If the starting temperature of the mixture is between T_a and T_b, both $\dot{q}_g - \dot{q}_l$ and $d(\dot{q}_g - \dot{q}_l)/dt$ are negative so that the mixture cools down to T_a at a continuously decreasing rate. Such cooling curves are shown in Figure 4.7(a) for three different values of the starting temperature.

If the starting temperature of the mixture is greater than T_b, both dT/dt and d^2T/dt^2 are positive so that the temperature of the reacting gases increases at an accelerating rate as shown by four curves in Figure 4.7(a). It is quite clear from this discussion that the two equilibrium points a and b are of different nature from one another. a is a stable point whereas b is metastable. That is, if the temperature of the gases is equal to T_a, the system would preserve this temperature even if perturbations are caused. The same comment does not hold for the reactor temperature T_b. T_b is an equilibrium point but any downward perturbation on the reactor temperature would

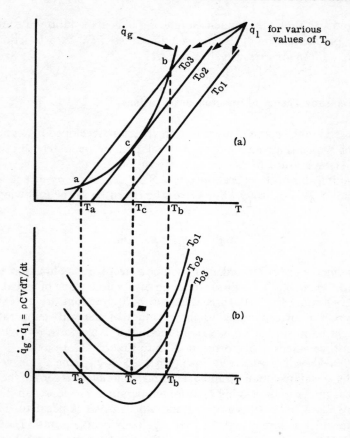

Figure 4.6 Semenov's thermal balance at ignition. (a) thermal balance, (b) excess energy

cause cooling to T_a and any upward perturbation would cause an "indefinite" accelerative heating. In order to indicate that b is a metastable equilibrium point, the state of the reactor at $T_b(t)$ is shown with a broken line in Figure 4.7(a).

Wall temperature relatively high: When $T_0 = T_{01}$, the $\dot{q}_g$ curve and $\dot{q}_l$ line never intersect. Thus $\dot{q}_g - \dot{q}_l$ is always positive. As shown in Figure 4.7(c), the temperature of the gases increases acceleratively.

Wall temperature moderate: As the wall temperatures progressively higher than T_{03} are considered, the points a and b approach one another when ultimately they coincide at the point c corresponding to a critical wall temperature T_{02} in Figure 4.6(a). The heat balance curve for this situation is shown in 4.6(b). The heating curves are shown in Figure 4.7(b).

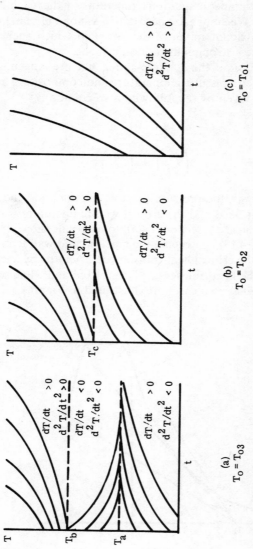

Figure 4.7 Temperature-time variations in a combustible system in (a) subcritical (b) critical and (c) supercritical regimes.

Implicit in the preceding discussion is an important criterion concerning spontaneous ignition. The wall temperature T_{02} is a limiting one beyond which the reaction progressively accelerates. The corresponding temperature* T_c is called spontaneous ignition temperature of the reactant gas mixture in the given vessel. We emphasize here that T_c is not a fundamental property of the given fuel/oxidant mixture; the vessel in which such a mixture is contained has quite a strong influence on it.

At the critical point, c, the $\dot{q}_g$ curve and $\dot{q}_l$ line are tangential. The interrelationship between pressure, temperature and composition at the ignition threshold hence is given by the following two equations.

$$(\dot{q}_g)_c = (\dot{q}_l)_c \tag{4.18}$$

$$\left(\frac{d\dot{q}_g}{dT}\right)_c = \left(\frac{d\dot{q}_l}{dT}\right)_c \tag{4.19}$$

In the above analysis, the pressure (i.e. the reaction) and the heat transfer coefficient are kept fixed and the critical wall temperature is deduced. It is possible, as mentioned in Section 4.4(a), to arrive at the same conclusions (viz., Eqs. 4.18 and 4.19) by keeping the reaction and wall temperature fixed while seeking for the critical heat transfer coefficient or by keeping the wall temperature and the heat transfer coefficient fixed and while seeking for the critical reaction. These two cases are illustrated in Figures 4.8 and 4.9.

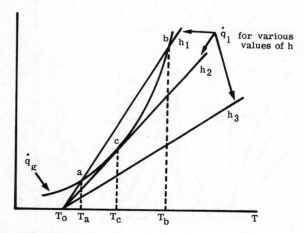

Figure 4.8 Second illustration of Semenov's thermal balance at ignition

---

*That T_c is approximately equal to T_{02} will be shown in the next section.

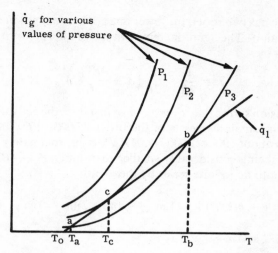

Figure 4.9 Third illustration of Semenov's thermal balance at ignition

4.6 Application of Semenov Theory to Predict Ignition Range

If we try to ignite a gaseous mixture in a vessel spontaneously, we will discover that the critical (ignition) temperature T_c is a strong function of the pressure in the vessel. For instance, if a simple reactant mixture were confined in a vessel at 1 atm. pressure, the ignition temperature is, say, 100 °C. Intuitively, it can be expected that the same mixture in the same vessel but at a higher pressure, say, 2 atm. will be ignited at a much lower temperature. The qualitative dependency of the ignition temperature on other variables, such as the mixture composition, initial temperature, vessel size, etc., can also be predicted intuitively.

The critical point c, in Figure 4.6(a), marks the transition of a slow stable reaction into one that is explosive. Assuming Arrhenius type rate law, Eqs. 4.18 and 4.19 become

$$\Delta H V k_n C_{Ac}^n e^{-E/RT_c} = hS(T_c - T_{0c}) \quad (4.18a)$$

$$\Delta H V k_n C_{Ac}^n e^{-E/RT_c}\left(\frac{E}{RT_c^2}\right) = hS \quad (4.19a)$$

Eliminating $\Delta H V k_n C_{Ac}^n / hS$ from these two equations and with an assumption that the amount of reactant consumed in the ignition delay is negligible,

$$\frac{RT_c^2}{E} = (T_c - T_{0c}) \quad (4.20)$$

This quadratic has two roots; the lower one applies to ignition and the upper one to extinction. The ignition temperature of the mixture in the given vessel is given by

$$T_c = \frac{E}{2R} - \frac{E}{2R}\left(1 - \frac{4RT_{0c}}{E}\right)^{1/2} \quad (4.21)$$

This expression shows that $T_c \approx T_{0c}$. For instance, the activation energy for most hydrocarbon/air flames is of the order 25,000 to 60,000 calories per mole. If T_{0c} is about 500 to 1,000 °K, RT_{0c}/E is approximately equal to 0.05. This quantity being considerably smaller than unity, $(1 - 4RT_{0c}/E)^{1/2}$ may be expanded into an infinite series by the identity.

$$(1 + x)^m = 1 + mx + (m - 1)m\frac{x^2}{2!} + (m - 2)(m - 1)m\frac{x^3}{3!} + \cdots : |x| < 1$$

Setting $m = 1/2$ and $x = -4RT_{0c}/E$,

$$\left(1 - \frac{4RT_{0c}}{E}\right)^{1/2} \approx 1 - \frac{2RT_{0c}}{E} - 2\left(\frac{RT_{0c}}{E}\right)^2$$

Substituting in Eq. 4.21,

$$(T_c - T_{0c}) \approx \frac{RT_{0c}^2}{E} \quad (4.20a)$$

Comparison of this result with Eq. 4.20 shows that $T_c \approx T_{0c}$. For $E \approx$ 40,000 cal/mole and $T_{0c} \approx 1000$ °K, $(T_c - T_{0c})$ will only be 50 °K. Thus whether we define the critical (ignition) temperature as the wall temperature T_{0c} or as the "tangent" gas temperature T_c, the possible error will only be of the order of 5%.

Substituting Eq. 4.20 in Eq. 4.18(a) we obtain for a simple second order* thermal reaction

$$\Delta H V k_2 C_{Ac}^2 e^{-E/[R(T_c + (RT_c^2/E))]} = hSRT_c^2/E.$$

Since $RT_c/E \ll 1$,

$$\Delta H V k_2 C_{Ac}^2 e^{-E/RT_c} = hSRT_c^2/E \quad (4.22)$$

If the gases are assumed perfect and if P_c and P_A are respectively the total pressure and the species A partial pressure,

$$C_{Ac} \equiv \frac{P_{Ac}}{RT_c} \equiv \frac{X_{Ac} P_c}{RT_c}$$

where X_A is the mole fraction of the species A.

---

* Recall that most of the hydrocarbon/air reactions are of approximate order of 2.

Kinetically Controlled Combustion Phenomena

Equation 4.22 thus becomes,

$$\Delta H V k_2 (X_{Ac} P_c / RT_c)^2 e^{-E/RT_c} = hSRT_c^2/E \tag{4.23}$$

If the composition X_A is kept fixed, Eq. 4.23 relates the critical pressure with the critical temperature.

$$\frac{P_c^2}{T_c^4} = \left(\frac{hSR^3}{\Delta H V k_2 X_A^2 E}\right) e^{E/RT_c}$$

Logarithmically,

$$\ln\left(\frac{P_c}{T_c^2}\right) = \ln\left(\frac{hSR^3}{\Delta H V k_2 X_A^2 E}\right)^{1/2} + \frac{E}{2RT_c} \tag{4.24}$$

This equation* is known in the literature as *Semenov Equation*. Plotting $\ln(P_c/T_c^2)$ on the y-axis and $(1/T_c)$ on the x-axis, Eq. 4.24 gives a straight line with a slope of $E/2R$. (See Figure 4.10).

The study of ignition thus provides an ingenious method of measuring the activation energy for simple thermal reactions. If the activation energy determined by this method is consistent with that determined by Arrhenius'

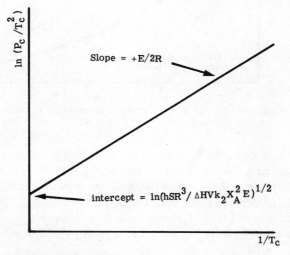

Figure 4.10 Critical pressure versus temperature

---

* In general, for an *n*th order thermal reaction, Eq. 4.24 has the form,

$$\ln\left(\frac{P_C}{T_c^{(n+2)/n}}\right) = \ln\left(\frac{hSR^{(n+1)}}{\Delta H V k_n X_A^n E}\right)^{1/n} + \frac{E}{nRT_c}$$

method (Section 2.8), one may infer that the assumptions† of Semenov Theory are valid.

If h, S, ΔH, V, k_2 and X_A^2 are known, Eq. 4.24 may also be plotted on a $P_c - T_c$ plane, as shown in Figure 4.11, to delineate the ignitable from non-ignitable conditions. Figure 4.11 shows that our intuitive speculations, made at the beginning of this section, are in general true for thermal reactions. At low pressures, very high temperatures are needed to accomplish ignition and vice versa.

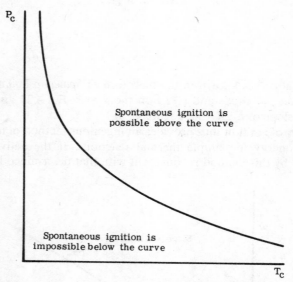

Figure 4.11 Critical pressure versus temperature on a linear plot to delineate spontaneous ignition zone from the no-ignition zone

Equation 4.24 can also be used to understand the ignition ranges on a $T - X_A$ plane at a fixed total pressure and on a $P - X_A$ plane at a fixed temperature. Figures 4.12 and 4.13 show these results schematically. In general, these graphs are U-shaped. The conditions lying inside the U result in an ignition whereas those lying outside, do not. Several inferences can be drawn from these figures. Consider the $T - X_A$ relation given by Figure 4.12.

Firstly, there exist a lower and an upper concentration limits for ignition; if the mixture is too fuel-lean or too fuel-rich, ignition is not possible no

---

† Two most important of these assumptions, to summarize here, are
  (a) the reaction is thermal.
  (b) the reactant consumption is negligible until ignition occurs.

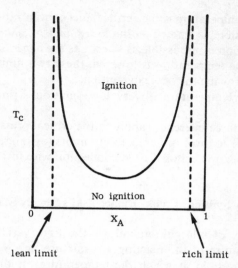

Figure 4.12 Flammability limits

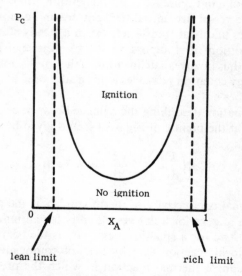

Figure 4.13 Flammability limits

matter what the temperature is. The critical fuel concentration, below which ignition is impossible, is known as the *lower limit of ignitability*; and that above which ignition is impossible, is known as the *upper limit*.

Secondly, as the temperature is lowered, these two limits approach one another, thus narrowing the range of ignition.

Thirdly, if the temperature is very low, ignition is impossible at any composition.

The preceding three inferences apply to the critical pressure-composition diagram as well. The topic of flammability limits (or range of ignition) will be further discussed in Section 4.10 in connection with forced ignition.

4.7 Spontaneous Ignition: Frank–Kamenetski's Steady State Analysis

Semenov's theory of self-ignition follows the time variation of spatially averaged temperature of the reacting gas mixture in a vessel in order to determine the threshold at which the temperature rapidly increases in a seemingly boundless manner.

Frank–Kamenetski considers the impossibility of the steady state represented by the point c in Figure 4.6(a) as the ignition criterion. Under certain conditions, of the vessel size, initial temperature and composition, reaction kinetics/energetics and heat transfer etc., such a steady state exists whereas under other conditions it is impossible. Frank–Kamenetski's spontaneous ignition analysis thus involves a delineation of the conditions under which the *steady* state energy equation possesses a physically valid solution from those under which it does not.

The energy equation describing the balance between conductive loss and reaction source (at the point c of Figure 4.6(a)) is given by

$$K\left(\frac{d^2T}{dx^2} + \frac{\beta}{x}\frac{dT}{dx}\right) + \dot{q}''' = 0 \qquad (4.25)$$

where K is thermal conductivity of the mixture, x is the position and β is a constant equal to zero for a flat vessel, unity for an infinitely long cylindrical vessel and two for a spherical vessel. $\dot{q}'''$ is the heat source obtained by multiplying local reaction rate by heat of combustion. Assuming an Arrhenius type simple thermal reaction in which the preignition reactant consumption is negligible*,

$$\dot{q}''' = \Delta H k_n C_{AO}^n e^{-E/RT}$$

---

* That this assumption is valid is substantiated by Frank-Kamenetski's work.

Equation 4.25 has to be solved under the following boundary conditions.

$$T = T_0 \quad \text{at} \quad x = \pm a$$
$$\frac{dT}{dx} = 0 \quad \text{at} \quad x = 0 \quad (4.26)$$

$2a$ is the thickness of the reacting gas body. Defining the following nondimensional temperature and position,

$$\theta \equiv \frac{E(T - T_0)}{RT_0^2}$$

$$\xi \equiv \frac{x}{a}$$

Eq. 4.25 is reduced to

$$\frac{d^2\theta}{d\xi^2} + \frac{\beta}{\xi}\frac{d\theta}{d\xi} + \frac{\Delta H k_n C_{AO}^n E a^2}{KRT_0^2} e^{-E/RT} = 0 \quad (4.27)$$

If $(T - T_0) \ll T_0$, the exponential in Eq. 4.27 can be simplified by the identity $(1 + z)^{-1} \approx (1 - z)$ when z is small.

$$e^{-E/RT} = e^{-E/R(T + T_0 - T_0)} = e^{-(E/RT_0)[1 + (T - T_0)/T_0]^{-1}}$$
$$\approx e^{-(E/RT_0)[1 - (T - T_0)/T_0]} = e^{\theta} \cdot e^{-E/RT_0}$$

Equation 4.27 thus reduces as following.

$$\frac{d^2\theta}{d\xi^2} + \frac{\beta}{\xi}\frac{d\theta}{d\xi} + \hat{\Delta} e^{\theta} = 0 \quad (4.28)$$

where

$$\hat{\Delta} \equiv \left[\frac{\Delta H k_n C_{AO}^n E a^2 e^{-E/RT_0}}{KRT_0^2}\right] \quad (4.29)$$

$\hat{\Delta}$ is a nondimensional property of the system relating the heat generated in the chemical reaction and the heat drained away by conduction.

$$\hat{\Delta} \equiv \frac{\Delta H k_n C_{AO}^n e^{-E/RT_0}}{KRT_0^2/Ea^2} [=] \frac{\text{cal/cm}^3/\text{sec}}{\text{cal/cm}^3/\text{sec}}$$

When the reaction is fast and highly exothermic, and when the conductive drain is small, $\hat{\Delta}$ attains a large value. For a weakly exothermic reaction in a small vessel, $\hat{\Delta}$ is small.

The boundary conditions in nondimensional form take the following simple form.

$$\theta = 0 \quad \text{at} \quad \xi = 1$$

$$\frac{d\theta}{d\xi} = 0 \quad \text{at} \quad \xi = 0$$

Equation 4.28 can now be solved with these boundary conditions to obtain $\theta(\xi)$ for various values of the parameter $\hat{\Delta}$. For $\hat{\Delta}$ greater than a critical value $\hat{\Delta}_c$, solutions will be found to be impossible. This is because when $\hat{\Delta}$ is large, the heat generation rate exceeds the heat loss rate; the system heats up as a consequence and the steady state equation is physically invalid. The critical value of $\hat{\Delta}$ thus provides us with the conditions provoking the onset of an unstable, progressively accelerating reaction (i.e., ignition). Frank–Kamenetski's solutions yield $\hat{\Delta}_c$ as 0.88 for a flat vessel (i.e., a vessel formed by two infinite parallel plates separated by a distance $2a$); 2.0 for an infinitely long cylindrical vessel (i.e. a tube) and 3.32 for a spherical vessel.

The maximum temperature occurs at the center of the vessel when $\hat{\Delta} = \hat{\Delta}_c$.

$$\theta_{\max} \equiv \frac{E(T_{\max} - T_0)}{RT_0^2} = \theta(0, \hat{\Delta}_c)$$

Frank–Kamenetski's solutions give $\theta(0, \hat{\Delta}_c) = 1.2$ for a flat vessel, 1.37 for a cylindrical vessel and 1.6 for a spherical vessel. These numbers* indicate that $(T_{\max} - T_0)$ is of the order of magnitude RT_0^2/E which for most hydrocarbon fuels burning in air is about 0.05. The assumption of $(T - T_0) \ll T_0$ made to simplify the exponential term in the energy equation is thus very good. For a second order reaction, the relation between the ignition conditions, as given by $\hat{\Delta}_c = $ constant, can be written in terms of the total pressure, mole fraction and temperature as

$$\hat{\Delta}_c = \frac{\Delta H k_2 (P_c X_{Ac}/RT_c)^2 e^{-E/RT_c} E a_c^2}{KRT_c^2}$$

$$= 0.88 \text{ for a slab}$$
$$= 2.00 \text{ for a cylinder} \quad (4.30)$$
$$= 3.32 \text{ for a sphere}$$

Equation 4.30 is equivalent to Semenov's criterion (given by Eq. 4.23) which can be rewritten as

$$\frac{\Delta H k_2 (P_c X_{Ac}/RT_c)^2 e^{-E/RT_c}}{h \dfrac{S}{V} \dfrac{RT_c^2}{E}} = 1$$

---

* Cross-check with Eq. 4.20(a) of Semenov Analysis.

The surface to volume ratio is $1/a$ for a flat vessel, $2/a$ for a cylindrical vessel and $3/a$ for a spherical one. If the heat loss is by conduction alone, the heat transfer coefficient is approximately (K/a) so that Semenov relation is given by

$$\frac{\Delta H k_2 (P_c X_{Ac}/RT_c)^2 e^{-E/RT_c} E a_c^2}{KRT_c^2} = 1 \text{ for a slab}$$
$$= 2 \text{ for a cylinder} \quad (4.31)$$
$$= 3 \text{ for a sphere}$$

The agreement between Eqs. 4.30 and 4.31 is extremely good. Equation 4.30 can be plotted in the same way as we did in Semenov analysis to illustrate ignition range graphically. For reactions whose kinetics and energetics are well-known, the ignition ranges can be predicted as in Figures 4.14 to 4.16 which are drawn on the basis of Frank–Kamenetski's report. Remembering that even minor errors in temperature measurements cause drastic errors in e^{-E/RT_c}, the agreement between the current theory and experiments can be noted as excellent.

Ideal ignition delay is calculated as below from Frank–Kamenetski's analysis. The amount of heat liberated in a unit volume in a time duration of

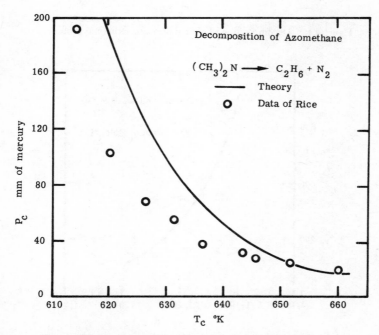

Figure 4.14 Decomposition of azomethane as a thermal process

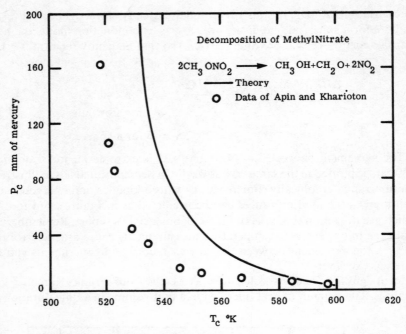

Figure 4.15 Decomposition of methylnitrate as a thermal process

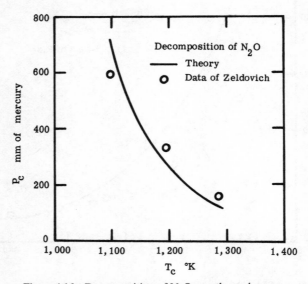

Figure 4.16 Decomposition of N_2O as a thermal process

Kinetically Controlled Combustion Phenomena

t_i is roughly equal to

$$t_i \, \Delta H k_n C_{AO}^n e^{-E/RT} \text{ cal/cm}^3$$

Neglecting losses, this heat is utilized to raise the temperature of the contents from T_0 to T_c.

$$\rho C(T_c - T_0) \approx t_i \, \Delta H k_n C_{AO}^n e^{-E/RT_c}$$

TABLE 4.1

Activation Energies (kcal/mole) for Combustion in Air*

Fuel	$E_{\text{activation}}$	Fuel	$E_{\text{activation}}$
Acetaldehyde	45.4	Ethyl bromide	42.0
Acetone	55.0	Ethyl carbonate	38.4
Acetylene	31.0	Furan	43.4
Acrolein	35.5	Furfuryl alcohol	47.9
Allyl alcohol	39.6	n-Heptane	60.5
n-Amyl alcohol	48.0	n-Hexane	50.7
Aniline	48.0	Hydrogen	57.0
Benzaldehyde	40.0	Kerosene	46.0
Benzene	47.2	Methane	29.0
Benzyl acetate	36.2	Methyl acetate	34.8
Benzyl alcohol	44.0	Methanol	41.3
Bromobenzene	49.5	Methylene chloride	41.9
Buta gas	50.0	Methylethyl ketone	49.3
n-Butyl acetate	36.5	Methyl formate	30.7
n-Butyl alcohol	48.4	Nitroethane	42.6
Butyl phthalate	43.4	Nitromethane	39.2
Calor gas	52.0	iso-Octane	32.4
Carbon disulfate	27.2	paraldehyde	45.0
Carbon monoxide	78.0	40–60° Petroleum ether	45.2
Cumene	46.7	80–100° Petroleum ether	44.8†
Cyclo-hexane	46.4	100–120° Petroleum ether	42.3†
Cyclo-hexene	43.2	n-Propyl acetate	32.3
trans-Decalin	47.4	n-Propyl alcohol	45.3
n-Decyl alcohol	51.7	iso-Propyl alcohol	37.8
Diethylamine	37.0	Propylene oxide	46.4
Diethyl ether	52.5	Pyridine	31.7
iso-Dodecane	51.0	Styrene	31.8
Ethane	49.0	Toluene	41.0
Ethyl acetate	35.8	Tri-iso-Butylenes	34.8
Ethanol	42.2	Turpentine	40.6
Ethylbenzene	45.7	Xylene	34.3

* This table is a summary of Mullins' data: from Mullins, B. P. and Penner, S. S., *Explosions, Detonations, Flammability and Ignition*, pp. 201–202, AGARDograph No. 31, Pergamon Press (1959).
† Aromatic-free.

Resolving for t_i and recalling that $T_c - T_0 \approx RT_c^2/E$,

$$t_i \approx \frac{\rho CRT_c^2}{\Delta H k_n C_{AO}^n E} e^{E/RT_c} \tag{4.32}$$

Comparison of Eqs. 4.32 and 4.17(a) indicates that the influence of various properties on ignition delay is consistent. Jackson and Brokaw's and Mullins' measurements of the ignition delay for various hydrocarbon/air flames confirm the correctness of Eq. 4.32. Differentiation of Eq. 4.32 with the neglect of T_c relative to E/R provides a new method of obtaining the activation energy.

$$\frac{d(\ln t_i)}{d(1/T_c)} \approx \frac{E}{R}$$

The activation energy thus measured is presented in Table 4.1 (adapted from Mullins' work).

Experiments show that there is a marked effect of the fuel type and fuel additive on the ignition delay of fuels mixed with air. This is in contrast to the absence of a marked effect on the propagation rate of a flame in a homogeneous mixture. (This topic is considered in Chapter 8). This contrast has lead to the conclusion that the global reaction has an activation energy which changes as the temperature goes from intermediate values (where spontaneous ignition effects are determined) to higher values (where flame propagation effects are determined). At intermediate temperatures, the activation energy is higher for smaller and compact fuel molecules. This leads to longer delay times. The activation energy becomes smaller as the complexity of the fuel molecule increases. Addition of knock suppressant agents, flame inhibitors and fire retardants again is expected to increase the apparent low temperature global activation energy.

4.8 Fowe Comments on Spontaneous Ignition

As indicated by Eqs. 4.24 and 4.31, the self-ignition temperature (pressure and composition) is not a fundamental property of the given fuel/oxidant mixture, but depends upon the apparatus and the method of measurement. This is clear from the appearance of the system volume and surface area in Eqs. 4.24 and 4.31 and the heat transfer coefficient which depends upon not only the method of mixing and introduction of the gas into the vessel, but also upon the cleanliness and thermal properties of the bounding walls.

The second comment of pertinence here concerns with the assumption of neglecting the preignition reactant consumption. In reality, a measurable but small amount of the reactant is depleted during the induction period. The effect of this depletion is to reduce the reaction rate slightly. Corrections can be made for such a reduction. For most of the ignition situations encountered in practice, however, these corrections are quite small.

Thirdly, let us briefly explore the experimental methods available to measure self-ignition temperature.

Le Chatelier's Method: Fuel/oxidant mixture of a known composition is introduced into a vacuum vessel located in an oven at a fixed temperature. The combustible mixture is heated by heat transfer from the hot vessel walls. The mixture temperature is continuously monitored with time. That particular vessel wall temperature which results in an abrupt rise in the mixture temperature is taken to be the ignition temperature. Le Chatelier's method suffers with the disadvantage of the need to heat the *cold* mixture *in* the vessel. This heating takes a definite time which is included in the measured induction time. The main advantage of this method lies in the fact that control of pressure and composition can be accomplished to a high degree of accuracy.

Compression Method: The combustible mixture is heated by compression in a cylinder-piston arrangement. The degree of compression at which ignition occurs is recorded. The critical pressure and temperature are obtained from the compression ratio by the polytropic formulae.

$$P_c = P_0(V_0/V_c)^{k^*}$$
$$T_c = T_0(V_0/V_c)^{k^*-1}$$

P_0, V_0 and T_0 are respectively the initial pressure, volume and temperature and P_c, V_c, T_c are their critical values. k^* is the polytropic constant which is equal to the ratio of the specific heats when the compression is strictly adiabatic. The compression method, like any other closed vessel method, possesses the advantage of good control of the mixture composition. However, in order to be able to stop the compression precisely at the instant when self-ignition occurs, the piston has to travel at an extremely slow rate. Such a slow compression means increased losses to the cylinder walls and consequent non-adiabatic heating.

Dixon's Concentric Tube Method: In this method, the fuel and the oxidant flow *separately* in heated concentric tubes. When the two heated streams mix (primarily by diffusion and turbulence) at the exit of the tubes, they ignite if the temperature is high enough. That temperature of the tube walls which yields ignition is considered to be the ignition temperature. This method offers an elegant technique to eliminate the mixture preheating but it has the drawback of uncertainty regarding the mixture composition.

Droplet Method: Liquid fuel is supplied drop by drop from a burette into a heated crucible through which the oxidant gas flows at a metered rate. The temperature of the crucible at which the flame appears is called ignition temperature. Simple as it is, the droplet method suffers with the same disadvantages as the concentric tube method.

The fourth comment deals with the assumption that the reaction is thermal. With the exception of a handful of reactions mentioned by Frank–Kamenetski, most of the combustion reactions are complex. The flammability limits measured for these complex reactions may not be as sharply defined as the thermal theory predicts. Several intermediate reactions occurring in the system may change their role of dominance from one range of temperature and pressure to another. Chain reaction theory enables study of flammability of such complex situations. (This is done in Section 4.12.)

Most of the combustion reactions are neither purely thermal nor purely chain-carrier type. Thermal and chain theories of ignition thus only offer an understanding of the behavior of the limiting systems. Such an understanding is useful to interpret many of the real-life ignition processes.

4.9 Forced Ignition

(a) *Some Preliminary Concepts*

When a cold reactant mixture is rapidly and *locally* heated by a heat source such as an incandescent solid particle, a heated electrical filament or a spark, a pocket of hot gas, a pilot flame, etc., a flame can be initiated in the vicinity of the heat source and propagated into the rest of the cold mixture. Such an *initiation* of a *propagating* flame is defined as forced ignition. It is clear from this definition that local heating is much less favorable for ignition than bulk heating.

Both spontaneous and forced ignition share the common virtue of self-acceleration and auto-catalytic behavior motivated by thermal and/or chain branching reactions. Initial provocation by an external source of energy is necessary for both types of ignition.

The phenomenon of forced ignition is considered in this section in a thermal view-point in order to systematically develop some simple relationships between the critical physico-chemical properties of the system.

Let us examine the definition of forced ignition by a hot particle in an attempt to establish the concepts of initiation and propagation of a flame. Consider as shown in Figure 4.17 an incandescently hot metal particle (whose temperature is T_w) located in an infinite combustible mixture (whose temperature T_0 is smaller than T_w). Due to the temperature difference, the particle loses heat to the adjacent mixture; the rate of heat flow is a function of the flow and thermal properties of the mixture. If T_w is moderate, a steady state temperature profile $T = T(x)$ is established in such a way that the steepest temperature gradients are confined to a thin boundary layer around the particle. Figure 4.17 shows this profile (a) when the mixture is noncombustible and (b) when it is combustible. The difference between these two profiles is due to the heat release in chemical reaction. From the temperature gradients at the wall, one can note that the heat flow *from* the wall to the mixture is lower when the gases react exothermically than when they are inert.

If the particle temperature were chosen higher than before, the difference between the reactive and inert profiles will be more pronounced. The higher the particle temperature, the lower the heat flux from the wall. At a critical particle temperature T_c, as shown in Figure 4.18, the heat flux from the wall to the reactive mixture is zero. If the particle temperature is even slightly greater than T_c, the enhanced reaction in the mixture shows a maxima of temperature a small distance away from the particle surface. Heat flows then partly *to* the particle and mostly to the reservoir of gases. A steady state temperature profile under these conditions becomes impossible because the temperature maxima continuously moves away from the particle wall. When

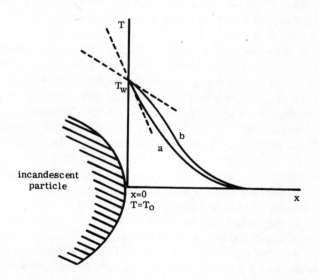

Figure 4.17 Temperature profile in the vicinity of an incandescent particle located in (a) a noncombustible medium, (b) a combustible medium

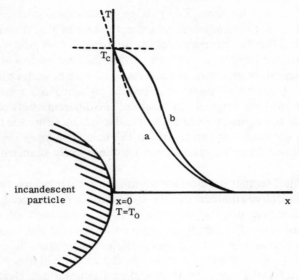

Figure 4.18 Same as Figure 4.17 but the particle is at a higher temperature

the temperature gradient at and normal to the igniting particle suface is zero, the reaction layer of gas (i.e., the flame) begins to propagate into the unburnt mixture. The onset of this propagation is conveniently considered as the criterion for forced ignition.

It should be noted here that temperature gradients in a reacting mixture are always accompanied by composition gradients. In the immediate vicinity of the particle, the concentration of the reactants will be the lowest and that of the products will be the highest; far away from the particle, the product concentration will be zero and the reactant concentration will be equal to the initial value. As a result of the concentration gradients, the products diffuse away from the particle whereas the reactants diffuse towards the particle to replenish the depleted mixture with fresh new reactants. A rigorous treatment of the ignition problem thus requires consideration of the influences of species diffusion. For example, the introduction of an incandescent particle into a combustible mixture need not necessarily cause ignition because in the vicinity of the particle, where the temperature is most favorable for a high reaction rate, the reactant concentration is least favorable. This fact was first discovered by Davy in 1812 in England in connection with the inability of a glowing piece of charcoal to initiate a propagating flame in a combustible mixture.

Mallard and Le Chatelier discovered that in order to cause ignition, the temperature of the incandescent particle has to be greater when the particle

Kinetically Controlled Combustion Phenomena

is smaller. This is mainly because the temperature and concentrations drop very abruptly when the particle is small. Another example to illustrate the role of the decrease of the reactant concentration is provided by the difficulty one encounters when one tries to ignite a mixture with a catalytic hot wire (such as platinum). The intense catalytic surface reactions deplete the reactants even more than when the wire is non-catalytic. The temperature of the surface rises considerably but the gas phase composition and temperature are less favorable to ignition.

Let us now examine the propagation of the flame from the ignition source. We will, here, only deal with the rudiments of the theory of flame propagation since this topic is to be thoroughly discussed in Chapter 8. The propagation of a flame is often studied by considering the steady propagation of a fixed temperature (or composition) profile into the unburnt mixture as shown in Figure 4.19.

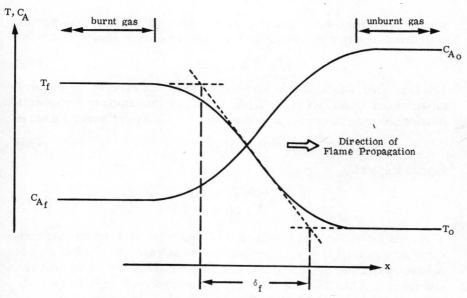

Figure 4.19 Phenomenology of a propagating flame front

The *fundamental flame speed* is defined as the speed at which the flame front travels in a direction normal to itself and relative to the speed of the unburnt mixture. We denote it by u_0. In Figure 4.19, we show how the temperature across the flame front varies from the flame temperature T_f behind the front to the initial mixture temperature T_0 ahead of the front. The flame thickness δ_f is conveniently defined as the ratio of the maximum temperature difference and the maximum temperature gradient.

$$\delta_f \equiv \frac{(T_f - T_0)}{(dT/dx)_{\max}} \tag{4.33}$$

Figure 4.19 also shows the fuel concentration profile across the flame front.

Phenomenologically, the picture is as following. Heat is generated by combustion in a layer of gas mixture of thickness δ_f. Due to the temperature gradient, the heat so generated is transferred by conduction to the unburnt gases. Utilizing this heat, the unburnt mixture heats up so that the combustion front progresses forward at a velocity u_0.

If $\dot{W}'''$ is the rate of combustion averaged over the flame thickness δ_f, the heat generation rate is given by

$$\dot{q}_g \approx \Delta H \cdot \dot{W}''' \cdot \delta_f \cdot A \tag{4.34}$$

where $\delta_f \cdot A$ is the volume of the flame front, A being the area of cross section considered. The rate of conduction heat transfer is given approximately by

$$\dot{q}_K \approx KA(T_f - T_0)/\delta_f \tag{4.35}$$

The heat needed to raise the temperature of a unit mass of the unburnt mixture from T_0 to T_f is $C(T_f - T_0)$ cal/gm. If ρ_0 is the density of the unburnt mixture, the mass heated per unit time is $\rho_0 u_0 A$. The rate of heating, hence is

$$\dot{q}_h \approx \rho_0 u_0 AC(T_f - T_0) \tag{4.36}$$

Equating $\dot{q}_g$ and $\dot{q}_K$,

$$\delta_f \approx \left(\frac{K(T_f - T_0)}{\Delta H \dot{W}'''}\right)^{1/2} \tag{4.37}$$

This equation shows that the flame thickness is larger if the mixture conductivity is larger, $(T_f - T_0)$ is greater and the average heat production rate is lower. Equating $\dot{q}_K$ and $\dot{q}_h$ and recalling the definition of thermal diffusivity $\alpha_0 \equiv K/\rho_0 C$, we obtain an alternative expression for δ_f to show the influence of specific heat and the flame speed

$$\delta_f \approx \frac{\alpha_0}{u_0} \tag{4.38}$$

Equating $\dot{q}_g$ and $\dot{q}_h$,

$$\delta_f \approx \frac{\rho_0 u_0 C(T_f - T_0)}{\Delta H \dot{W}'''} \tag{4.39}$$

Kinetically Controlled Combustion Phenomena

Eliminating δ_f from Eqs. 4.37 and 4.39,

$$u_0 \approx \frac{1}{\rho_0 C}\left(\frac{K \Delta H \dot{W}'''}{T_f - T_0}\right)^{1/2} \quad (4.40)$$

That is, the flame speed is greater if the mixture conductivity is greater, mean heat production rate $\Delta H \dot{W}'''$ is greater, the volumetric specific heat of the original mixture is lower and $(T_f - T_0)$ is lower. Eliminating ΔH,

$$u_0 \approx \left(\frac{K \dot{W}'''}{\rho_0^2 C C_{AO}}\right)^{1/2} \quad (4.40\text{a})$$

In Eq. 4.40(a), it is assumed that the concentration of the fuel in the completely burnt mixture is zero.

It is important to note the influence of pressure on the flame speed. Noting that the conductivity, specific heat, flame temperature and heat of combustion are essentially independent of total pressure, Eq. 4.40 shows that $u_0 \propto (\dot{W}''')^{1/2}/\rho_0$. For an nth order reaction the mean reaction rate is proportional (refer to Chapter 2) to the total pressure raised to n. Furthermore, ρ_0 is directly proportional to P. Hence

$$u_0 \propto P^{(n-2)/2} \quad (4.41)$$

For most hydrocarbon fuels burning in air or oxygen, n is roughly equal to 2 so that the fundamental flame speed is nearly independent of the total pressure. For further details of this dependency, refer to chapter 8.

(b) Ignition by Heated Spheres and Rods

When spheres of quartz or platinum are shot into a combustible gas mixture, ignition is found to be possible if the sphere temperature T_w is greater than a critical value T_c. Experiments show that the critical sphere temperature is dependent upon such variables as the sphere size, sphere catalytic properties, shooting velocity, mixture thermal, chemical and kinetic properties, etc. Considering that ignition is assured if the heat generated by the chemical reaction in a layer of the combustible mixture adjacent to the hot sphere is in excess of the heat lost by the same layer, the critical criterion can be postulated as below.

Let the temperature fall linearly from T_w at the sphere wall to T_0 in the mixture in a thin layer of thickness τ around the sphere. The thickness τ naturally depends upon the sphere velocity, sphere diameter $2a$, fluid viscosity and thermal properties. The volume* of the gas layer of thickness τ

---

* The volume of a sphere of radius r is given by $\frac{4}{3}\pi r^3$. The increment in this volume due to a small increment in r is $4\pi r^2\, dr$. If $r = a$ and $dr = \tau$, $\Delta V = 4\pi a^2 \tau$.

in which the reaction occurs is, approximately, $4\pi a^2 \tau$. The area of the shell through which heat is lost to the expanse of combustible mixture is, approximately $4\pi a^2$. Assuming that heat loss occurs by conduction and that the reaction rate law is of Arrhenius type, ignition is expected if

$$\Delta H 4\pi a^2 \tau k_n C_{AO}^n e^{-E/RT_w} \geq 4\pi a^2 K(T_w - T_0)/\tau$$

Here, the reactant consumption is neglected even though such a neglect can be serious as mentioned in the preceding discussion. Simplifying,

$$\frac{(T_w - T_0)}{\tau} \leq \frac{\Delta H \tau k_n C_{AO}^n e^{-E/RT_w}}{K} \qquad (4.42)$$

The equality sign holds when T_w is equal to the critical temperature T_c. Equation 4.42 indicates that the temperature gradient in the reaction layer is an important factor governing the possibility of ignition.

From Chapter 3, we can recall that when the Reynolds number of flow around the sphere and the Prandtl number are high, the thermal boundary layer thickness is small. Consequently the thermal gradients in the vicinity of the sphere wall will be large. Hence, with a given wall temperature, ignition is harder to accomplish at higher velocities of flow. Recalling the definitions of Nusselt number and heat transfer coefficient, the ignition criterion can be written as

$$\frac{(T_c - T_0)}{a} = \frac{2a\,\Delta H k_n C_{AO}^n e^{-E/RT_c}}{\text{Nu}\ K} \qquad (4.42a)$$

Explicitly, the ignition temperature depends upon the sphere size a, according to

$$a^2 = \frac{\text{Nu}\ K(T_c - T_0)e^{E/RT_c}}{2\,\Delta H k_n C_{AO}^n} \qquad (4.42b)$$

This equation is verified by Silver's experiments which show that the critical temperature is smaller when the sphere size is larger and when the velocity of shooting is lower. In fact, plotting $\ln(T_c - T_0)/a$ vs. $(1/T_c)$ when the velocities are low gives a straight line whose slope is $(-E/R)$. The activation energy determined by this technique by Silver for the combustion of pentane, illuminating gas and hydrogen with air are quite close to the values determined by other techniques.

An analysis similar to the preceding one can also be made for ignition of gaseous mixtures flowing across heated rods.

(c) *Ignition by Flames*

The energy required to ignite a combustible mixture may be provided by a pilot flame. The possibility of ignition is determined by such properties as the mixture composition, time duration of contact between the pilot flame

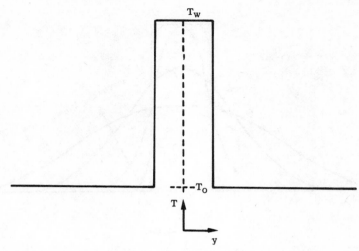

Figure 4.20 Temperature profile when a hot slab pilot flame is introduced into a combustible mixture

and the mixture, the size and temperature of the pilot flame, intensity of mixing, etc. Consider as shown in Figure 4.20 an infinite slab of a pilot flame whose temperature is T_w and whose thickness is $2a$. (In reality, pilot flames are finite in size and three dimensional. Here, the choice of an infinite slab, of thickness $2a$, makes the mathematical analysis conveniently unidimensional. The nature of the results can be extrapolated for two and three dimensional pilot flames quite easily.) Let the pilot flame slab be placed at time $t = 0$ in an infinite reservoir of a combustible mixture whose temperature is T_0. Assuming that convective currents are absent, the transient energy conservation equation is given by

$$\rho C \frac{\partial T}{\partial t} = K \frac{\partial^2 T}{\partial y^2} + \Delta H k_n C_A^n e^{-E/RT}. \tag{4.43}$$

Governed by this equation, the temperature varies with time and location in the mixture with the following initial and boundary conditions.

$$t \leq 0 \quad T = T_w \quad \text{in} \quad 0 < |y| < a$$
$$T = T_0 \quad \text{in} \quad a < |y| < \infty$$

$$t > 0 \quad \frac{\partial T}{\partial y} = 0 \quad \text{at} \quad y = 0$$

$$\frac{\partial T}{\partial y} = 0 \quad \text{at} \quad |y| = \infty$$

$$T = T_0 \quad \text{at} \quad |y| = \infty$$

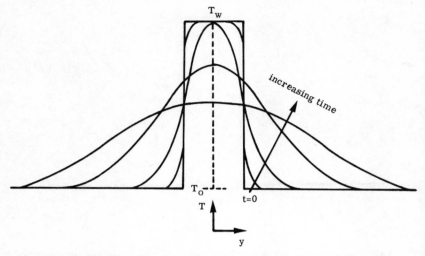

Figure 4.21 Temperature decay of the pilot flame slab as time increases

Analytical solution* of Eq. 4.43 with these conditions can be obtained if the chemical source term is absent. Such a solution describes the gradual spread and decay of temperature with time as shown in Figure 4.21.

If the chemical reaction term is retained in Eq. 4.43, a closed form solution is not possible. Numerical or graphical solutions, however, can be obtained. These solutions show that initially the chemical reaction term indeed is negligible and, as a result, the temperature profile tends to decay with time, as indicated in Figure 4.21. Two cases now arise. Firstly, if the initial pilot flame slab is thinner than a certain critical size, the temperature profile continues to decay to eventually extinguish the pilot flame. It is intuitively clear that the extinguishment is a consequence of the interplay between excessive conductive drain of energy and moderate or low generation rate.

Secondly, if the slab is thicker than the critical size, the exothermic reaction reverses the decaying trend of the temperature profile and enables it to

---

*
$$\left(\frac{T-T_o}{T_w-T_o}\right) = \frac{1}{2}\left(\text{erf}\left[\frac{a-y}{2\sqrt{\alpha t}}\right] + \text{erf}\left[\frac{a+y}{2\sqrt{\alpha t}}\right]\right)$$

where α is thermal diffusivity ($\equiv K/\rho C$) and $\text{erf}(\Psi)$ is error function defined as

$$\text{erf}(\Psi) \equiv \frac{2}{\sqrt{\pi}}\int_0^{\Psi} e^{-\eta^2}\, d\eta$$

It can be found tabulated in C.R.C. *Handbook of Physics and Chemistry* and in Jahnke-Emde *Tables of Functions*.

propagate. The critical thickness of the igniting slab is of the order of twice the magnitude of a steadily propagating flame thickness.

$$a_c \approx \delta_f \tag{4.44}$$

From Eq. 4.38,

$$a_c \approx \frac{\alpha}{u_0} \approx \left(\frac{K(T_f - T_0)}{\Delta H \dot{W}'''}\right)^{1/2} \tag{4.45}$$

Consistent with our intuition, this equation shows that a thicker pilot flame is necessary to ignite a mixture whose conductivity is high and flame temperature is high. If the mean heat release rate is high, the critical pilot flame thickness is small. Furthermore, since thermal diffusivity is inversely proportional to density,

$$a_c \propto \frac{1}{\rho_0 u_0} \propto \frac{1}{P^{n/2}} \tag{4.46}$$

The critical slab thickness thus is smaller at higher pressures. If the order of the reaction is 2, $a_c \propto P^{-1}$.

There exists an alternative method of ignition using a flame. In this method, the ignition source is kept in communication with the combustible mixture through a constriction or a "peep-hole." If the flame passes through the constriction, ignition may be assured. Whether the flame would pass through the constriction or not is determined by the considerations pertaining to flame quenching by solid walls and narrow passages, the topic of concern for the next subsection.

(d) Ignition by Sparks

Ignition of combustible mixtures can be accomplished by striking an electrical spark* between two electrodes placed in the mixture as shown in Figure 4.22. The electrodes may either be flanged or unflanged.

The spark is caused either by capacitance discharges (which are relatively fast ≈ 0.01 μsec) or by inductance discharges. A capacitance discharge is produced by quickly discharging a condenser whereas an inductance discharge is produced by opening a circuit which involves transformers, ignition coils and magnetos.

If C_1 is the capacitance of a condenser and V_1 and V_2 are the voltages on the condenser respectively before and after the spark, the energy discharged is given by

$$E = \tfrac{1}{2}C_1(V_1^2 - V_2^2)$$

---

* Parenthetically, ignition by electrically heated wires is known to be very inefficient compared with ignition by spark discharges.

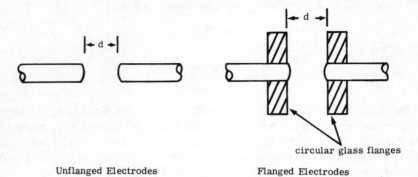

Figure 4.22 Geometric arrangement of flanged and unflanged electrodes for study of ignition by sparks

Given a mixture of a known composition in which the electrodes are located with a separation distance d, experiments show that ignition is possible only if the energy discharged is greater than some threshold value known as ignition energy. It varies with d in the fashion shown in Figure. 4.23(a).

At small values of d, the electrodes remove excessive quantities of heat from the incipient flame, thus disabling it to propagate; initiation of a propagating flame then requires a very high amount of energy discharge. In fact, when d is smaller than a certain value d_q, ignition of the mixture is impossible with any physically possible energy. d_q is called quenching distance.

As d is increased from d_q, the ignition energy continuously decreases first rapidly and then slowly. It reaches a minimum, beyond which it increases with further increase in d. This later behavior is due to the fact that when d is large, the spark loses much heat to the mixture.*

The *minimum ignition energy*, $E_{\min}$, can be defined as the smallest quantity of the spark energy which can initiate a propagating flame in the given mixture. The *quenching distance*, d_q, can be defined as the smallest gap between two solid walls through which a flame can propagate. $E_{\min}$ and d_q depend primarily upon the mixture physico-chemical properties, pressure, velocity and temperature and, rather weakly, on the electrode geometry.

---

* It is imperative that ignition is only possible in the area lying above the E-d curve of Figure 4.23(a). If the electrodes are flanged, this ignition area is reduced by increased flange diameter as shown in Figure 4.23(b). To summarize, if the spark energy is less than $E_{\min}$, ignition of the mixture is impossible for any electrode separation and if the electrode separation is less than d_q, ignition is impossible for any spark energy.

Kinetically Controlled Combustion Phenomena

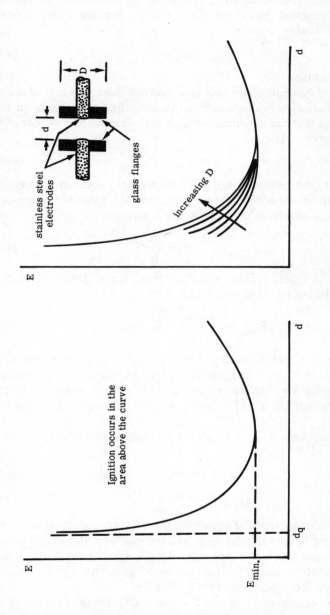

Figure 4.23 Ignition energy versus electrode separation (a) minimum ignition energy and quenching distance, (b) effect of flange diameter

E_{min} and d_q are often used to characterize the ignitability of different mixtures. Experimental measurements indicate that they are mutually related by the following relation

$$E_{min} = \kappa d_q^2 \qquad (4.47)$$

where κ is a constant.

The concept of quenching distance is somewhat similar to that of critical pilot flame slab thickness discussed in the preceding subsection. In fact, experience shows that the magnitude of d_q is of the same order as that of the critical slab thickness. That is,

$$d_q \approx 2a_c \qquad (4.48)$$

Postulating that ignition is achieved when the spark provides enough energy to raise the temperature of the mixture contained in a sphere of diameter d_q from T_0 to T_f, the minimum ignition energy is given by

$$E_{min} \approx \frac{\pi}{6} d_q^3 \cdot \rho_0 C(T_f - T_0) \qquad (4.49)$$

where $\pi d_q^3/6$ is the sphere volume and C, as usual, is the specific heat of the mixture. Substituting Eqs. 4.48 and 4.45 in 4.49,

$$E_{min} \approx \frac{\pi}{3} \frac{K(T_f - T_0)}{u_0} d_q^2 \qquad (4.50)$$

u_0 is the fundamental flame speed. If the order of the combustion reaction is 2, the factor $K(T_f - T_0)/u_0$ is independent of total pressure. The experimental finding given by Eq. 4.47 can thus be explained by simple energetic analysis. Table 4.2 shows that for a wide variety of hydrocarbon/air mixtures $\kappa \approx 0.0017$ cal/cm^2.

Since $d_q \approx 2a_c$, Eqs. 4.41, 4.46 and 4.50 yield the following pressure dependencies.

$$d_q \propto P^{-(n/2)} \qquad (4.51a)$$

$$E_{min} \propto P^{(2-3n)/2} \qquad (4.51b)$$

Equation 4.51 indicates that the higher the pressure, the narrower is the gap through which the flame can propagate and the smaller is the energy needed to initiate the flame. For hydrocarbon/air flames, $d_q \propto P^{-1}$ and $E_{min} \propto P^{-2}$.

These considerations are of extreme use in designing the ignition systems for propulsion devices operating at high altitudes.

The minimum ignition energy and the quenching distance vary with the fuel/oxidant proportions in a mixture in such a way that they are the smallest for a mixture near stoichiometric composition. A plot of E_{min} (or d_q) vs. %

TABLE 4.2

Quenching Distance and Ignition Energy for Stoichiometric Mixtures at Room Temperature and 1 atm.

Fuel	Oxidant	d_q (mm)	E_{min} ($10^{-5} \times$ cal)	$-\dfrac{d \ln d_q}{d \ln P}$
Hydrogen	45% Bromine	3.63	—	0.87–0.98
Hydrogen	Air	0.64	0.48	1.14
Hydrogen	Oxygen	0.25	0.10†	1.08
Methane	Air	2.55	7.90	—
Methane	Oxygen	0.30	0.15†	—
Acetylene	Air	0.76	0.72	—
Acetylene	Oxygen	0.09	0.01†	—
Ethylene	Air	1.25	2.65†	—
Ethylene	Oxygen	0.19	0.06†	—
Propane	Air	2.03	7.29	0.88
Propane	Argon "air"	1.04	1.84†	—
Propane	Helium "air"	2.53	10.83†	—
Propane	Oxygen	0.24	0.10†	—
1-3 Butadiene	Air	1.25	5.62	—
iso-Butane	Air	2.20	8.22†	—
n-Pentane	Air	3.30	19.60	—
Benzene	Air	2.79	13.15	—
Cyclohexene	Air	3.30	20.55	—
Cyclohexane	Air	4.06	32.98	—
n-Hexane	Air	3.56	22.71	—
1-Hexane	Air	1.87	5.24†	—
n-Heptane	Air	3.81	27.49	—
iso-Octane	Air	2.84	13.71†	—
n-Decane	Air	2.06	7.21†	0.89
1-Decane	Air	1.97	6.60†	—
n-Butyl benzene	Air	2.28	8.84†	—
Ethylene oxide	Air	1.27	2.51	—
Proplyene oxide	Air	1.30	4.54	—
Methyl formate	Air	1.65	14.82	—
Diethyl ether	Air	2.54	11.71	—
Carbon disulfide	Air	0.51	0.36	—

† Estimated. The rest, compiled from the data contained in NACA 1300 and in the article of Potter, A. E., Flame Quenching, pp. 145–181, *Progress in Combustion Science and Technology: I*, Ducarme, J., et al. (eds.), Pergamon (1960).

fuel in the mixture is usually U-shaped. We will consider the influence of the mixture composition on ignition in section 4.10 in connection with flammability limits.

When the combustible mixture is in motion, the minimum ignition energy increases nearly linearly with the flow velocity. In turbulent flow, the intensity of turbulence is known to increase the E_{min}. These effects are mainly caused by the increased heat losses. In order to visualize the influence of other variables on E_{min}, let Eqs. 4.45 and 4.40 be substituted in Eq. 4.50.

$$E_{min} \approx \frac{4\pi}{3} \rho_0 C \left(\frac{K}{\Delta H \dot{W}'''}\right)^{3/2} (T_f - T_0)^{5/2} \qquad (4.52)$$

When the nitrogen in a methane/air mixture is replaced by helium, E_{min} is experimentally found to be higher whereas when replaced by argon, it is found to be lower. This observation is consistent with Eq. 4.52 which shows that E_{min} is high if the mixture thermal conductivity is high. Addition of inert diluents to the mixture to alter ρ, C, K, T_f and $\dot{W}'''$ is a very well-known experimental technique.

Summarizing, the most conducive conditions for ignition are provided by

(i) a low flame temperature,
(ii) a high initial temperature,
(iii) a high heat of combustion,
(iv) a high mean reaction rate,
(v) a low volumetric heat capacity,
(vi) a low thermal conductivity,*
(vii) a high total pressure,
(viii) a mixture whose composition is nearly stoichiometric,
(ix) a low gas velocity,
(x) a low intensity of turbulence, if the flow is turbulent, and
(xi) an electrode separation distance near d_q.

4.10 Range of Ignition: (or Flammability Limits)

In the design of combustion engines and furnaces which can be ignited easily and reliably and in learning to prevent the initiation of unwanted fires and explosions, an understanding of the so-called *flammability limits* is essential. Flammability limits are usually given by specifying the range of conditions (i.e. temperature, pressure and composition) under which ignition is possible.

---

* See footnote on page 129.

Kinetically Controlled Combustion Phenomena

Figures 4.11, 4.12 and 4.13 delineate the range in which spontaneous ignition occurs from the range in which it does not occur. These figures are deduced from Semenov's theory. In section 4.9, similarly, we have obtained the forced ignition criteria which relate the conditions (of P, T and X_A) under which propagable flame can be initiated with an ignition source such as a spark. We discovered that unless a given fuel/oxidant mixture is provided with a minimum amount of energy, ignition is not possible. The minimum ignition energy depends not only on the nature of the fuel and the oxidant, but also on their proportions.

As mentioned in Section 4.9, if one determines the minimum ignition energy for mixtures of various proportions of a given fuel and oxidant, a U-shaped plot is obtained as shown in Figure 4.24. This figure indicates that E_{min} is the least for a mixture whose composition is stoichiometric (or nearly stoichiometric). If the mixture gets "leaner" or "richer," the E_{min} increases first gradually and then abruptly. The abrupt rise of E_{min} suggests that when the mixture is too "lean" (i.e. if the fuel content is less than $l\%$ in Figure 4.24) or too "rich" (i.e. if the fuel content is more than $r\%$ in Figure 4.24) ignition is possible only if "infinite" amount of energy is supplied through the igniter.

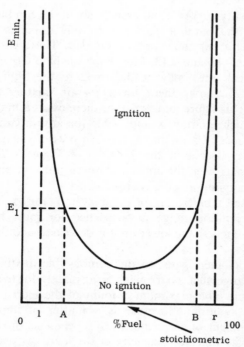

Figure 4.24 Effect of the mixture ratio on the minimum ignition energy. (Flammability limits)

Since the energy any real system can supply is finite, this means ignition of very lean and very rich mixtures is impossible.

Several comments are imminent.

(i) All the energies and compositions corresponding to the area lying inside the U in Figure 4.24 result in an ignition whereas those lying outside, do not.

(ii) Ignition is impossible if the mixture is too lean or too rich. The upper and lower limits of flammability (corresponding to r and l in Figure 4.24) are characteristic of the nature of fuel/oxidant combination. Table 4.3, abstracted mainly from N.A.C.A. 1300, and Mullins' papers, gives the flammability limits for various fuel/oxidant mixtures. This table also furnishes much other valuable combustion data.

(iii) For any given energy E_1, all mixtures, lying between an upper (B) and a lower (A) limit of composition, can be ignited. All the mixtures lying outside these limits cannot be ignited.

(iv) The range of ignition—AB—is narrowed if E_1 is chosen smaller. If E_1 is smaller than the least E_{min}, ignition is impossible for any mixture.

(v) Flammability curves found extensively in the combustion literature delineate the ignitible and nonignitible domains on a pressure (or temperature) and composition plot. These curves are similar to Figures 4.12 and 4.13. The shape of these curves is not always symmetrical; they may be deformed to the right or to the left. Figure 4.25 shows them for H_2 + air mixture and $CO + O_2$ mixture. This figure indicates that there exists a minimum pressure below which ignition is impossible for any mixture composition. The existence of a minimum ignition pressure suggests that as the pressure is decreased the U-curve of Figure 4.24 progressively becomes narrower, the upper and lower limits ultimately coinciding as the minimum pressure is approached.

(vi) If the initial temperature of the mixture is raised, it is well-known that the ignition range is broadened for most hydrocarbon/air mixtures. That is, the lower limit is decreased and the upper limit is raised.

(vii) Addition of inert gases to the combustible mixtures narrows the ignition range, such a narrowing being mainly due to a decrease in the upper limit. The maximum amount of the inert gas necessary to eliminate ignition altogether is a matter of importance in the extinguishment of accidental fires. It seems as though the thermal conductivity and heat capacity of the inert gas play an important role in determining its effectiveness as an extinguishing agent. It is

Kinetically Controlled Combustion Phenomena

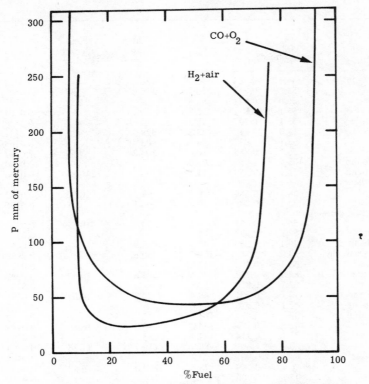

Figure 4.25 Flammability limits for two different mixtures

clear from Eq. 4.40 that while conductivity encourages flame propagation, specific heat inhibits it.* A good fire extinguishing agent, hence, can be expected to have a low conductivity and a high specific heat. In fact, experience indicates that agents with high C/K are better extinguishers.†

It is also well-known that the amount of an extinguishing agent required to prevent forced ignition is several times larger than that required to prevent spontaneous ignition.

(viii) A gradual replacement of the fuel A in an A/oxidant mixture by another fuel B will gradually shift the flammability curves from those of the A/oxidant mixture to those of the B/oxidant mixture.

---

* Note from Section 4.9 that a larger conductivity makes ignition harder (Eq. 4.52) and propagation easier (Eq. 4.40).

† Polyatomic heavy organic compounds such as ethyl, ethylene and methylene bromides, ethyl and methyl iodides, bromoform, chloroform, carbon tetrachloride, etc.

TABLE 4

Fuel	Mol. wt.	Spec. grav.	T_{Boil} (°C)	Heat of vap. (cal/gm)	Heat of comb. (kcal/gm)	Stoichiometry % Vol.	f
Acetaldehyde	44.1	0.783	−56.7	136.1	—	0.0772	0.1280
Acetone	58.1	0.792	56.7	125.0	7.36	0.0497	0.1054
Acetylene	26.0	0.621	−83.9	—	11.52	0.0772	0.0755
Acrolein	56.1	0.841	52.8	—	—	0.0564	0.1163
Acrylonitrile	53.1	0.797	78.3	—	—	0.0528	0.1028
Ammonia	17.0	0.817	−33.3	328.3	—	0.2181	0.1645
Aniline	93.1	1.022	184.4	103.4	—	0.0263	0.0872
Benzene	78.1	0.885	80.0	103.2	9.56	0.0277	0.0755
Benzyl alcohol	108.1	1.050	205.0	—	—	0.0240	0.0923
1,2-Butadiene (methylallene)	54.1	0.658	11.1	—	10.87	0.0366	0.0714
n-Butane	58.1	0.584	−0.5	92.2	10.92	0.0312	0.0649
Butanone (methylethyl ketone)	72.1	0.805	79.4	106.1	—	0.0366	0.0951
1-Butene	56.1	0.601	−6.1	93.3	10.82	0.0377	0.0678
d-Camphor	152.2	0.990	203.4	—	—	0.0153	0.0818
Carbon disulfide	76.1	1.263	46.1	83.9	—	0.0652	0.1841
Carbon monoxide	28.0	—	−190.0	50.6	—	0.2950	0.4064
Cyclobutane	56.1	0.703	12.8	—	—	0.0377	0.0678
Cyclohexane	84.2	0.783	80.6	85.6	10.47	0.0227	0.0678
Cyclohexene	82.1	0.810	82.8	—	—	0.0240	0.0701
Cyclopentane	70.1	0.751	49.4	92.8	10.56	0.0271	0.0678
Cyclopropane	42.1	0.720	−34.4	—	—	0.0444	0.0678
trans-Decalin	138.2	0.874	187.2	—	—	0.0142	0.0692
n-Decane	142.3	0.734	174.0	86.0	10.56	0.0133	0.0666
Diethyl ether	74.1	0.714	34.4	83.9	—	0.0337	0.0896
Ethane	30.1	—	−88.9	116.7	11.34	0.0564	0.0624
Ethyl acetate	88.1	0.901	77.2	—	—	0.0402	0.1279
Ethanol	46.1	0.789	78.5	200.0	6.40	0.0652	0.1115
Ethylamine	45.1	0.706	16.7	146.1	—	0.0528	0.0873
Ethylene oxide	44.1	1.965	10.6	138.9	—	0.0772	0.1280
Furan	68.1	0.936	32.2	95.6	—	0.0444	0.1098
n-Heptane	100.2	0.688	98.5	87.1	10.62	0.0187	0.0661
n-Hexane	86.2	0.664	68.0	87.1	10.69	0.0216	0.0659
Hydrogen	2.0	—	−252.7	107.8	28.65	0.2950	0.0290
iso-Propanol	60.1	0.785	82.2	158.9	—	0.0444	0.0969
Kerosene	154.0	0.825	250.0	69.5	10.30	—	—
Methane	16.0	—	−161.7	121.7	11.95	0.0947	0.0581
Methanol	32.0	0.793	64.5	263.0	4.74	0.1224	0.1548
Methyl formate	60.1	0.975	31.7	112.8	—	0.0947	0.2181
n-Nonane	128.3	0.772	150.6	68.9	10.67	0.0147	0.0665
n-Octane	114.2	0.707	125.6	71.7	10.70	0.0165	0.0633
n-Pentane	72.1	0.631	36.0	87.1	10.82	0.0255	0.0654
1-Pentene	70.1	0.646	30.0	—	10.75	0.0271	0.0678
Propane	44.1	0.508	−42.2	101.7	11.07	0.0402	0.0640
Propene	42.1	0.522	−47.7	104.5	10.94	0.0444	0.0678
n-Propanol	60.1	0.804	97.2	163.9	—	0.0444	0.0969
Toluene	92.1	0.872	110.6	86.7	9.78	0.0227	0.0743
Triethylamine	101.2	0.723	89.4	—	—	0.0210	0.0753
Turpentine	—	—	—	—	—	—	—
Xylene	106.0	0.870	130.0	80.0	10.30	—	—
Gasoline 73 octane	120.0	0.720	155.0	81.0	10.54	—	—
Gasoline 100 octane	—	—	—	—	—	—	—
Jet fuel JP1	150.0	0.810	—	—	10.28	0.0130	0.0680
JP3	112.0	0.760	—	—	10.39	0.0170	0.0680
JP4	126.0	0.780	—	—	10.39	0.0150	0.0680
JP5	170.0	0.830	—	—	10.28	0.0110	0.0690

* Compiled from data presented in NACA 1300 and estimated from various miscellaneous sources.

Kinetically Controlled Combustion Phenomena

Physical and Combustion Properties of Selected Fuels in Air*

Flammability limits (% stoichio.)		Spont. ign. temp. (°C)	Fuel for max. flame speed (% stoichio.)	Max. flame speed (cm/sec)	Flame temp. at max. fl. speed °K	Ign. energy (10⁻⁵ cal.)		Quenching dist. mm	
Lean	Rich					Stoich.	Min.	Stoich.	Min.
—	—	—	—	—	—	8.99	—	2.29	—
59	233	561.1	131	50.18	2,121	27.48	—	3.81	—
31	—	305.0	133	155.25	—	0.72	—	0.76	—
48	752	277.8	100	61.75	—	4.18	—	1.52	—
87	—	481.1	105	46.75	2,461	8.60	3.82	2.29	1.52
—	—	651.1	—	—	2,600	—	—	—	—
—	—	593.3	—	—	—	—	—	—	—
43	336	591.7	108	44.60	2,365	13.15	5.38	2.79	1.78
—	—	427.8	—	—	—	—	—	—	—
—	—	—	117	63.90	2,419	5.60	—	1.30	—
54	330	430.6	113	41.60	2,256	18.16	6.21	3.05	1.78
—	—	—	100	39.45	—	12.67	6.69	2.54	2.03
53	353	443.3	116	47.60	2,319	—	—	—	—
—	—	466.1	—	—	—	—	—	—	—
18	1,120	120.0	102	54.46	—	0.36	—	0.51	—
34	676	608.9	170	42.88	—	—	—	—	—
—	—	—	115	62.18	2,308	—	—	—	—
48	401	270.0	117	42.46	2,250	32.98	5.33	4.06	1.78
—	—	—	—	44.17	—	20.55	—	3.30	—
—	—	385.0	117	41.17	2,264	19.84	—	3.30	—
58	276	497.8	113	52.32	2,328	5.74	5.50	1.78	1.78
—	—	271.7	109	33.88	2,222	—	—	—	—
45	356	231.7	105	40.31	2,286	—	—	2.06	—
55	2,640	185.6	115	43.74	2,253	11.71	6.69	2.54	2.03
50	272	472.2	112	44.17	2,244	10.04	5.74	2.29	1.78
61	236	486.1	100	35.59	—	33.94	11.47	4.32	2.54
—	—	392.2	—	—	—	—	—	—	—
—	—	—	—	—	—	57.36	—	5.33	—
—	—	428.9	125	100.35	2,411	2.51	1.48	1.27	1.02
—	—	—	—	—	—	5.40	—	1.78	—
53	450	247.2	122	42.46	2,214	27.49	5.74	3.81	1.78
51	400	260.6	117	42.46	2,239	22.71	5.50	3.56	1.78
—	—	571.1	170	291.19	2,380	0.36	0.36	0.51	0.51
—	—	455.6	100	38.16	—	15.54	—	2.79	—
46	164	632.2	106	37.31	2,236	7.89	6.93	2.54	2.03
48	408	470.0	101	52.32	—	5.14	3.35	1.78	1.52
—	—	—	—	—	—	14.82	—	2.79	—
47	434	238.9	—	—	—	—	—	—	—
51	425	240.0	—	—	2,251	—	—	—	—
54	359	284.4	115	42.46	2,250	19.60	5.26	3.30	1.78
47	370	298.3	114	46.75	2,314	—	—	—	—
51	283	504.4	114	42.89	2,250	7.29	—	2.03	1.78
48	272	557.8	114	48.03	2,339	6.74	—	2.03	—
—	—	433.3	—	—	—	—	—	—	—
43	322	567.8	105	38.60	2,344	—	—	—	—
—	—	—	—	—	—	27.48	—	3.81	—
—	—	252.2	—	—	—	—	—	—	—
—	—	298.9	—	—	—	—	—	—	—
—	—	468.3	106	37.74	—	—	—	—	—
—	—	248.9	107	36.88	—	—	—	—	—
—	—	261.1	107	38.17	—	—	—	—	—
—	—	242.2	—	—	—	—	—	—	—

4.11 Auto-Catalytic Ignition

The concept of ignition has been developed in the preceding sections from a purely thermal viewpoint by considering the progressively accelerating rate of a simple, exothermic Arrhenius type reaction. Heating is held to be the sole reason for the progressive acceleration of the rate. The only aspect which continuously resists such an accelerating reaction rate is the concentration term, C_A^n, in the rate law. As the reactant concentration continuously falls (and hence the product concentration continuously increases), an isothermal reaction rate continuously decreases with time as discussed in Section 4.2.

Many combustion reactions often exhibit a peculiar behavior in which the reaction rate initially increases with increasing product concentration even if the temperature is kept constant. Such reactions are known as *auto-catalytic* or self-catalyzing. The rate law for such reactions is typically,

$$\frac{dC_A}{dt} = -k_n C_A^{n-j} C_P^j e^{-E/RT} \qquad (4.53)$$

where C_P is the product concentration. Initially $C_P = 0$ and therefore the reaction rate is zero. In order to start an auto-catalytic reaction, it is necessary to generate an initial amount of the catalyzing product by some external means. Usually, this is furnished by a slow thermal reaction which ceases when it generates enough catalyzing product. Once initiated, the catalytic reaction rate of an exothermic reaction accelerates under the combined influence of the calalytic (C_P^j term) and thermal ($e^{-E/RT}$ term) effects.

The critical conditions for auto-catalytic ignition are determined by the same criterion as for purely thermal ignition—viz. at ignition, the heat generation rate due to the reaction exceeds the heat loss rate. An auto-catalytic ignition is often recognized by a relatively longer induction time.

Sometimes thermal ignition is preceded by auto-catalytic heating. Ignition of ethane/oxygen mixture contaminated with methyl nitrite offers an example for such a situation. The main part of the induction time for this mixture is taken up by the auto-acceleration of the reaction and thus does not involve steep temperature rise. During the final portion of the induction time, however, experiments show a very abrupt temperature rise which indicates that the ignition indeed is thermal.

4.12 Chemical Chain Ignition and Explosions

In Chapter 2 (Section 2.5) we have introduced the concept of chain reactions which indeed belong to the broader spectrum of auto-catalytic reactions. The rate of a chain reaction is greatly influenced not only by the final product concentration but also by some intermediate unstable species concentration. In the reaction between hydrogen and bromine, for example, hydrogen and bromine atoms are the active catalyzing intermediate species. Once these chain carriers are produced by supplying energy from an external means, the propagation of chain continues until and unless the chain carriers are suppressed by third body collisions. Third body collisions can be broadly classified into two classes—those in which the vessel walls act as the third body and those in which the molecules, or radicals and atoms of any species in the vessel act as the third body. When two chain carriers collide simultaneously with a third body, they recombine to form an inactive molecule, thus terminating the chain.

A chain reaction may either be straight or branching. Straight chains are those in which one chain carrier is consumed to generate another. Branching chains are those in which for every chain carrier consumed two or more new chain carriers are generated. The examples of $H_2 + Br_2$, $H_2 + Cl_2$, propane decomposition and acetaldehyde decomposition reactions mentioned in Section 2.5 illustrate straight chains. A spectacular example to illustrate branching chains is that of $H_2 + O_2$ reaction in which for every H atom consumed, three H atoms are generated.

As shown in Section 2.6, if k_i is the specific forward reaction rate constant for the chain initiation reaction, k_{ii} is that for the chain branching reaction in which n' chain carriers are generated for every one consumed, and k_t is that for chain termination, the steady state radical concentration is given by Eq. 2.14 as

$$C_R = k_i C_A C_B C_C \cdots /(k_t - [n' - 1]k_{ii} C_A C_B C_C \cdots) \qquad (2.14)$$

This equation implies several important points concerning chain reactions. If $n' = 1$, the chain is straight. C_R then will attain a steady state value $k_i C_A C_B C_C \ldots /k_t$; the reaction rate attains a limiting value. A straight chain reaction is therefore never explosive. For branching chain reactions (i.e. $n' \geq 2$) if the denominator of the right hand side of Eq. 2.14 vanishes, and the numerator is nonzero, C_R tends to infinity. This means when the probabilities of branching and termination are equal, the radical concentration and consequently the reaction rate become enormously large (even under isothermal conditions). An explosion is the result. These type of explosions are known as *chemical chain explosions* or chain carrier explosions.

Now we can explain why the critical pressure and temperature for $H_2 + O_2$ mixture ignition behave in the complexly strange manner shown in Figure 2.2. Recalling what has been said in Section 2.6, the ignition threshold pressure-temperature curve is S-shaped instead of monotonically decreasing as suggested by thermal theory of ignition. (cf. Figure 4.11). At a very low temperature the system can explode only at a very high pressure. At a very high temperature explosion occurs even at very low pressures. When the temperature is moderate, as the pressure is increased the lower threshold boundary is crossed and explosion occurs. Further increase in pressure beyond the middle boundary will result in a nonexplosive situation. However, continued elevation of the pressure beyond the upper boundary ensures explosion once again.

The lower boundary is greatly influenced by the dimensions, material and the nature of the surface of the vessel. Addition of inert gases to $H_2 + O_2$ mixture lowers the lower boundary.

The dimensions, material and surface cleanliness do not have noticeable effect on the middle boundary. The uppermost boundary follows the laws of thermal ignition quite satisfactorily. The question we ask ourselves therefore is this. What are the physical and chemical phenomena that cause the ignition peninsula?

This question can be answered by the chain reaction theory as below. It is reasonable to postulate that chemical chain ignition occurs when the chain branching rate exceeds the chain termination rate.

The lower boundary is explained by considering chain termination due to collisions involving the vessel walls. The rate of such termination depends upon the nature of the vessel wall and the rate at which the chain carriers diffuse to the wall. The diffusion rate is governed by the coefficient of diffusion D (a function of pressure, temperature and the molecular constants of the mixture), the size and shape of the vessel. Mainly because the diffusion coefficient is inversely proportional to the pressure, lower pressure makes diffusional transport of the chain carriers to the wall easier. As they hit the wall, the chain is terminated. Therefore the critical temperature is higher for lower pressures. Semenov derives the following equation to relate the critical pressure and temperature for the lower boundary,

$$P_C = Ae^{B/T_c} \qquad (4.54)$$

A and B are constants which depend upon the nature of the chain carrier, other participating reactants and inert additives, dimensions and nature of the walls, etc.

The middle boundary is explained by considering chain termination due to collisions involving the molecules, radicals and atoms in the bulk of the mixture in the vessel. Based upon the premise that chain branching rate is a weaker function of pressure than chain breaking, a gradual increase in pressure from the lower boundary would yield at some point a situation in which chain breaking dominates chain branching. The frequency of chain breaking collisions in the volume depends upon the composition of the mixture at any instant and the molecular constants. A formula for the critical

Kinetically Controlled Combustion Phenomena

conditions on the middle boundary can be derived as

$$P_c = A' e^{-B'/T_c} \tag{4.55}$$

Equations 4.54 and 4.55 are found to predict the ignition peninsula for $H_2 + O_2$ mixtures as well as for $CO + O_2$ mixtures.

Ignition conditions measured for various hydrocarbon-oxygen mixtures show irregularly shaped S-curves on a P-T graph. Such irregularities indicate that perhaps chain carrier and thermal effects play a combined role simultaneously and that the thermal and chain carrier ignition hypotheses postulated in this chapter, being ideal, can truly represent only a few real reactions.

4.13 Longwell's Well-Stirred Reactor: (Space Heating Rate)

In their quest to obtain a flame dominated by chemical kinetics J. P. Longwell and his associates developed a combustor in which the mixing rate is much faster than the reaction rate. Figure 4.26 shows this combustor schematically. The spherical chamber is lined with insulating fire brick to minimize heat losses. Premixed fuel and air enter at sonic velocity through six jets aimed at the center and hot combustion products leave through pores in the brick.

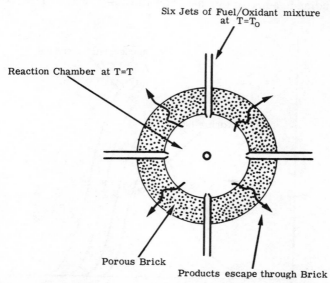

Figure 4.26 Longwell's well-stirred reactor (adapted with the permission of American Chemical Society, from: J. P. Longwell and M. A. Weiss, *Ind. and Engr. Chemistry*, **47**, p.1634 (1955))

At low pressures excellent mixing is accomplished. The combustion reaction maintains a high temperature in the reactor. By thus eliminating the time lags associated with mixing and evaporation, Longwell's homogeneous reactor simulates a typical turbojet combustor.

If $\dot{W}'''$ is the uniform reaction rate (gms of fuel A per unit volume per unit time), ΔH is the heat of combustion (cal per gm of A), V is volume of the reactor, the amount of heat generated by combustion is equal to $\dot{W}'''\Delta H \cdot V$ cal/sec. This amount of heat is used to increase the temperature of the flow ($\dot{m}$ gm per sec) from T_0 to T, where T is the reactor uniform temperature. If C is the mean specific heat of the combustion gases, since heat loss is negligible, energy balance is given by

$$\dot{m}C(T - T_0) = \dot{W}'''V \Delta H \qquad (4.56)$$

Rearranging,

$$\frac{\dot{m}}{V} = \frac{\dot{W}'''\Delta H}{C(T - T_0)} \qquad (4.57)$$

Figure 4.27 shows Eq. 4.57 which implies that an increase in the mass input rate per unit volume of the reactor causes a decrease in the reactor temperature. Longwell's experiments show that there exists a maximum allowable rate of reaction $\dot{W}'''$ (and hence heat generation rate $\dot{W}'''\Delta H$) corresponding to a maximum mass flow rate ($\dot{m}/V$) which, if exceeded, blows

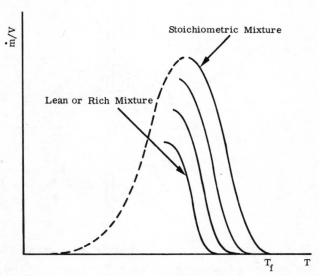

Figure 4.27 Blow-off in Longwell reactor

out the flame. Solving the equations of energy and species conservation this behavior is theoretically predictable.

The maximum space heating rate $(\Delta H \dot{W}'''_{max}) = (\dot{m}C/V)(RT^2/E)$ is an important limiting factor in the design of propulsion and fire extinguishment systems. Consider, for example, the fact that the combustor will be 100% efficient if the reactor maximum temperature is equal to the adiabatic flame temperature of the fuel/air mixture. But to accomplish $T_{max} = T_f$, one needs an infinitely large combustion chamber.

In reality, an infinite combustion chamber is not possible. The engineer therefore should use his judgment to compromise between the combustor efficiency and its weight. If the order of the reaction is 2 (i.e., hydrocarbon-air flames), a small combustor operated at high pressures will have as much efficiency as a large combustor operated at lower pressures. The maximum space heating rate, which occurs at the blow-out, represents limiting efficiency of the combustor/fuel/oxidant system. The great strides of progress in the design of combustors can be recognized by comparing the maximum space heating rate (SHR) of a domestic boiler ($0.02 \cdot 10^6$ BTU/ft^3/Hr) with that of a modern propulsion combustor motor ($2 \cdot 10^{17}$ BTU/ft^3/Hr). Table 4.4 gives some values of SHR of use in evaluation and design of combustors for propulsion purposes and in the estimation of combustibility of various fuel/oxidant systems.

4.14 Conclusions

The field of chemical kinetics is still very much an active frontier of science. The work outlined in this chapter is an engineering approach that borrows from the kineticist some of his methods and serves to give a basis for predicting the behavior of combustion systems. Study of flames is greatly facilitated by the systematic categorization present in Section 4.1. The fact that reactions can be classified into two types—simple and complex—is elaborated. A simple reaction is that whose rate is influenced only by the temperature and concentration of the initial reactants. Simple reactions are also sometimes called thermal reactions. Complex reactions are those in which the concentrations of the final and intermediate products influence the reaction rate.

Defining ignition as a state in which the reactant mixture continues to react at a progressively accelerating rate, two types of ignition are introduced. In spontaneous ignition, the mixture is heated in a bulk by enclosing it in a closed vessel or by subjecting it to compression. Ignition occurs when heat is produced at a rate which exceeds the heat loss rate. In forced ignition, the mixture is heated intensely and locally by an energy source such as a spark,

TABLE 4.4*
Combustion Kinetics, Flame Temperatures and Space Heating Rates

1 BTU/ft³/hr = 2.4912 × 10⁻⁶ cal/cm³-sec.

Fuel	Oxidant	Pressure atm.	Initial temp. °K	Activation energy kcal/mole	Preexponential	Overall order	Flame temp. °K	Average reaction rate moles/cm³/sec	Maximum reaction rate moles/cm³/sec	SHR max. BTU/ft³/hr
Propane	Air	—	378	31.0	2×10^{14}	1.56	2,250	—	—	—
Propane	Air	—	400	31.0	—	1.60	2,313	1.04	3.27	$6.44 \cdot 10^8$
Propane	Air	—	483	—	—	—	2,351	—	—	—
Propane	Air	—	558	—	—	—	2,384	—	—	—
Propane	Helium "air"	—	—	—	—	1.63	2,557	2.94	8.40	$1.66 \cdot 10^9$
Propane	Argon "air"	—	—	—	—	1.71	2,557	3.82	11.30	$2.24 \cdot 10^9$
Propane	0.17 O_2 + N_2	—	313	—	—	—	2,051	—	—	—
Propane	0.21 O_2 + N_2	—	—	—	—	1.90	2,253	—	—	—
Propane	0.30 O_2 + N_2	—	—	—	—	—	2,558	—	—	—
Propane	0.50 O_2 + N_2	—	—	—	—	—	2,844	—	—	—
Propane	0.70 O_2 + N_2	—	—	—	—	—	2,970	—	—	—
Methane	Air	—	—	29.0	—	1.60	2,236	1.60	5.07	$3.90 \cdot 10^8$
Methane	O_2	—	—	—	—	—	3,020	—	—	—
Ethane	N_2O	—	—	49.0	—	—	2,740	—	—	—
Ethane	NO	—	—	—	—	—	2,840	—	—	—
Butane	O_2	—	—	21.0	5.4×10^{13}	2.00	2,256	—	—	—
Hexane (n)	Air	—	—	51.0	—	—	2,239	—	—	—

Decane (*n*)	Air	—	—	51.0	—	1.62	2,286	—	—	—
Octane (*iso*)	Air	—	—	32.4	—	1.52	2,302	0.24	0.72	$3.6 \cdot 10^{7}$
Octane (*n*)	Air	1	400	40.0	—	1.80	2,251	—	—	$1.0 \cdot 10^{9}$
Octane (*n*)	Air	15	400	40.0	—	1.80	2,316	—	—	$7.3 \cdot 10^{10}$
Octane (*n*)	O_2†	1	400	40.0	—	1.80	3,056	—	—	$4.5 \cdot 10^{11}$
Octane (*n*)	O_2†	15	400	40.0	—	1.80	3,422	—	—	$2.7 \cdot 10^{14}$
Hexene	Air	—	—	43.2	—	—	2,324	—	—	—
Decene	Air	—	—	—	—	—	2,322	—	—	—
Benzene	Air	—	—	47.2	—	—	2,365	—	—	—
Acetylene	N_2O	—	298	31.0	—	—	2,940	—	—	—
Acetylene	NO	—	—	—	—	—	3,080	—	—	—
Ethylene	Air	—	298	—	—	1.60	2,362	5.00	15.10	$1.93 \cdot 10^{9}$
Ethylene	N_2O	—	—	—	—	—	2,820	—	—	—
Ethylene	NO	—	—	—	—	—	2,920	—	—	—
Ethylene oxide decomposition		—	298	—	—	—	1,217	—	—	—
Ammonia	O_2†	1	400	49.5	—	1.70	2,782	—	—	$1.50 \cdot 10^{10}$
Ammonia	O_2†	15	400	49.5	—	1.70	3,041	—	—	$3.00 \cdot 10^{12}$
Hydrogen	Air	—	313	15.0	1.6×10^{12}	2.17	2,380	1.69	4.63	$1.10 \cdot 10^{9}$
Hydrogen	O_2†	1	400	18.0	—	2.00	3,018	—	—	$4.60 \cdot 10^{11}$
Hydrogen	O_2†	15	400	18.0	—	2.00	3,384	—	—	$1.10 \cdot 10^{15}$
Hydrogen	Br_2(40%)	—	323	—	—	1.50	1,490	—	—	—
Hydrogen	F†	1	400	50.0	—	2.00	3,962	—	—	$7.00 \cdot 10^{13}$
Hydrogen	F†	15	400	50.0	—	2.00	4,546	—	—	$1.60 \cdot 10^{17}$

* Compiled from a variety of sources.

† Stoichiometric mixtures.

an incandescent particle, etc. A flame is initiated in the vicinity of the source and propagated into the rest of the mixture.

Simple reactions and straight chain reactions occurring at constant temperature never accelerate to ignition. Simple reactions occuring under adiabatic conditions always attain ignition when the heat loss rate is smaller than generation rate. Branching chain reactions, likewise, lead to ignition when the chain carrier generation rate exceeds their termination rate.

The concepts of flammability, quenching and space heating rate are also introduced in this chapter.

In every biennial international combustion symposium, a dominant place in the program is occupied by research in the chemical kinetics of combustion processes. A continuous review of progress in this area usually appears in the bimonthly journal of *Combustion and Flame*. A very instructive review of spontaneous ignition is found in AGARDograph 4 by B. P. Mullins. This review introduces the reader to the wealth of literature in this field. For a more concise account and experimental results, the book by Jost contains an excellent summary.

For ignition delay in rocket propellants, Penner's book, *Chemistry Problems in Jet Propulsion*, is recommended along with his chapter on this subject in the Princeton Series Volume on Combustion. An elaborate but excellent summary of various combustion phenomena (including ignition and flammability) related to hydrocarbon/air mixtures is given by N.A.C.A. Report 1300. The reader is referred to Longwell's paper for a discussion on the homogeneous reactor.

Semenov's and Frank–Kamenetski's books deal with ignition and explosions very thoroughly.

References

Apin, Todes, Kariton, *Zhur. Fiz. Khim.*, **8**, p. 866 (1936).
Frank-Kamenetski, *Diffusion and Heat Exchange in Chemical Kinetics*. Translated from Russian by N. Thon. Princeton University Press, Princeton (1955).
Gaydon, A. G. and Wolfhard, H. G., *Flames*, Chapman and Hall, London (1960).
Jackson, J. L., and Brokaw, R. S., N.A.C.A., R.M. E54B19 (1954).
Jost, W., *Explosion and Combustion Processes in Gases*, McGraw-Hill, New York (1946).
Knitrin, L. N., *Physics of Combustion and Explosion*, National Science Foundation, translation from Russian (1962) (Israel program for Scientific Translations, Jerusalem).
Lewis, B. and von Elbe, G., *Combustion, Flames and Explosion of Gases*, Academic, New York (1951). Longwell, J. P. and Weiss, M. A., *Ind. Eng. Chem.*, **47**, p. 1634 (1955).
Mullins, B. P., *Fuel*, **32**, p. 327, 363, and 451 (1953).
Mullins, B. P., AGARDograph 4, Advisory Group for Aeronautical Research and Development Organization, N.A.T.O.
Mullins, B. P. and Penner, S. S., *Explosions, Detonations, Flammability and Ignition*, AGARDograph No. 31, Pergamon Press, New York (1959).

N.A.C.A. Report 1300, *Basic Considerations in the Combustion of Hydrocarbon Fuels in Air*, National Advisory Committee for Aeronautics (1957).

Penner, S. S., *Chemistry Problems in Jet Propulsion*, Pergamon Press, New York (1957).

Penner, S. S., *Princeton High Speed Aerodynamics and Jet Propulsion Series*, Volume on Combustion.

Potter, A. E., Flame Quenching, pp 145–181, *Progress in Combustion Science and Technology, I*, Ducarme, J., et al. (eds.), Pergamon, New York (1960).

Rice, *Journ. Am. Chem. Soc.*, **57**, p. 310, 1044, 2212 (1935); *Journ. Chem. Phys.*, **7**, p. 701 (1939).

Semenov, N. N., *Chemical Kinetics and Chain Reactions*, Oxford University Press, London (1935).

Spalding, D. B., *Some Fundamentals of Combustion*, Butterworths, London (1955).

Vulis, L. A., *Thermal Regimes of Combustion* (translated from Russian by M. D. Friedman), McGraw-Hill, New York (1961).

Williams, F. A., *Combustion Theory*, Addison-Wesley, Reading (1964).

Zeldovich and Yakovlev, *Doklady Akad. Nauk. S.S.S.R.*, **19**, p. 699 (1938).

CHAPTER 5

Diffusion Flames in Liquid Fuel Combustion

Following the categorization of flames introduced in Section 4.1, diffusion flames are those in which the chemical reaction rate* is very high compared to the rate of flow, diffusion and mixing. Diffusion flames are known to man since the dawn of history. Unwanted natural fires are mostly composed of diffusion flames. Flames in piles of wooden sticks used by the cave man and oil wick torches and candle flames used by medieval man are some primitive examples of diffusion flames. Burning of a fuel droplet in a ramjet, turbojet, or a diesel engine, of a gasoline spill on an aircraft carrier deck, of a gaseous fuel jet in a kitchen stove, of a carbon particle in a furnace etc., are some contemporary examples.

An ideal diffusion flame consists of an infinitesimally thin exothermic reaction zone which separates the fuel stream from the oxidant stream; that is, in ideal diffusion flames the fuel and oxidant are separated by an interface. They diffuse towards the reaction zone due to the composition gradients which are strongly governed by the flow and mixing. Because diffusion, flow, mixing and heat transfer play an important role, study of diffusion flames logically involves boundary layer type heat and mass transfer analyses.†

Broadly, diffusion flames are either homogeneous or heterogeneous. In homogeneous diffusion flames, the fuel as well as the oxidant are in gaseous state, such as in the case of combustion of a jet of natural gas issuing into air. Most commonly, heterogeneous diffusion flames occur when a solid or liquid fuel burns in an oxidizing gaseous atmosphere. A complete account

---

* If the reaction mechanism involves several intermediate steps, a diffusion flame is observed when the slowest of the reaction steps is much faster than diffusion.

† It is precisely for this reason, one of the most important factors influencing diffusion flames is the surface to volume ratio of the burning body of the fuel. A smaller particle of fuel, for example, receives more heat and consequently burns faster. The use of solid fuels in pulverized form and liquid fuels in the form of a spray of fine droplets is thus eminently justified.

Diffusion Flames in Liquid Fuel Combustion

of diffusion flames consists, therefore, of three separate sections—liquid fuel flames, solid fuel flames and gaseous fuel flames.

5.1 Preliminary Remarks on Solid and Liquid Fuels

Liquid and solid fuels* are often considered to be members of a single family as described below. The gasification (or vaporization) temperature of these fuels is usually lower than the self-ignition temperature of the fuel vapor/air mixture. Consequently, combustion occurs mainly (and sometimes, only) in the gas phase. The gasification of light weight liquid fuels is purely a physical process whereas that of heavy liquids and solids involves a chemical break-up, sometimes known as destructive distillation. The surface temperature of a steadily burning light weight liquid fuel is close to, but slightly less than, the boiling temperature. Heavy fuels, on the other hand, very seldom have a unique boiling temperature. They burn with a rather high surface temperature at which the liquid disintegrates into light weight combustible vapors and carbonaceous residue. Same is true with most of the solids. Heavy liquid fuels such as crude oil, heavy fuel oil, etc. thus offer a link between simple light weight liquids and complex, charring solids. Conversely, subliming and melting solid fuels burn in the same way as simple liquids.

Light liquid fuels can be classified into two types—those that are easily vaporizable and those that are not. Easily vaporizable liquids such as gasoline are first vaporized and mixed with air in a "carburetor" before introducing the combustible mixture into the combustion chamber. Ignition and flame propagation in such a mixture are governed by the laws of "premixed flames" (i.e. Chapters 4 and 8). The liquids which are "hard to vaporize" are evaporated *as they burn*. In general, these fuels are moderately heavy. Combustion of a diesel fuel spray in hot compressed air and combustion in open trays are two examples of this type.

5.2 Flash Point and Fire (or Ignition) Point Temperatures

Consider a test tube containing a liquid fuel at temperature T_o. Corresponding to T_o, the liquid vaporizes at a small rate and mixes with the air. If now a small pilot flame is brought into contact with the vapor/air mixture above the liquid surface, a flame will flash through the mixture if T_o is high enough.

---

* For a clarification on the types of solid fuels, see Chapter 6.

TABLE 5.1

Flash and Fire Points of Liquid Fuels*

Fuel	T_{flash} °C	T_{fire} °C	T_B °C	$(T_{fire}-T_{flash})$ °C
Ethyl Alcohol	10	69–76	78	59–66
n-butyl Alcohol	34	105	117	71
Acetone	−20	55	56	75
Petroleum	30	—	—	—
Solar oil	148	—	—	—
Machine oil	196	—	—	—
Cylinder oil	215	—	—	—
Gasoline	10	16	100–150	6
Lube oil	285	344	—	59

* Partly from Blinov and Khudyakov: *Diffusive Burning of Liquids*. Pergamon translation from Russian, (1960).

TABLE 5.2

Thermal and Thermochemical Properties of Liquids*

	M	ρ_l gm/cm³	L cal/gm	C_l cal/gm/°C	C_{vap} cal/gm/°C	T_B °C	ΔH kcal/gm	f
n-Pentane	72	0.631	87.1	0.557	0.397	36.0	10.82	0.314
n-Hexane	86	0.664	87.1	0.536	0.398	68.0	10.69	0.314
n-Heptane	100	0.688	87.1	0.525	0.399	98.5	10.62	0.314
n-Octane	114	0.707	86.5	0.526	0.400	125.0	10.60	0.316
iso-octane	114	0.702	78.4	0.515	0.400	125.0	10.59	0.316
n-decane	142	0.734	86.0	0.523	0.400	174.0	10.56	0.317
n-deodane	170	0.753	85.5	0.521	0.400	200.0†	10.54	0.319
Octene	112	0.710	80.5	0.525	0.400	121.0	10.59	0.322
Benzene	78	0.884	103.2	0.411	0.277	80.0	9.56	0.359
Methanol	32	0.796	263.0	0.566	0.410	64.5	4.74	0.726
Ethanol	46	0.794	200.0	0.560†	0.460	78.5	6.40	0.528
Gasoline	120†	0.720†	81.0	0.490	0.400†	155.0	10.54	0.318
Kerosene	154	0.825	69.5	0.460	0.400†	150.0	10.30	0.316
Light Diesel	170	0.876	63.9	0.450	0.400†	250.0†	10.12	0.316
Medium Diesel	184	0.920	58.4	0.430	0.400†	260.0†	10.00	0.315
Heavy Diesel	198	0.960	55.5	0.420	0.400†	270.0†	9.88	0.318
Acetone	58	0.791	125.0	0.506	0.340†	56.7	7.36	0.453
Toluene	92	0.870	84.0	0.386	0.400†	110.6	10.16	0.320
Xylene	106	0.870	80.0	0.411	0.400†	130.0	10.30	0.319

* Extracted from various Data Books.
† Estimated.

Diffusion Flames in Liquid Fuel Combustion

The least value of the liquid temperature T_O, at which the mixture is marginally capable of flashing with a pilot flame, is known as the *flash point* of the liquid fuel. At the flash point, the ignition source can *only flash* the mixture; but the mixture will be too weak to sustain a continued self-supporting flame. In order to accomplish such a sustainment, the liquid temperature usually has to be raised far above the flash point. The lowest liquid temperature at which the mixture, once ignited by a pilot flame, continues to burn by itself is known as the *ignition* or *fire point* of the liquid fuel.

Following ignition, the surface evaporation and the gas phase combustion continue one another; heat is received from the flame by the liquid surface which in turn provides more vapor to burn. When a steady state is established, the rate of vaporization is equal to the rate of burning. The surface temperature of the liquid then is higher than the fire point and close to, but slightly less than the boiling point.

Table 5.1 indicates the flash, fire and boiling points for various common fuels. It may be noted that the ignition point is considerably higher than the flash point particularly when the latter is high.

Tables 4.3 and 5.2 provide vaporization and combustion properties of several liquid fuels of pertinence to this study.

5.3 Scope of This Chapter

Postponing solid fuel combustion to Chapter 6 and gaseous diffusion flames to Chapter 7, this chapter is devoted to liquid fuel diffusion flames. In particular, the topic of steady state evaporation and combustion of a liquid fuel droplet is considered.

Specifically, Section 5.4 deals with the techniques of manufacturing liquid fuel sprays. Spalding's theory of mass transfer is presented in Sections 5.5 to 5.7 to calculate the rate of vaporization, *without* combustion, of a droplet. It is then extended to solve the problem *with* combustion in Sections 5.8 and 5.9. The concepts of evaporation/burning constant and evaporation/burning time of a droplet are also introduced. Evaporation and combustion in sprays (i.e. clouds of droplets) are considered in Section 5.10.

The mass transfer equation derived from the droplet analysis is shown to be useful to predict the vaporization and/or burning rates of *fuel surfaces of various geometries and flow conditions*. Examples of such geometries are offered by a vertical fuel-soaked wick burning in air, combustion of a fuel-spill on a flat surface, burning of the inner lining of a duct, etc. Several numerical examples are worked out in the end of the chapter to illustrate the mass transfer calculations.

5.4 Atomization of Liquids

As mentioned in the introduction of this chapter, in order to intensify combustion it is desirable to introduce the fuel in the form of a spray of finely divided droplets so that the required rapid vaporization and uniform distribution of the fuel is accomplished. The process of manufacture and dispersion of the droplets is known as *atomization*. Atomization involves stretching of the liquid into sheets and filaments by accelerating it through a nozzle orifice, formation of ripples and disturbances in the filaments due to air drag and turbulence, breakage of the filaments which collapse into droplets due to surface tension and further break up of large drops.

Atomization is frequently accomplished by the following techniques.

(a) *Pressure Atomization*

The injector is simply an orifice through which the fuel is forced with a high velocity, frequently with a swirl, into a relatively quiescent air. A very unstable hollow conical film of fuel is thereby formed. Severe holes and waves in this film initiate the break up into drops and droplets. This sort of atomization technique is widely used in gas turbines and spray drying. The nozzle on a garden hose offers a familiar example of a pressure atomizer.

(b) *Air (Gas or Pneumatic) Atomization*

A high velocity air stream flows by a relatively low velocity fuel jet so that the resulting surface drag distorts and breaks up the fuel jet into fine droplets. An every day example of a pneumatic atomizer is given by an insect spray gun or a paint spray gun.

Air atomization is known to produce smaller average droplets than pressure atomization. This technique is often used in diesel engine combustion chambers.

(c) *Impinging Jets*

Two jets of the liquid fuel are made to collide head-on to form an unstable sheet at the stagnation point. The liquid sheet containing holes and waves disintegrates into droplets in much the same way as the conical sheet in pressure atomization. This technique finds wide application in liquid fuel rocket combustors.

An interesting variation of the impinging jet method is obtained by impinging a single jet on a solid wall. The phenomenon of droplet formation is equivalent to that when two jets of equal strength impinge on one another. Perhaps a familiar example of this case would be a glass of milk spilling on the floor.

Diffusion Flames in Liquid Fuel Combustion

Lord Rayleigh showed that a rod of liquid undergoes surface contortions before breaking up into drops; the rod disintegrates into drops when the wave length of these surface disturbances exceeds the circumference of the undisturbed rod. Such a criterion may also be extended to sheets of liquid.

Weber applied Rayleigh's concepts to viscous fluids to show that the critical wavelength for a jet of a given diameter increases with increasing liquid viscosity and decreasing surface tension. It is, furthermore, clear that the break up is easier if the relative velocity between the liquid and air streams is higher.

Break up of a bulk of liquid into droplets requires work in order to increase the surface area against the surface tension resistance. The work done against the surface tension is only a small fraction of the total pumping work. This fraction is known as the efficiency of atomization. Most of the pump work is expended to accelerate the liquid to a high velocity. The efficiency of atomization depends upon the viscosity and surface tension of the liquid and the method of atomization. It is usually between 0.1 and 1%. Air atomization and impinging jet atomization are known to be more efficient than pressure atomization.

The drop size distribution in a spray is usually described either in "cumulative" or in "differential" count. Cumulative count (Figure 5.1) shows the fractional number of drops whose size is greater than a certain size d as a function of d. If $d = 0$, it is clear that *all* the droplets are larger than d. As the value of d increases, the number of drops bigger than d rapidly falls to zero.

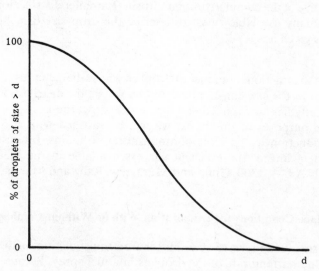

Figure 5.1 Cumulative size distribution of droplets in a spray

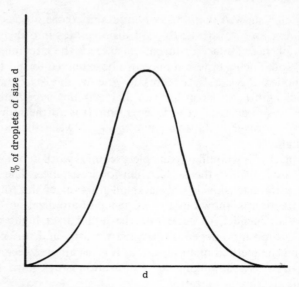

Figure 5.2 Differential size distribution of droplets in a spray

On the other hand, the differential count (Figure 5.2) shows the fractional number of drops whose size is equal to d. For example, there are usually few very large or very small droplets in a spray; the majority of droplets are of size about a mean diameter. It is clear that the differential count is simply the derivative of the cumulative count. Rosin–Rammler distribution function is one of many available rules to describe the drop size distribution in a spray. It is given as

$$\psi = e^{-qd^r}$$

where ψ is the fractional number of droplets whose diameter is larger than d. q is known as the size constant and r is known as the distribution constant which usually lies between 2 and 5 for most hydrocarbons.

For the purposes of this book, we will be satisfied with the preceding brief introduction to the topic of atomization of liquids. Interested reader is referred to any of the excellent surveys available in the literature (for example, N.A.C.A. 1300, Giffin and Muraszew, Ranz and Marshall, etc.).

5.5 Surface Conditions for Evaporation With or Without Combustion

The performance and size of a combustor are strong functions of the vaporization and the burning rate of the droplets in a fuel spray. Because the interaction of neighboring droplets in a spray is not yet clearly understood,

Diffusion Flames in Liquid Fuel Combustion

(a) the detailed studies on isolated drops are extrapolated and (b) empirical rules for such interactions are developed in practice. In this chapter we consider the vaporization of a droplet with and without combustion in detail and demonstrate the possible avenues of extrapolation to the sprays.

In the prevention of accidental fires, the rate of evaporation and combustion on various shapes of the fuel surface in various flow situations (i.e. laminar, or turbulent, free or forced convection) is of importance. We will keep this in mind all through this chapter and work out some example problems in the end. The rate of vaporization of a droplet depends mainly on the type of the fuel, the free stream temperature and composition, and the size of the droplet. In order to determine this rate, it is necessary to know the profiles of temperature and composition in both the liquid and the gas phases. Solution of the energy and species conservation equations to obtain these profiles warrants enough boundary conditions to evaluate all the integration constants. Under steady state vaporization conditions, the temperature and composition profiles in the liquid phase become inconsequential. In this section, we will derive the surface flux conditions for the energy and species conservation in steady state vaporization (and subsequent combustion, if any).

Consider, as shown in Figure 5.3, a small area of the evaporating surface (which is assumed to be flat for the present purpose). Let the surface be exposed to a hot air stream. Let T be the temperature and Y_i be the mass fraction of the ith species. Furthermore, let the subscripts ∞, W and R res-

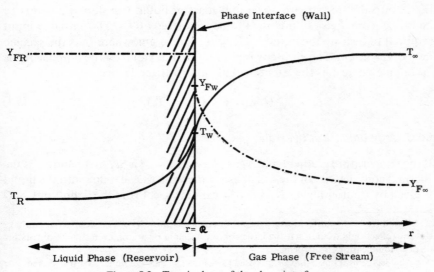

Figure 5.3 Terminology of the phase interface

pectively refer to the free stream, the gas phase immediately adjacent to the gas-liquid interface and the reservoir (i.e. deep beneath the surface in the liquid phase). If the air is fuel-free, the free stream fuel mass fraction $Y_{F\infty}$ will be zero. Similarly, if the liquid fuel is pure and does not contain any dissolved gases, the fuel mass fraction in the reservoir Y_{FR} will be unity.

It will be shown later that if the free stream temperature T_∞ is much greater than the fuel boiling point T_B, the wall temperature T_W may be assumed equal to or slightly less than the boiling point. The mass fraction of the fuel at the surface in the gas phase Y_{FW} is usually much smaller than Y_{FR}. T_W and Y_{FW} are generally unknown. Solution of the problem, as we will see later, partly deals with determining these wall conditions.

Let r be the coordinate fixed normal to the surface (i.e. radially) with $r = 0$ at the droplet center and $r = \mathscr{R}$ at the drop surface. The steady state temperature and fuel species mass fraction profiles are then expected to be as shown in Figure 5.3.

As a consequence of the temperature difference $(T_\infty - T_W)$, heat is transferred from the air stream to the fuel surface.* As a consequence of the composition difference $(Y_{FW} - Y_{F\infty})$, the fuel vapor is diffused from the fuel surface to the free stream. The net mass flow $\dot{W}''_W$ gms/cm^2/sec normal to the surface is equal to the vaporization rate, calculation of which is the primary goal of this work. The equation of continuity (refer to Chapter 3) implies that the total rate of mass transfer $(\dot{W}''' 4\pi r^2)$ gm/sec. remains constant and equal to $(\dot{W}''_W 4\pi \mathscr{R}^2)$.

Let Q be the amount of heat required to transfer a unit mass of the reservoir liquid into the free stream. Each unit mass of liquid transferred has first to be raised from T_R to T_W and then to be vaporized at T_W. The enthalpy input required to heat up the liquid is $C_l(T_W - T_R)$ cal/gm, where C_l is the specific heat of the liquid. The enthalpy input required to vaporize the liquid at T_W is L cal/gm, where L is the latent heat of vaporization. Hence,

$$Q \equiv L + C_l(T_W - T_R) \tag{5.1}$$

(a) *Energy Balance at the Wall*

If the temperature gradient in the gas phase at $r = \mathscr{R}$ is $(dT/dr)_W$ and K_g is the thermal conductivity of the gas phase at the wall, the heat flux into the liquid is given by Fourier law as $K_g(dT/dr)_W$ cal/cm^2/sec. From the definition† of Q

---

* Radiative heat transfer is neglected all through this work. If one so desires one can take this into effect only to find that the basic concepts remain invariant, even though the mathematical nonlinearities overwhelm.

† That is, the heat flux $K_g(dT/dr)_W$ is used to heat just so much of liquid as is transferred. For further elaboration on this point see Section 5.12.

Diffusion Flames in Liquid Fuel Combustion

(the heat needed to transfer a unit mass of the liquid into the free stream), the heat flux *into* the liquid surface and the mass flux *out of* it are then related as below.

$$\dot{W}_W'' Q = K_g \left.\frac{dT}{dr}\right|_W \tag{5.2}$$

Recalling the definition of thermal diffusivity as $\alpha_g \equiv K_g/\rho_g C_g$, Eq. 5.2 can be rearranged to

$$\dot{W}_W'' = \rho_g \alpha_g \left.\frac{d}{dr}\left(\frac{C_g T}{Q}\right)\right|_W \tag{5.3}$$

Defining the nondimensional temperature as

$$b_T \equiv \frac{C_g(T - T_\infty)}{Q}, \tag{5.4}$$

since T_∞ is a constant, Eq. 5.3 transforms to

$$\dot{W}_W'' = \rho_g \alpha_g \left.\frac{db_T}{dr}\right|_W \tag{5.5}$$

This equation gives the mass transfer rate provided the temperature gradient at the wall is known. However, since this is not known at present, Eq. 5.5 is useless to calculate the mass transfer rate. It simply relates two unknowns $\dot{W}_W''$ and $(db_T/dr)_W$ and thus serves as a boundary condition for the energy conservation equation in the gas phase.

(a) *Species F Balance at the Wall*

In every $\dot{W}_W''$ gms of the reservoir liquid lost through a unit surface area per unit time, the reservoir is deprived of ($\dot{W}_W'' \cdot Y_{FR}$) gms of the fuel species F. This much of the fuel reaching the surface from the reservoir, leaves into the free stream under the combined influence of diffusion (by virtue of the non-zero concentration gradient at the surface) and convection (by virtue of the net fluid flow at the wall at a rate of $\dot{W}_W''$ gm/cm²/sec). Fick's law gives the diffusion rate as $-\rho_g D_F (dY_F/dr)_W$. Here D_F is the diffusion coefficient of the fuel vapor in the fuel/air mixture. The convective rate is simply ($\dot{W}_W'' \cdot Y_{FW}$). In steady state, the fuel species is conserved at the surface if

$$\dot{W}_W'' Y_{FR} = \dot{W}_W'' Y_{FW} + \left(-\rho_g D_F \left.\frac{dY_F}{dr}\right|_W\right) \tag{5.6}$$

Rearrangement to resolve for $\dot{W}_W''$ gives

$$\dot{W}_W'' = \rho_g D_F \left.\frac{d}{dr}\left(\frac{Y_F}{Y_{FW} - Y_{FR}}\right)\right|_W \tag{5.7}$$

Defining the normalized mass fraction as

$$b_D \equiv \left(\frac{Y_F - Y_{F\infty}}{Y_{FW} - Y_{FR}}\right) \tag{5.8}$$

Eq. 5.7 reduces to the same form as Eq. 5.5.

$$\dot{W}'''_W = \rho_g D_F \frac{db_D}{dr}\bigg|_W \tag{5.9}$$

This equation, relating the unknown mass transfer rate at the wall $\dot{W}'''_W$ with the unknown mass fraction gradient at the wall, serves as a boundary condition for the fuel species conservation equation in the gas phase.

5.6 Simple Steady State Vaporization of a Droplet Without Combustion

(a) *Equations*

Equations 3.66 and 3.67 give the conservation equations for energy and species in rectangular coordinates. These equations written in spherical coordinates for the case of steady state, constant properties and noncombustion situation will be as following.

$$\frac{d}{dr}\left(K_g 4\pi r^2 \frac{dT}{dr}\right) - \frac{d}{dr}([\dot{W}''' 4\pi r^2] C_g T) = 0 \tag{5.10}$$

$$\frac{d}{dr}\left(\rho_g D_F 4\pi r^2 \frac{dY_F}{dr}\right) - \frac{d}{dr}([\dot{W}''' 4\pi r^2] Y_F) = 0 \tag{5.11}*$$

The first terms of Eqs. 5.10 and 5.11 represent the surplus conductive and diffusive fluxes across a thin spherical shell of thickness dr. The second terms represent the surplus convective fluxes across the same shell. The square brackets around the factor $[\dot{W}''' 4\pi r^2]$ is to indicate that this product is invariant with r, due to the continuity equation.

Dividing both sides of Eq. 5.10 by the constant Q, and Eq. 5.11 by the constant $(Y_{FW} - Y_{FR})$ and introducing the definitions of the normalized

---

* Equations 5.10 and 5.11 indicate that at any r in the gas phase, the sum of diffusive and convective fluxes remains constant. This is clear by writing Eq. 5.11, for example, as below.

$$\frac{d}{dr}\left(-4\pi r^2 \rho_g D_F \frac{dY_F}{dr} + 4\pi r^2 \dot{W}''' Y_F\right) = 0$$

Diffusion Flames in Liquid Fuel Combustion

temperature and fuel mass fraction, Eqs. 5.10 and 5.11 reduce to the following analogous form.

$$\rho_g \alpha_g \frac{d}{dr}\left(r^2 \frac{db_T}{dr}\right) - [\dot{W}''_W \mathscr{R}^2] \frac{db_T}{dr} = 0 \quad (5.12)$$

$$\rho_g D_F \frac{d}{dr}\left(r^2 \frac{db_D}{dr}\right) - [\dot{W}''_W \mathscr{R}^2] \frac{db_D}{dr} = 0 \quad (5.13)$$

The boundary conditions are given by Eqs. 5.5 and 5.9 along with the following. Corresponding to the wall condition $r = \mathscr{R}$, $T = T_W$ and $Y_F = Y_{FW}$, we obtain $b_T = C_g(T_W - T_\infty)/Q \equiv b_{TW}$ and $b_D = (Y_{FW} - Y_{F\infty})/(Y_{FW} - Y_{FR}) \equiv b_{DW}$. Corresponding to the free stream condition $r \to \infty$ (i.e. far away from the droplet surface), $T = T_\infty$ and $Y_F = Y_{F\infty}$ we obtain $b_{T\infty} = 0$ and $b_{D\infty} = 0$. Hence, to summarize,

$$r = \mathscr{R} \quad \left.\begin{array}{l} b_T = b_{TW} \\ b_D = b_{DW} \end{array}\right\} \quad (5.14)$$

$$\left.\begin{array}{l} \dot{W}''_W = \rho_g \alpha_g \dfrac{db_T}{dr}\bigg|_W \\ \\ \dot{W}''_W = \rho_g D_F \dfrac{db_D}{dr}\bigg|_W \end{array}\right\} \quad (5.15)$$

$$r \to \infty \quad \left.\begin{array}{l} b_T = b_{T\infty} = 0 \\ b_D = b_{D\infty} = 0 \end{array}\right\} \quad (5.16)$$

Not only the conservation equations but also their boundary conditions exhibit a striking analogy between the heat and mass transfer processes. Therefore, the solution of either the energy or the species equation yields that of the other readily. In fact, if the Lewis number is unity, (i.e. $\alpha_g = D_F$), the energy and species solutions become one and the same. The assumption of unit Lewis number is very commonly made in combustion analyses.

Any variable b governed by a differential equation of the type of Eqs. 5.12 and 5.13 in which the eigenvalue is related to b by Eq. 5.15* is known in mass transfer literature as a *conserved variable*. The solution of a simple second order differential equation requires two boundary conditions. However, Eqs. 5.12 and 5.13 make three conditions available. This seems to be an over-specification at first sight. But a closer look at Eqs. 5.12 to 5.16 reveals that

---

* For the sake of completeness, a conserved variable b may be defined as below. The flux of b is spatially conserved under the combined influence of transport by diffusion and convection. The slope of the conserved variable at the wall is linearly related to the eigenvalue mass flux at the wall, $\dot{W}''_W$.

$\dot{W}''_W$ and T_W are two extra unknowns, determination of which requires two more conditions than usual; that is, a total of four. The fourth condition is obtained from thermodynamics of saturation vapor/liquid phase equilibrium.

(b) *Solution*

Assuming $\alpha_g = D_F$, Eq. 5.12 (or 5.13) can be solved with the boundary conditions as below. We will ignore the subscripts T and D since they are now synonymous. Integrating once, Eq. 5.12 becomes

$$\rho_g \alpha_g r^2 \frac{db}{dr} - [\dot{W}''_W \mathscr{R}^2] b = \text{constant independent of } r \quad (5.17)$$

The constant of integration is evaluated by applying the wall condition equations. It is,

$$\rho_g \alpha_g \mathscr{R}^2 \frac{db}{dr}\bigg|_W - [\dot{W}''_W \mathscr{R}^2] b_W = [\dot{W}''_W \mathscr{R}^2](1 - b_W)$$

Equation 5.17, therefore, reduces to the following.

$$\rho_g \alpha_g r^2 \frac{db}{dr} - [\dot{W}''_W \mathscr{R}^2](b - b_W + 1) = 0$$

Separating the variables,

$$\frac{db}{b - b_W + 1} = \frac{[\dot{W}''_W \mathscr{R}^2]}{\rho_g \alpha_g} \frac{dr}{r^2}$$

Integration yields,

$$\ln(b - b_W + 1) = -\frac{[\dot{W}''_W \mathscr{R}^2]}{\rho_g \alpha_g} \cdot \frac{1}{r} + \text{constant}$$

Applying Eq. 5.16, the constant is evaluated* as $\ln(b_\infty - b_W + 1)$. The profile of $b(r)$ is, thereby,

$$\ln\left[\frac{(b_\infty - b_W + 1)}{(b - b_W + 1)}\right] = \frac{[\dot{W}''_W \mathscr{R}^2]}{\rho_g \alpha_g} \cdot \frac{1}{r} \quad (5.18)$$

The mass transfer rate is obtained by setting $b = b_W$ when $r = \mathscr{R}$.

$$\dot{W}''_W = \frac{\rho_g \alpha_g}{\mathscr{R}} \ln(b_\infty - b_W + 1) \quad (5.19)$$

---

* Even though $b_\infty = 0$, we will carry out the derivation as though it were not. At the end we substitute $b_\infty = 0$.

Diffusion Flames in Liquid Fuel Combustion

This equation shows that smaller droplets evaporate at a faster rate than larger ones.

Eliminating $\dot{W}'''_W$ from Eqs. 5.18 and 5.19, the following alternative profile equation is obtained.

$$(b - b_W + 1) = (b_\infty - b_W + 1)^{(1 - \mathcal{R}/r)} \tag{5.20}$$

A plot of $\ln(b - b_W + 1)$ versus $(1 - \mathcal{R}/r)$ for various values of $(b_\infty - b_W)$ is shown in Figure 5.4.

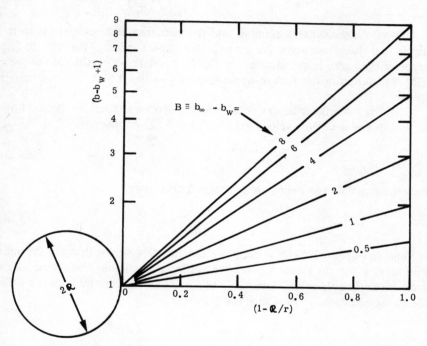

Figure 5.4 Profiles of the conserved variable b in droplet vaporization

The difference between the free stream and wall values of b is known as the mass transfer driving force or simply *mass transfer number* and is denoted by B. Because, according to our definitions of b_T and b_D, $b_{T\infty} = b_{D\infty} = 0$, $B \equiv b_\infty - b_W = -b_W$.

$$B \equiv \frac{C_g(T_\infty - T_W)}{L + C_l(T_W - T_R)} \equiv \frac{(Y_{F\infty} - Y_{FW})}{(Y_{FW} - Y_{FR})} \tag{5.21}$$

The second equality of Eq. 5.21 arises from the fact that $b_T \equiv b_D$ when $\alpha_g = D_F$.

(c) *Discussion of the Vaporization Rate Equation (5.19)*

When a droplet of diameter d is suspended in an infinite hot gas medium, the heat transfer problem is described (Chapter 3. Table 3.4) by the Nusselt number dependency on the flow Reynolds number and Prandtl number. If u is the flow velocity of the medium whose conductivity, kinematic viscosity and thermal diffusivity are respectively denoted by K_g, v_g and α_g,

$$\overline{\mathrm{Nu}}_d^0 \equiv \frac{\bar{h}^0 d}{K_g} = 2 + 0.6 \left(\frac{ud}{v_g}\right)^{1/2} \left(\frac{v_g}{\alpha_g}\right)^{1/3} \tag{5.22}$$

The bar on the Nusselt number and the heat transfer coefficient is to indicate that the values are averaged over the whole surface of the sphere. The superscripted zero is to indicate that Eq. 5.22 is derived for the case of pure heat transfer from the stream to an imperveous (inert and nonvaporizing) sphere.

If the drop diameter is very small and the drop is carried by the gas flow so that the relative flow velocity is small, Eq. 5.22 simplifies to

$$\frac{\bar{h}^0 d}{K_g} \approx 2 \tag{5.23}$$

Rearranging with the definition of thermal diffusivity,

$$\frac{\bar{h}^0}{C_g} \approx \frac{2\rho_g \alpha_g}{d} \tag{5.24}$$

Since the drop diameter is equal to $2\mathcal{R}$, Eq. 5.24 shows that the thermal boundary layer thickness around the sphere is such that the temperature gradient at the sphere wall is equal to $(T_\infty - T_W)/\mathcal{R}$. This point is clear by the following equation for heat flux and Figure 5.5.

$$\dot{q}'' \equiv \bar{h}^0(T_\infty - T_W) \equiv K_g(T_\infty - T_W)/\mathcal{R}$$

Substituting Eq. 5.24 in 5.19 we obtain the following general mass transfer equation which is applicable to any arbitrary fuel surface geometry and flow conditions.

$$\dot{W}_W'' = \frac{\bar{h}^0}{C_g} \ln(B + 1) \tag{5.25}$$

The vaporization problem is separated, as shown by Eq. 5.25, into two aspects—the heat transfer aspect and the thermodynamic aspect. The factor $\bar{h}^0/C_g$ accounts for the heat transfer aspect whereas the factor B accounts for the thermodynamic aspect. The fact that these two aspects can be separated from one another, considerably simplifies the mass transfer calculations. The

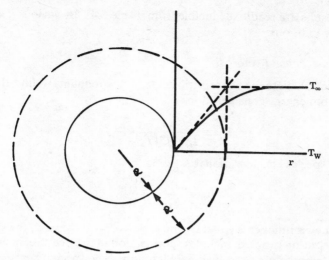

Figure 5.5 The boundary layer about a small spherical droplet in a nearly quiescent atmosphere.

method to find $\bar{h}^0/C_g$ and B to calculate the vaporization rate by Eq. 5.25 for various geometries of the fuel body is described below.

(d) Heat Transfer Factor

The heat transfer coefficient $\bar{h}^0$ is obtained as

$$\bar{h}^0 \equiv \frac{\rho_g C_g \alpha_g}{l} \cdot \overline{\mathrm{Nu}}_l^0$$

where l is the characteristic dimension of the vaporizing body. The Nusselt number is a function of the flow Reynolds number (or Grashof number) and the Prandtl number as given in Table 3.4. Consider the following two examples for illustration.

A pan of fuel (of diameter l) evaporating in a gentle gust of air (of velocity u) may be considered as a flat plate in laminar flow, for which Table 3.4 gives the average heat transfer as $\overline{\mathrm{Nu}}_l^0 = 0.664 \, \mathrm{Re}_l^{1/2} \, \mathrm{Pr}_r^{1/3}$. Hence,

$$\frac{\bar{h}^0}{C_g} = \frac{\rho_g \alpha_g}{l} \cdot 0.664 \left(\frac{ul}{v_g}\right)^{1/2} \left(\frac{v_g}{\alpha_g}\right)^{1/3}$$

For a vertical fuel soaked flat wick (of height l) the laminar free convective heat transfer is given by $\overline{\mathrm{Nu}}_l^0 = 0.59 (\mathrm{Gr}_l \, \mathrm{Pr})^{1/4}$. Hence,

$$\frac{\bar{h}^0}{C_g} = \frac{\rho_g \alpha_g}{l} \cdot 0.59 \left(\frac{gl^3 \beta \Delta T}{v_g^2}\right)^{1/4} \left(\frac{v_g}{\alpha_g}\right)^{1/4}$$

Similar results are readily deducible from Table 3.4* for various geometries and flow conditions.

(e) *Thermodynamic Factor*: B

In order to calculate the vaporization rate, it now remains to find B which is defined from energy considerations as

$$B_T \equiv \frac{C_g(T_\infty - T_W)}{L + C_l(T_W - T_R)} \tag{5.26}$$

and from species considerations as

$$B_D \equiv \frac{Y_{F\infty} - Y_{FW}}{Y_{FW} - Y_{FR}} \tag{5.27}$$

When Lewis number is equal to unity, $B_D = B_T$. In general, the statement of a vaporization problem includes specification of the free stream and liquid reservoir temperatures and fuel mass fractions. The specific heats C_g and C_l and the fuel latent heat L can be obtained from property tables.†

But because neither the wall temperature T_W nor the fuel mass fraction Y_{FW} are known, the computation of B_T or B_D is not possible at this stage. Finding T_W and Y_{FW} is part and parcel of any mass transfer problem. The equation $B_T = B_D$ gives one of the two conditions necessary to solve for the two variables T_W and Y_{FW}. The second necessary condition is obtained from the vapor/liquid phase equilibrium thermodynamics.

Assuming that the liquid and vapor phases are in equilibrium at the evaporating fuel surface, Clausius-Clapeyron equation gives the saturation vapor pressure P_F of the fuel as a function of the temperature.

$$P_F = C_1 e^{C_2/T} \tag{5.28}$$

C_1 and C_2 are characteristic constants for a given liquid. Table 5.3 and Figures 5.6 and 5.7 give Eq. 5.28 for several hydrocarbon liquids. Extensive data similar to those given in Table 5.3 are available in the C.R.C. Handbook of Physics and Chemistry.

Consider from Appendix A, the following derivation relating the partial pressure with the mass fraction. For a binary mixture of species 1 and 2,

---

* Table 3.4 is only a short dictionary of heat transfer results. The student is referred to the abundant heat transfer literature for more extensive results.

† Tables 4.3, 5.2 and such handbooks as the following:
  International Critical Tables,
  C.R.C. Handbook of Physics and Chemistry,
  Mark's Mechanical Engineer's Handbook,
  Perry's Chemical Engineer's Handbook.

TABLE 5.3
Saturation Vapor Pressures

Name	Formula	\multicolumn{6}{c}{$T\,°C$ when P_F is equal to}

Name	Formula	1 mm	10 mm	40 mm	100 mm	400 mm	760 mm
Methanol	CH_3OH	−44.0	−16.0	5.0	21.2	49.9	64.7
Acetic Acid	$C_2H_4O_2$	−17.0	17.5	43.0	63.0	99.0	118.0
Ethanol	C_2H_6O	−31.0	−2.3	19.0	35.0	64.0	78.4
Acetone	C_3H_6O	−59.0	−31.0	−9.4	7.7	39.5	56.5
1-Propanol	C_3H_8O	−15.0	14.7	36.4	52.8	82.0	97.8
Glycerol	$C_3H_8O_3$	125.5	167.2	198.0	220.1	263.0	290.0
Butyl Alcohol	$C_4H_{10}O$	−1.2	30.2	53.0	70.0	100.0	117.0
Benzene	C_6H_6	−36.7	−11.5	7.6	26.1	60.6	80.1
Cyclohexane	C_6H_{12}	−45.3	−15.9	6.7	25.5	60.8	80.7
Toluene	C_7H_8	−26.7	6.4	31.8	51.9	89.5	110.6
Heptane	C_7H_{16}	−34.0	−2.1	22.3	41.8	78.0	98.4
2-Xylene	C_8H_{10}	−3.8	32.1	59.5	81.3	121.7	144.4
Octane	C_8H_{18}	−14.0	19.2	45.0	65.7	104.0	125.6

the sum of partial densities is equal to the total density.

$$\rho_1 + \rho_2 = \rho$$

Assuming that air and the fuel vapor behave as ideal gases, $\rho_i = P_i M_i / RT$, $i = 1, 2$. P_i is the partial pressure of the ith species and M_i is its molecular weight. Substituting this result,

$$\frac{P_1 M_1}{RT} + \frac{P_2 M_2}{RT} = \frac{PM}{RT}$$

where P and M respectively are the total mixture pressure and the molecular weight.

The mass fraction of species 1 is defined as the ratio of the partial density of species 1 to the mixture density.

$$Y_1 \equiv \rho_1/\rho = \frac{\rho_1}{\rho_1 + \rho_2} = \frac{1}{1 + \dfrac{P_2 M_2}{P_1 M_1}} \tag{5.29}$$

By Dalton's law, $P = P_1 + P_2$. Hence

$$Y_1 = 1 \bigg/ \left[1 + \left(\frac{P}{P_1} - 1\right)\frac{M_2}{M_1}\right]$$

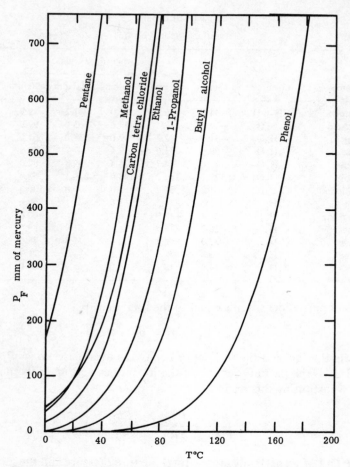

Figure 5.6 Vapor pressure versus temperature for various liquid fuels

Letting species 1 to be the fuel and species 2 to be the air, substitution of the saturation $P_F(T)$ relation Eq. 5.28 gives at the liquid fuel surface,

$$Y_{FW} \equiv Y_F(T_W) = 1 \bigg/ \left[1 + \left(\frac{P}{C_1 e^{C_2/T_W}} - 1 \right) \frac{M_g}{M_F} \right] \quad (5.30)$$

Substitution in Eq. 5.27 yields,

$$B_D = \frac{Y_{F\infty} - Y_F(T_W)}{Y_F(T_W) - Y_{FR}} \quad (5.31)$$

Diffusion Flames in Liquid Fuel Combustion

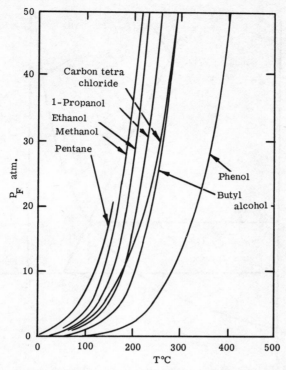

Figure 5.7 Vapor pressure versus temperature for various liquid fuels

We thus have Eq. 5.26 giving B_T as a function of T_W and Eq. 5.31 giving B_D as a function of T_W. B_T and B_D are equal to one another when $Le = 1$. The surface conditions can hence be most conveniently obtained by plotting $B_T(T_W)$ and $B_D(T_W)$ against T_W and finding the point of intersection. Note that when $L \gg C_l(T_W - T_R)$, $B_T(T_W)$ is a linear function of T_W with a negative slope. The straight line intersects the x-axis at $T_W = T_\infty$.

When T_W is zero, P_{FW} is zero (by Eq. 5.29) so that Y_{FW} is zero (by Eq. 5.30). Then $B_D = -Y_{F\infty}/Y_{FR}$. When T_W is equal to the boiling point T_B, P_{FW} is equal to the total pressure so that $Y_{FW} = Y_{FR}$. From Eq. 5.31, then* $B_D = \infty$. B_D as a function of T_W, thus, is a steeply rising function.

Figure 5.8 shows $B_T(T_W)$ and $B_D(T_W)$ functions for a given set of the conditions $Y_{F\infty}$, Y_{FR}, P, M_g, M_F, C_1, C_2, T_∞, T_R, C_g and L. Inclusion of the $C_l(T_W - T_R)$ term in Eq. 5.26 introduces a slight curvature in the function $B_T(T_W)$. The point of intersection of $B_T(T_W)$ and $B_D(T_W)$ gives B and T_W.

---

* Since infinite driving force is physically unreal, this discussion implies that the surface of a vaporizing liquid never reaches the boiling temperature T_B.

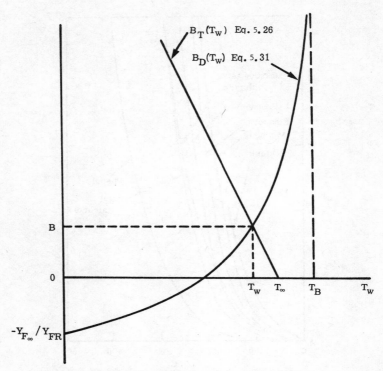

Figure 5.8 Determination of the wall condition

Two extremes are apparent in Figure 5.8.

Firstly, when T_∞ is much greater than the fuel boiling point temperature T_B, the B_T and B_D functions intersect at a wall temperature T_W which is nearly equal to but slightly less than T_B. Then B may be calculated by assuming that T_W is equal to T_B. Figure 5.9 indicates this situation.

$$B = B_T \approx \frac{C_g(T_\infty - T_B)}{L + C_l(T_B - T_R)} \qquad (5.32)$$

Secondly, when the free stream temperature T_∞ is much lower than the boiling point temperature T_B, the intersection occurs, as shown in Figure 5.10, very close to $B = 0$. B may then be calculated by Eq. 5.31 by assuming that the wall mass fraction is that corresponding to the saturation vapor pressure at T_∞. By Eq. 5.30 then $Y_{FW} = Y_F(T_\infty)$ is calculated to find

$$B = B_D \approx \frac{Y_{F\infty} - Y_F(T_\infty)}{Y_F(T_\infty) - Y_{FR}} \qquad (5.33)$$

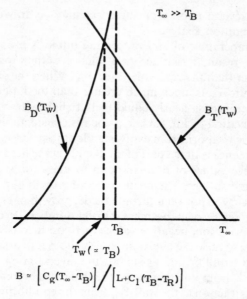

Figure 5.9 Approximation of the wall condition when the ambient temperature is much greater than the fuel boiling point

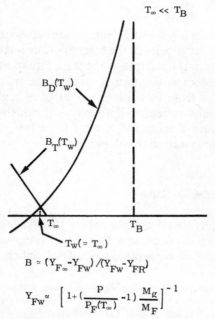

Figure 5.10 Approximation of the wall condition when the ambient temperature is much lower than the fuel boiling point

The following several important inferences may be drawn from the preceding surface condition analysis.

The surface temperature of an evaporating liquid is always less than its boiling point. As a result, the surface mass fraction of the vapor Y_{FW} is always considerably lower than the reservoir value Y_{FR}. When the ambient (i.e. free stream) temperature T_∞ is much much greater than the liquid boiling point, the surface temperature is nearly equal to but slightly less than T_B as discussed in the derivation of Eq. 5.32. Y_{FW} then is considerably closer to Y_{FR}. The concept of flash vaporization evolves under these circumstances.*

The second inference is that a part of the $B_D(T_W)$ curve lies in the negative-B quadrant. Negative values of B correspond to mass transfer from the free stream to the liquid surface. The phenomena of growth of a rain drop, condensation of mercury vapor on a turbine blade, growth of crystals in a solution etc. are described by negative mass transfer numbers. In connection with our study of vaporization, negative values of B are inconsequential.

It is proper here to raise the question "how valid is it to assume that vapor/liquid saturation equilibrium exists at the evaporating fuel surface?" Spalding estimates, from kinetic theory, the departure of real-life nonequilibrium vapor mass fraction at the wall Y_{FW} from the equilibrium mass fraction Y_{FWe} as

$$(Y_{FWe} - Y_{FW}) \approx 3\dot{W}''_W / \mathscr{S} \rho_W a_W \tag{5.34}$$

where $\mathscr{S}$ is called sticking (or accommodation) coefficient which is of the order of unity, ρ_W is the density of the fuel vapor/air mixture at the wall and a_W is the sonic speed in the mixture at the wall. Equation 5.34 indicates that unless the mass transfer rate is extremely high (of the order of sonic blowing), the departure of Y_{AW} from its equilibrium value is negligibly small. If we exclude microcapillary systems whose physical dimensions are minutely small, our equilibrium assumption is valid for a broad spectrum of technically important evaporation problems.

5.7 Droplet Vaporization Time

In Section 5.6 we described the procedure to calculate the heat transfer coefficient and mass transfer number B from which the mass transfer rate can be estimated. The rate of vaporization of a small spherical fuel droplet in a

---

* It is a common experience that when a liquid drop falls on an extremely hot metallic pan surface, the droplet evaporates vigorously and moves rapidly around on a cushion of hot vapors. This phenomenon was first analyzed by Leidenfrost and is now known as the Leidenfrost effect.

Diffusion Flames in Liquid Fuel Combustion

relatively quiescent atmosphere is given by Eq. 5.19.

$$\dot{W}''_W = \frac{2\rho_g \alpha_g}{d} \ln(B + 1) \qquad (5.35)$$

Using Eq. 5.35, we can calculate the time taken by a droplet of a given diameter to completely evaporate. This time is known as the life-time or *evaporation time* t_v of the drop. Droplet evaporation time is an important parameter in combustion chamber design. Consider the following examples to establish the pertinence of this statement.

The evaporation time of the largest droplet in a spray determines the minimum time the droplet must be allowed to reside in a furnace or a combustion chamber. The residence time depends upon such factors as the velocity of the air stream through the combustor, the velocity of the spray injection, the length and width of the combustor, etc. Water droplet evaporation rate during its fall through the atmosphere is of interest to the meteorologist in predicting the probability of precipitation. Aerosol spray behavior governs many technical problems of the modern civilization. The combustion in a diesel engine is another example in which droplet evaporation is a crucial factor. When a spray of water is delivered from a fire extinguishant sprinkler to attack a fire, the fate and effectiveness of the spray are mainly determined by the adverse and hot fire gas flow. Warranted by these and other examples, we will now undertake the task of calculating the evaporation time of a droplet.

As the droplet vaporizes, its size diminishes; the size decrement is related to the evaporation rate as below. If d is the droplet diameter at any time t, and Δd is the decrease in diameter due to evaporation in a short time Δt, the decrease in volume is given by $\pi d^2 \Delta d/2$. Decrease in the weight of the droplet then is simply $\rho_l \pi d^2 \Delta d/2$. The rate of weight decrease, hence, is $\rho_l \pi d^2 \Delta d/2 \Delta t$ gms per sec. This rate of weight decrease is equal to the rate at which the fuel is lost by the droplet through its entire surface area. This later rate is given by the product of $\dot{W}''_W$ and the surface area πd^2. Assuming quasi-steady state,

$$- \rho_l \frac{\pi d^2}{2} \frac{\Delta d}{\Delta t} = \dot{W}''_W \pi d^2$$

Substituting Eq. 5.35 for $\dot{W}''_W$, the rate of disappearence of the droplet follow

$$\frac{\Delta d}{\Delta t} = - \frac{4\rho_g \alpha_g}{\rho_l d} \ln(B + 1) \qquad (5.36)$$

It is apparent that the rate of disappearance is greater if the droplet is smaller.

If d_0 is the initial diameter of the droplet, integration of Eq. 5.36 gives the drop diameter as a function of time.

$$d^2 = d_0^2 - \left[\frac{8\rho_g \alpha_g}{\rho_l} \ln(B + 1)\right] \cdot t \qquad (5.37)$$

Equation 5.37 suggests that a plot of d^2 versus t is a straight line with a negative slope of

$$\left[\frac{8\rho_g \alpha_g}{\rho_l} \ln(B + 1)\right].$$

This is illustrated in Figure 5.11.

The slope is known in the literature as the *evaporation constant*, λ_v.

$$\lambda_v \equiv \left[\frac{8\rho_g \alpha_g}{\rho_l} \ln(B + 1)\right] \text{ cm}^2/\text{sec.} \qquad (5.38)$$

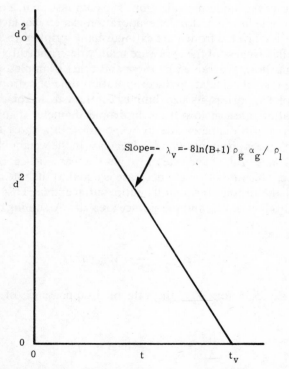

Figure 5.11 The square law of droplet vaporization and the definition of the evaporation constant

Diffusion Flames in Liquid Fuel Combustion

The evaporation time of the droplet is thereby equal to

$$t_v = \frac{d_0^2}{\lambda_v} = \frac{\rho_l d_0^2}{8\rho_g \alpha_g \ln(B+1)} \quad (5.39)*$$

It is obvious from Eqs. 5.38 and 5.39 that t_v is longer for larger droplets if all other conditions are kept unaltered. Lighter fuels evaporate faster. A high gas conductivity and low gas specific heat provide high evaporation constant.

5.8 Evaporation Followed by Combustion

So far in this chapter, we have considered the problem of evaporation without any combustion. The theoretical concepts developed in Sections 5.5 to 5.7 can readily be adopted for the case when the evaporation is followed by combustion by simply modifying the definitions of the conserved variables.

Combustion reactions introduce sources of energy and products and sinks of the fuel and oxidant into the gas phase conservation equations. Let D_i be the coefficient of diffusion of species i into the mixture. Constant property unidimensional conservation equations in cartesian coordinates are given as below.

Energy:

$$K_g \frac{d^2 T}{dy^2} - [\dot{W}''']C_g \frac{dT}{dy} + \dot{q}''' = 0 \quad (5.40a)$$

Fuel:

$$\rho_g D_F \frac{d^2 Y_F}{dy^2} - [\dot{W}'''] \frac{dY_F}{dy} + \dot{W}_F''' = 0 \quad (5.41a)$$

Oxygen:

$$\rho_g D_O \frac{d^2 Y_O}{dy^2} - [\dot{W}'''] \frac{dY_O}{dy} + \dot{W}_O''' = 0 \quad (5.42a)$$

Products:

$$\rho_g D_P \frac{d^2 Y_P}{dy^2} - [\dot{W}'''] \frac{dY_P}{dy} + \dot{W}_P''' = 0 \quad (5.43a)$$

---

* Alternatively, if W_O is the initial weight of the droplet ($= \rho_l \pi d_0^3/6$) and $\dot{W}_W$ is the rate of weight loss ($= \dot{W}_W'' \pi d^2$), the life time of the droplet can be estimated as $t_v \approx W_O/\dot{W}_W$. Since $\dot{W}_W$ continuously increases with decreasing diameter, it has to be evaluated at some diameter d between O and d_O. Assuming $d = 2d_0/3$, t_v is obtained as $\rho_l d_0^2/8\rho_g \alpha_g \ln(B+1)$.

Overall Continuity:

$$\dot{W}'' = \text{constant for all } y \qquad (5.44a)$$

In spherical coordinates, assuming circumferential symmetry, the preceding equations take the following form.

Energy:

$$K_g \frac{d}{dr} r^2 \frac{dT}{dr} - C_g \frac{d}{dr}[\dot{W}''r^2]T + r^2 \dot{q}''' = 0 \qquad (5.40)$$

Fuel:

$$\rho_g D_F \frac{d}{dr} r^2 \frac{dY_F}{dr} - \frac{d}{dr}[\dot{W}''r^2]Y_F + r^2 \dot{W}_F''' = 0 \qquad (5.41)$$

Oxygen:

$$\rho_g D_O \frac{d}{dr} r^2 \frac{dY_O}{dr} - \frac{d}{dr}[\dot{W}''r^2]Y_O + r^2 \dot{W}_O''' = 0 \qquad (5.42)$$

Products:

$$\rho_g D_P \frac{d}{dr} r^2 \frac{dY_P}{dr} - \frac{d}{dr}[\dot{W}''r^2]Y_P + r^2 \dot{W}_P''' = 0 \qquad (5.43)$$

Overall Continuity:

$$\dot{W}''r^2 = \text{constant for all } r \qquad (5.44)$$

Since droplet combustion is of utmost importance for us, we deal with Eqs. 5.40 to 5.44 here. The results of the spherical analysis are later modified for application to a multitude of liquid fuel diffusion flames. As before, the square brackets around $[\dot{W}''r^2]$ in Eqs. 5.40 to 5.43 are to emphasize the meaning of the continuity equation. $\dot{q}'''$ is the energy released due to the reaction in a unit volume in a unit time. $\dot{W}_i'''$ is the mass of i consumed or released in a unit volume in a unit time.

In a diffusion flame, as discussed earlier, the chemical reactions occur infinitely fast. As shown in Figure 4.1, a distinct and infinitesimally thin reaction zone separates the fuel from the oxygen. This means that no oxidant is present on the fuel side of the flame and no fuel, on the oxidant side. While the fuel and oxidant concentrations are zero at the "flame shell," the product concentration and temperature are maximum.

Solution of the mass transfer problem involves the determination of the vaporization rate, the location of the flame shell relative to the liquid fuel surface and the concentration and temperature profiles under the given conditions of geometry and flow.

Diffusion Flames in Liquid Fuel Combustion

Let us assume a single, simple, infinitely fast combustion reaction scheme, the stoichiometric equation for which is as given below.

$$f \text{ gms of fuel, } F + 1 \text{ gm of oxygen, } O \longrightarrow (1+f) \text{ gms of}$$
$$\text{products, } P + f \cdot \Delta H \text{ cals. of heat.} \quad (5.45)$$

Here f is the stoichiometric fuel/oxygen ratio (grams of fuel per gram of oxygen) and ΔH is the heat of combustion (cal per gm of fuel). This reaction stoichiometry relates the source-sink terms as following.*

$$-\frac{\dot{W}_F'''}{f} = -\dot{W}_O''' = +\frac{\dot{W}_P'''}{1+f} = +\frac{\dot{q}'''}{f\Delta H} \quad (5.46)$$

Consider the energy and fuel conservation equations. Equation 5.46 indicates that

$$\dot{q}''' + \Delta H \cdot \dot{W}_F''' = 0 \quad (5.47)$$

Multiplying Eq. 5.41 by ΔH and adding to Eq. 5.40, one obtains the following.

$$\rho_g \alpha_g \frac{d}{dr} r^2 \frac{dC_g T}{dr} - [\dot{W}_W''' \mathscr{R}^2] \frac{dC_g T}{dr} + \rho_g D_F \frac{d}{dr} r^2 \frac{d\Delta H Y_F}{dr}$$
$$- [\dot{W}_W''' \mathscr{R}^2] \frac{d\Delta H Y_F}{dr} + r^2(\dot{q}''' + \Delta H \dot{W}_F''') = 0$$

Applying Eq. 5.47 and assuming $\alpha_g = D_F$,

$$\rho_g \alpha_g \frac{d}{dr} r^2 \frac{d(C_g T + \Delta H Y_F)}{dr} - [\dot{W}_W''' \mathscr{R}^2] \frac{d(C_g T + \Delta H Y_F)}{dr} = 0$$

Dividing this equation through by a constant† $Q + \Delta H(Y_{FW} - Y_{FR})$, it reduces to the form of Eqs. 5.12 and 5.13.

$$\rho_g \alpha_g \frac{d}{dr} r^2 \frac{db_{FT}}{dr} - [\dot{W}_W''' \mathscr{R}^2] \frac{db_{FT}}{dr} = 0 \quad (5.48)$$

b_{FT} is defined as below

$$b_{FT} \equiv \frac{C_g(T - T_\infty) + \Delta H(Y_F - Y_{F\infty})}{Q + \Delta H(Y_{FW} - Y_{FR})} \quad (5.49)$$

---

* Recall from Chapter 2 that consumption of the reactants and production of the products in the reaction $aA + bB + \cdots \rightarrow mM + nN + \cdots$ are mutually related according to

$$-\frac{1}{a}\frac{dC_A}{dt} = -\frac{1}{b}\frac{dC_B}{dt} = \cdots = +\frac{1}{m}\frac{dC_M}{dt} = +\frac{1}{n}\frac{dC_N}{dt}$$

where C_i is the concentration of the species i.
† Recall from Eq. 5.1 that $Q \approx L + C_l(T_W - T_R)$.

$Y_{F\infty}$ is, of course, zero for a diffusion flame. It is retained in Eq. 5.49 in order to preserve the generality of the definition. Seeking for the boundary conditions of Eq. 5.48, we resort to Eqs. 5.2 and 5.7 which are valid even when combustion occurs.

$$\dot{W}''_W = \rho_g \alpha_g \frac{d}{dr}\left(\frac{C_g(T - T_\infty)}{Q}\right)\bigg|_W \tag{5.2}$$

$$\dot{W}''_W = \rho_g D_F \frac{d}{dr}\left(\frac{Y_F - Y_{F\infty}}{Y_{FW} - Y_{FR}}\right)\bigg|_W \tag{5.7}$$

Multiply Eq. 5.7 by ΔH. Rearrangement yields,

$$\dot{W}''_W Q = \rho_g \alpha_g \frac{dC_g T}{dr}\bigg|_W \tag{5.2a}$$

$$\dot{W}''_W \Delta H(Y_{FW} - Y_{FR}) = \rho_g D_F \frac{d\Delta H Y_F}{dr}\bigg|_W \tag{5.7a}$$

Adding these two equations and recalling that we assumed $\alpha_g = D_F$ in the derivation of Eq. 5.48, we obtain the following boundary condition for b_{FT}.

$$\dot{W}''_W = \rho_g \alpha_g \frac{db_{FT}}{dr}\bigg|_W \tag{5.50}$$

That is, b_{FT} is a "conserved variable" by our definition. The student may verify that when $\alpha_g = D_O$, the energy and oxygen conservation equations reduce to an equation of the form of Eq. 5.48 (with a boundary condition of the form of Eq. 5.50) if the conserved variable b_{OT} is defined as following.

$$b_{OT} \equiv \frac{C_g(T - T_\infty) + \Delta H f(Y_O - Y_{O\infty})}{Q} \tag{5.51}$$

When $D_F = D_O$, the oxygen and fuel equations combine, similarly, with a third conserved variable defined as

$$b_{FO} \equiv \frac{(Y_F - Y_{F\infty}) - f(Y_O - Y_{O\infty})}{(Y_{FW} - Y_{FR})} \tag{5.52}$$

Hence, b_{FT}, b_{OT} and b_{FO} satisfy the following equations

$$\rho_g \alpha_g \frac{d}{dr} r^2 \frac{db}{dr} - [\dot{W}''_W \mathcal{R}^2] \frac{db}{dr} = 0$$

$$\dot{W}''_W = \rho_g \alpha_g \frac{db}{dr}\bigg|_W$$

Diffusion Flames in Liquid Fuel Combustion

The solution follows immediately from Section 5.6.
The b-profile is given by,

$$\ln\left(\frac{B+1}{b-b_W+1}\right) = \frac{[\dot{W}_W'' \mathcal{R}^2]}{\rho_g \alpha_g} \cdot \frac{1}{r} \quad (5.53)$$

Or, eliminating $[\dot{W}_W'' \mathcal{R}]/\rho_g \alpha_g$,

$$(b - b_W + 1) = (B+1)^{(1-\mathcal{R}/r)} \quad (5.54)$$

The mass transfer rate is given by substituting $r = \mathcal{R}$ and $b = b_W$ in Eq. 5.53.

$$\dot{W}_W'' = \frac{\rho_g \alpha_g}{\mathcal{R}} \ln(B+1) \quad (5.55)$$

The mass transfer number B is defined by the following three mutually equivalent formulae.

$$B \equiv B_{OT} = B_{FO} = B_{FT} \quad \text{since} \quad \alpha_g = D_F = D_O$$

$$B_{OT} \equiv b_{OT\infty} - b_{OTW} = \frac{\Delta H f Y_{O\infty} + C_g(T_\infty - T_W)}{L + C_l(T_W - T_R)} \quad (5.56)$$

$$B_{FO} \equiv b_{FO\infty} - b_{FOW} = \frac{f Y_{O\infty} + Y_{FW}}{Y_{FR} - Y_{FW}} \quad (5.57)$$

$$B_{FT} \equiv b_{FT\infty} - b_{FTW} = \frac{-\Delta H Y_{FW} + C_g(T_\infty - T_W)}{L + C_l(T_W - T_R) + \Delta H(Y_{FW} - Y_{FR})} \quad (5.58)$$

The elimination of the nonlinear source-sink terms in the conservation equations, thus, exceedingly simplifies the diffusion flame problem. This elimination technique is known in the combustion literature as the *Schwab-Zeldovich transformation*.

Application of the preceding equations to find the mass transfer rate, flame location and the profiles of temperature and composition is presented below.

(a) Calculation of the Burning Rate

If the mass transfer number B is known, the burning rate of the droplet may be calculated by Eq. 5.55. The calculation of B, however, requires a knowledge of either T_W or Y_{FW}. Equating B_{FO} and B_{OT},

$$\frac{f Y_{O\infty} + Y_{FW}}{Y_{FR} - Y_{FW}} = \frac{\Delta H f Y_{O\infty} + C_g(T_\infty - T_W)}{L + C_l(T_W - T_R)}$$

This equation, along with the saturation vapor pressure equation of the liquid fuel under consideration, enables us to find the wall temperature and B. It is customary and convenient to assume that $T_W \approx T_B$ and calculate B_{OT} for B.

$$B \approx \frac{\Delta H f Y_{O\infty} + C_g(T_\infty - T_B)}{L + C_l(T_B - T_R)} \qquad (5.59)$$

Table 5.4 gives the B-values calculated by Eq. 5.59 (and property values of Table 5.2) for various hydrocarbon/air diffusion flames. Also presented in this table are values of B calculated by neglecting the $C_g(T_\infty - T_B)$ and $C_l(T_B - T_R)$ terms. Such a neglect yields the following approximate definition of B.

$$B \approx \frac{\Delta H}{L} f Y_{O\infty}$$

This approximation, as seen in Table 5.4, causes errors as high as 50% in the transfer number. Since $\ln(B + 1)$ is a very slowly varying function of B, such large errors are suppressed to within 20% when the mass transfer rate is considered.

From Eqs. 5.55 and 5.59, it is clear that the droplet combustion rate is increased by reducing the drop size, increasing K_g/C_g and increasing B. Combustion rate is higher if the ambient atmosphere is hot and rich in oxygen content, and if the fuel is preheated. A fuel with a high ratio of heat of combustion and latent heat of evaporation is expected to burn faster.

(b) *Location of the Flame*

At the flame, the fuel and oxidant concentrations occur to be in stoichiometric proportions so that reaction can take place instantaneously. Denoting the flame location by the subscript c,

$$Y_{FC} = f Y_{OC}$$

Substituting this in Eq. 5.53 and simplifying with Eqs. 5.52 and 5.57, the radius of the flame shell is obtained as

$$r_C = \frac{\dot{W}_W'' \mathscr{R}^2}{\rho_g \alpha_g} \cdot \frac{1}{\ln(1 + f Y_{O\infty}/Y_{FR})} \qquad (5.60)$$

If one wishes, $\dot{W}_W'' \mathscr{R}/\rho_g \alpha_g$ may be eliminated from Eq. 5.60 by using Eq. 5.55.

$$\frac{r_C}{\mathscr{R}} = \frac{\ln(B + 1)}{\ln\left(\dfrac{f Y_{O\infty}}{Y_{FR}} + 1\right)} \qquad (5.61)$$

Diffusion Flames in Liquid Fuel Combustion

TABLE 5.4
Mass Transfer Driving Force for Combustion of Various Liquids in Air

	$C_g(T_\infty - T_B)$*	$\Delta H f Y_{O\infty}$†	$C_l(T_B - T_R)$‡	L	B	B'	B''	$\ln(B+1)$	$\ln(B'+1)$	$\ln(B''+1)$	M	T_B
n-Pentane	−4.95	785	8.91	87.1	8.15	8.19	9.00	2.21	2.22	2.30	72	36.0
n-hexane	−14.8	770	25.80	87.1	6.70	6.82	8.83	2.04	2.06	2.29	86	68.0
n-heptane	−24.3	770	41.20	87.1	5.82	6.00	8.84	1.92	1.94	2.29	100	98.5
n-octane	−32.6	775	55.40	86.5	5.24	5.46	8.97	1.83	1.87	2.30	114	125.0
iso-octane	−32.6	770	54.00	78.4	5.56	5.82	9.84	1.88	1.92	2.38	114	125.0
n-decane	−47.7	770	80.50	86.0	4.34	4.62	8.95	1.68	1.73	2.30	142	174.0
n-deodane	−55.7	774	94.00	85.5	4.00	4.30	9.05	1.61	1.67	2.35	170	200.0
Octene	−31.0	783	53.00	80.5	5.64	5.86	9.72	1.89	1.93	2.37	112	121.0
Benzene	−18.6	790	24.70	103.2	6.05	6.18	4.65	1.95	1.97	2.16	78	80.0
Methanol	−13.8	792	25.20	263.0	2.70	2.74	3.00	1.31	1.32	1.39	32	64.5
Ethanol	−18.1	776	32.60	200.0	3.25	3.34	3.88	1.45	1.47	1.59	46	78.5
Gasoline	−41.8	774	66.00	81.0	4.98	5.25	9.55	1.79	1.83	2.35	120	155.0
Kerosene	−71.2	750	106.00	69.5	3.86	4.26	10.80	1.58	1.66	2.47	154	250.0
Light diesel	−71.2	735	103.50	63.9	3.96	4.40	11.50	1.60	1.69	2.52	170	250.0
Med. diesel	−74.3	725	107.50	58.4	3.94	4.38	12.45	1.60	1.68	2.59	184	260.0
Heavy diesel	−77.5	722	109.00	55.5	3.91	4.40	13.00	1.59	1.69	2.64	198	270.0
Acetone	−11.4	770	23.60	125.0	5.10	5.19	6.16	1.81	1.82	1.97	58	56.7
Toluene	−28.0	749	35.00	84.0	6.06	6.30	8.92	1.95	1.99	2.29	92	110.6
Xylene	−34.0	758	45.30	80.0	5.76	6.04	9.48	1.91	1.95	2.35	106	130.0

* Assumed $C_g \approx 0.31$ cal/gm/°C and $T_\infty = 20$ °C.
† $Y_{O\infty}$ assumed 0.232.
‡ T_R assumed 20 °C.

$$B = \frac{\Delta H f Y_{O\infty} + C_g(T_\infty - T_B)}{L + C_l(T_B - T_R)} \qquad B' = \frac{\Delta H f Y_{O\infty}}{L + C_l(T_B - T_R)} \qquad B'' = \frac{\Delta H}{L} f Y_{O\infty}$$

In all cases CO formation is neglected.

This equation shows that if the combustion rate is large, the flame will be farther away from the fuel surface. Furthermore, since B is always* greater than $fY_{O\infty}/Y_{FR}$, the liquid fuel diffusion flame never approaches the liquid surface; that is, $r_C/\mathscr{R}$ is always greater than unity. For hydrocarbon liquids the fuel oxidant ratio is about 0.32. If pure liquid ($Y_{FR} = 1$) is burning in standard air ($Y_{O\infty} = 0.232$), Eq. 5.61 can be written as

$$\frac{r_C}{\mathscr{R}} \approx 14 \frac{\dot{W}''_W \mathscr{R}}{\rho_g \alpha_g} = 14 \ln(B + 1)$$

It is timely here to comment on the effects of total pressure on the flame shell location and the burning rate. Experimental investigations show that an increase in pressure results in a slight reduction of the flame shell radius and a slight increase in the burning rate. The free convective heat transfer coefficient (to which the mass transfer is directly proportional), is known to increase directly proportional to the fourth root of the pressure. The radiative heat transfer from the flame to the fuel wall increases very strongly with pressure. These two factors influence $\dot{W}''_W$ through the factor $\bar{h}^0/C_g (= \rho_g \alpha_g/\mathscr{R}$ when these effects are neglected). On a secondary level, the slight decrease in the latent heat of evaporation with increased pressure results in a higher B and hence a higher mass transfer rate. It is obvious to the reader that our simple model yields the combustion rates only to the accuracy of an engineering need. Further detailed theoretical formulations are available in the literature for precise calculation of the burning rate of a droplet. (Williams' *Combustion Theory* is an example.)

(c) *The Temperature Profile*

In deducing the temperature profile, we make use of the fact that within the flame, (i.e., $\mathscr{R} < r < r_C$) there exists no oxygen and outside the flame, (i.e., $r_C < r < \infty$) no fuel exists. Thus the $b_{OT}(r)$ gives the temperature profile within the flame and $b_{FT}(r)$ gives it outside the flames.

$\mathscr{R} < r < r_C$:

Equation 5.53 gives $b_{OT}(r)$ as

$$\ln\left(\frac{b_{OT\infty} - b_{OTW} + 1}{b_{OT} - b_{OTW} + 1}\right) = \frac{[\dot{W}''\mathscr{R}^2]}{\rho_g \alpha_g} \cdot \frac{1}{r}$$

---

* According to Eq. 5.57, when B is extremely small, since $Y_{FW} \approx 0$, $B \to fY_{O\infty}/Y_{FR}$. The flame then is located *on the* fuel surface. This situation arises in solid fuel combustion as discussed in Chapter 6.

Diffusion Flames in Liquid Fuel Combustion

Substituting the definition of b_{OT} and setting $Y_O = 0$ within the flame,

$$\ln\left(\frac{C_g(T_\infty - T_W) + \Delta H f Y_{O\infty} + Q}{C_g(T - T_W) + Q}\right) = \frac{[\dot{W}''_W \mathcal{R}^2]}{\rho_g \alpha_g} \cdot \frac{1}{r}$$

Rearrangement yields

$$C_g(T - T_W) = [C_g(T_\infty - T_W) + \Delta H f Y_{O\infty} + Q] \cdot \exp[-(\dot{W}''_W \mathcal{R}^2)/\rho_g \alpha_g r] - Q \tag{5.62}$$

Or, alternatively, substituting Eqs. 5.59 and 5.55,

$$C_g(T - T_W) = Q[(B + 1) \cdot \exp[-(\mathcal{R}/r) \ln(B + 1)] - 1] \tag{5.63}$$
$$= Q[(B + 1)^{(1 - \mathcal{R}/r)} - 1] \tag{5.63a}$$

$r_C < r < \infty$:

Similarly, substituting the definition of b_{FT} in Eq. 5.53 and setting $Y_F = 0$ for $r > r_C$,

$$C_g(T - T_W) = [C_g(T_\infty - T_W) - \Delta H Y_{FR} + Q]$$
$$\times \exp[-(\dot{W}''_W \mathcal{R}^2)/\rho_g \alpha_g r] + \Delta H Y_{FR} - Q \tag{5.64}$$

Substituting Eqs. 5.55 and 5.58,

$$C_g(T - T_W) = [Q + \Delta H(Y_{FW} - Y_{FR})][(B + 1)^{(1 - \mathcal{R}/r)} - 1] + \Delta H Y_{FW} \tag{5.65}$$

Substitution of Eq. 5.60 in either Eq. 5.62 or 5.65, yields the flame temperature as

$$C_g(T_C - T_W) = \frac{C_g(T_\infty - T_W) + f Y_{O\infty}(\Delta H - Q/Y_{FR})}{1 + f Y_{O\infty}/Y_{FR}} \tag{5.66}$$

The assumptions to be remembered in using Eq. 5.66 are: (a) the gas specific heat C_g is independent of both temperature as well as composition, (b) ΔH is independent of temperature and (c) Lewis number is unity (i.e. $\alpha_g = D_F = D_O$). The first two assumptions may be relaxed but such a relaxation is not worthwhile until the Lewis number assumption is relaxed.

It is interesting to note that the flame temperature calculated by the techniques described in Appendix C is equal to that obtained from Eq. 5.66 if the dissociation effects are neglected. For example, if $f Y_{O\infty}$ grams of the fuel is mixed with a unit mass of air and allowed to react, the resulting flame temperature is given by an energy balance as below. The energy released by the combustion of $f Y_{O\infty}$ grams of fuel is $\Delta H f Y_{O\infty}$ calories. This energy is utilized to accomplish three missions: (a) to raise the temperature of a unit mass of air from T_∞ to T_C, (b) to raise the temperature of $f Y_{O\infty}/Y_{FR}$ gms of fuel vapor from T_W to T_C and (c) to provide for the preparation of the reservoir fuel to the gas phase at the wall. Thus, for energy conservation, assuming that

C_g is same for the fuel/air mixture of any composition and that ΔH is independent of temperature,

$$C_g(T_C - T_\infty) + \frac{fY_{O\infty}}{Y_{FR}} C_g(T_C - T_W) + \frac{fY_{O\infty}}{Y_{FR}} Q = fY_{O\infty}\Delta H$$

Resolving for $C_g(T_C - T_W)$,

$$C_g(T_C - T_W) = \frac{C_g(T_\infty - T_W) + fY_{O\infty}(\Delta H - Q/Y_{FR})}{1 + fY_{O\infty}/Y_{FR}}$$

This result is same as Eq. 5.66. This observation indicates that in spite of the drastic differences involved in the description of diffusion flames and premixed flames, the highest possible temperature of a diffusion flame is equal to that of a premixed flame.

The flame temperature is a function of B, $Y_{O\infty}$, Y_{FR}, f, T_∞, T_B, L and T_R. On the basis of Table 5.2, for hydrocarbons, ΔH and f may be assumed as constants respectively equal to 10,300 cal/gm and 0.296. Assuming these values and $T_\infty = 20$ °C, Spalding prepared a chart as shown in Figure 5.12 to quickly read out the temperature of a diffusion flame. Dissociation effects are neglected.

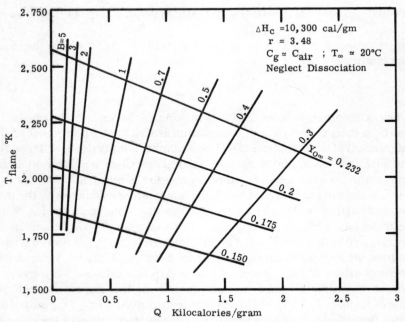

Figure 5.12 Spalding's flame temperature chart for some common hydrocarbon liquid fuels (r in the figure is $1/f$) (with the permission of The Combustion Institute, from: D. B. Spalding, p. 863, *4th International Symposium on Combustion*, (1953))

Diffusion Flames in Liquid Fuel Combustion

(d) *The Fuel and Oxygen Mass Fraction Profiles*

Substituting $b = b_{FO}$ in Eq. 5.53,

$$\frac{b_{FO} - b_{FOW} + 1}{b_{FO\infty} - b_{FOW} + 1} = \exp[-(\dot{W}_W'' \mathcal{R}^2)/\rho_g \alpha_g r] \tag{5.67}$$

$b_{FO}(r)$ gives Y_F profile in $\mathcal{R} < r < r_c$ and Y_O profile in $r_c < r < \infty$. Substituting $Y_{OW} = 0$ and $Y_{F\infty} = 0$, Eq. 5.67 reduces to the following.

$$\frac{f(Y_O - Y_{O\infty}) - Y_F}{Y_{FR} + fY_{O\infty}} = \exp[-(\dot{W}_W'' \mathcal{R}^2)/\rho_g \alpha_g r] - 1 \tag{5.68}$$

The fuel profile is obtained by substituting $Y_O = 0$ in Eq. 5.68.

$$Y_F(r) = (Y_{FR} + fY_{O\infty})\{1 - \exp[-(\dot{W}_W'' \mathcal{R}^2)/\rho_g \alpha_g r]\} - fY_{O\infty} \tag{5.69}$$

Substitution of $r = r_c$ from Eq. 5.60 shows that $Y_F = 0$ in $r > r_c$.

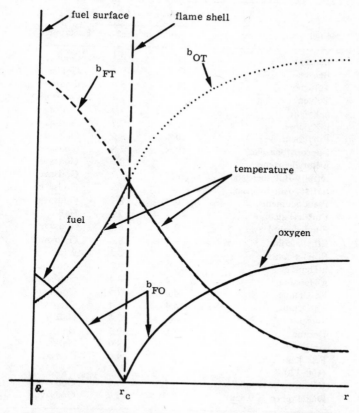

Figure 5.13 Species and temperature profiles for burning droplets

The oxygen profile is obtained by substituting $Y_F = 0$ in Eq. 5.68.

$$Y_O(r) = \frac{(Y_{FR} + fY_{O\infty})}{f} \{\exp[-(\dot{W}_W'' \mathscr{R}^2)/\rho_g \alpha_g r] - 1\} + Y_{O\infty} \quad (5.70)$$

Figure 5.13 qualitatively shows the profiles determined as above.

5.9 Droplet Burning Time

In a manner similar to the droplet vaporization time, the burning time can be calculated by assuming quasi-stationary state of burning. In fact the same derivation holds good if the mass transfer number B is defined by Eq. 5.59.

TABLE 5.5

Burning Constant for Various Hydrocarbons
(Oxidant: air at 20 °C and 1 atm.)

Fuel	$\lambda_{calc.}$	$\lambda_{meas.}$	Experimentor
	10^{-3} cm^2/sec		
Benzene	11.2	9.7	Godsave
Benzene	11.2	9.9	Goldsmith
Toluene	11.1	6.6	Godsave
Toluene	11.1	7.7	Goldsmith
o-Xylene	10.4	7.9	Godsave
p-Xylene	10.8	7.7	Godsave
Ethyl benzene	10.8	8.6	Godsave
isopropylbenzene	10.6	7.8	Godsave
n-Butylbenzene	—	8.6	Godsave
tertiary-Butylbenzene	10.4	7.7	Godsave
tertiary-Amylbenzene	—	7.8	Godsave
Pseudocumene	10.2	8.7	Godsave
Furfuryl alcohol	—	7.2	Hall
Ethyl alcohol	9.3	8.1	Godsave
Ethyl alcohol	9.3	8.6	Goldsmith
Ethyl alcohol	9.3	8.5	Wise
n-Heptane	14.2	9.7	Godsave
n-Heptane	14.2	8.4	Goldsmith
iso-Octane	14.4	9.5	Godsave
iso-Octane	14.4	11.4	Graves
Tetralin	—	7.6	Hall
Decane	11.6	10.1	Hall
Amyl acetate	—	8.0	Hall
Petroleum ether (100–120 °C)	—	9.9	Godsave
Kerosene ($\rho = 0.805$)	9.7	9.6	Godsave
Diesel oil ($\rho = 0.850$)	8.5	7.9	Godsave

Diffusion Flames in Liquid Fuel Combustion

The variation of the droplet diameter with time is given by

$$d^2 = d_O^2 - \lambda_b t \tag{5.71}$$

where the burning constant λ_b is defined by

$$\lambda_b \equiv \frac{8\rho_g \alpha_g}{\rho_l} \ln(B + 1) \tag{5.72}$$

The burning time is given by

$$t_b \equiv d_O^2 / \lambda_b \tag{5.73}$$

Table 5.5 presents some typical values of the burning constant calculated from Eq. 5.72 and the B-values given in Table 5.4. The experimental measurements of various workers are also shown. The agreement is excellent. It is to be pointed out that wide variations in the chemical structure of the fuel do not seem to systematically influence λ_b.

The general order of magnitude of λ_b for hydrocarbons burning in air is about 10^{-2} cm^2/sec.

5.10 Spray Combustion

So far in this chapter we have considered the evaporation and combustion of isolated droplets in an *infinite* medium of the oxidant stream. When the practically important problem of vaporization and combustion of an aggregate of droplets in a spray is considered, two major complications arise.

Firstly, a droplet under consideration is no longer in an infinite oxidant atmosphere. The influence of neighboring droplets becomes significant in their competition for oxygen, or alternatively, in their competition to *push away* the fuel vapor. Rex and his associates and Kanevsky experimented with a systematic model of this interference among neighboring droplets. Kanevsky arranged nine droplets (suspended on quartz filaments) spaced uniformly and measured the burning constant. His findings show that the burning time is still proportional to the square of the initial diameter of the droplets, but the burning constant is somewhat increased. Table 5.6 summarizes his data. In reference to this table, it can be noticed that λ_b is increased as much as 40%.

Secondly, in sprays the following question often arises. What is the rate of evaporation of the droplets relative to the rate of oxygen diffusion into the spray? The answer to this question is important in at least two respects.

If the droplet vaporization rate is much higher than the oxygen diffusion rate, in all probability, the droplets may vaporize completely before combustion occurs. The combustion in such a situation is governed by the laws of gaseous diffusion flames—the topic assigned to Chapter 7 of this book.

TABLE 5.6

Kanevsky's Data on Droplets Burning in an Array

```
|←S→|
 o  o  o  ↑
 o  o  o  S
 o  o  o  ↓
```

Fuel	S mm	λ_b cm²/sec
n-heptane	∞	0.97×10^{-2}
	9.5	1.28×10^{-2}
	8.5	1.16×10^{-2}
	7.5	1.23×10^{-2}
	5.8	1.28×10^{-2}*
	3.6	0.78×10^{-2}†
Methanol	8.7	1.04×10^{-2}
	7.5	1.09×10^{-2}
	5.8	1.08×10^{-2}*
	3.6	0.64×10^{-2}†

* Flames are partially merged.
† Flames are completely merged.
Uncommented flames are separate.
From Kanevsky, J., *Jet Propulsion*, **26**, p. 758 (1956).

The length of the spray is controlled by either the droplet vaporization process or the oxygen diffusion process. This problem obviously depends upon the combustion behavior of fuel droplets relative to fuel jets and hence we will postpone it to Chapter 7, as well.

In spray combustion, it is reasonable to expect the burning constant to depend upon the injector design and initial drop size distribution. If t_r is the residence time of the spray in the combustion chamber, and $\bar{d}_0$ is some average droplet diameter in the spray cross section near the injector, Probert uses the Rosin-Rammler size distribution to predict the percentage of unburnt fuel at the exit as a function of the nondimensional* factor $\lambda_b t_r / \bar{d}_0^2$. Obviously if $t_r = 0$, 100% of the fuel remains unburnt whereas if $t_r \approx \bar{d}_0^2 / \lambda_b$ practically all the fuel is burnt. The Rosin-Rammler size distribution constant tampers with this statement very little.

---

* Recognizing $\bar{d}_0^2 / \lambda_b$ as the burning time of the mean droplet, $t_{b_{\text{mean}}}$, this nondimensional factor can be seen as the ratio of residence time to $t_{b_{\text{mean}}}$.

Diffusion Flames in Liquid Fuel Combustion

5.11 Some Examples

In this section the procedure to calculate vaporization and combustion rate of droplets (and other geometries) is illustrated by working out several examples. The following steps are recommended in such calculations.

(a) Examine the geometry and flow conditions to choose a suitable heat transfer relation from Table 3.4. Deduce the average heat transfer coefficient.

$$\frac{\bar{h}^0}{C_g} = \frac{\rho_g \alpha_g}{l} \cdot \overline{\mathrm{Nu}}_l^0$$

(b) If T_W or Y_{FW} is given calculate B by Eq. 5.26 or 5.27 (if the vaporization is simple) or by Eq. 5.59 (if the vaporization is followed by combustion). If $T_\infty \gg T_B$ assume $T_W \approx T_B$. This assumption is particularly acceptable when flame occurs. For simple vaporization when $T_\infty \ll T_B$, first find Y_{FW} from the vapor pressure data at $T_W \approx T_\infty$ and calculate B using Eq. 5.27. Go to step (d).

(c) If T_∞ is neither very high nor very low when compared to T_B, plot $B(T_W)$ vs. T_W according to Eq. 5.26 (if vaporization is not followed by combustion) or Eq. 5.56 (if vaporization is followed by combustion). Then read P_{FW} for various values of T_W from the vapor pressure data. Knowing this relation and total pressure P calculate $Y_{FW}(T_W)$. Compute and plot $B(T_W)$ vs. T_W from Eq. 5.27 or Eq. 5.57 depending upon whether vaporization is simple or is followed by combustion. The intersection point of these two curves gives B.

(d) Calculate vaporization rate from

$$\dot{W}_W'' = \frac{\bar{h}^0}{C_g} \ln(B + 1)$$

Example 1

A 2 cm dia. naphthalene ball whose uniform interior temperature is 24 °C evaporates into still air at 28 °C. Calculate the mass-loss rate in grams per minute assuming that the specific heat of the vapor/air mixture is 0.25 cal/gm/°C, thermal conductivity is 0.025 cal/cm/°C/sec and the latent heat of evaporation is 100 cal/gm.

Solution

For a sphere in still air, Table 3.4 gives

$$\overline{\mathrm{Nu}}_d^0 = 2$$

Therefore,

$$\frac{\bar{h}^0}{C_g} = \frac{\rho_g \alpha_g}{d/2} = \frac{2K_g}{C_g d} = \frac{2 \cdot 0.025}{0.25 \cdot 2} = 0.1 \text{ gm/cm}^2/\text{sec}.$$

The mass transfer number B is given by $C_g(T_\infty - T_W)/L = 0.25(28 - 24)/100 = 0.01$. The mass transfer rate therefore, is

$$\dot{W}_W'' = \frac{\bar{h}^0}{C_g} \ln(B + 1) = 0.1 \ln 1.01 \approx 0.001 \text{ gm/cm}^2/\text{sec}.$$

The total mass loss rate is $\dot{W}_W'' \times$ surface area of the sphere.

$$\dot{W}_W'' = \dot{W}_W \pi d^2 = 0.754 \text{ gm/min}$$

Example 2

A 2 cm high wick soaked in ethyl alcohol evaporates into the room air free convectively. If the room temperature is 200 °C and the "reservoir" temperature is 20 °C, calculate the evaporation rate.

Solution

Grashof number is defined by $gl^3 \beta \Delta T / v_g^2$ where g is acceleration due to gravity, 981 cm/sec^2, l is the height of the body, 2 cm, β is the temperature coefficient of volumetric expansion of air ($\approx 1/T\,^\circ$K for a perfect gas), ΔT is the characteristic temperature differential, $(200 - 20)\,^\circ$C, and v_g is the kinematic viscosity of the vapor/air mixture estimated at some mean temperature to be about 0.15 cm^2/sec.

For ethanol, Table 5.2 gives $L = 200$ cal/gm, $C_l = 0.56$ cal/gm/°C, and $T_B \approx 78.5$ °C. Let us assume $K_g = 0.025$ cal/cm/°C/sec.

The Grashof number is to be calculated first of all.

$$\text{Gr}_l = \frac{gl^3 \beta \Delta T}{v_g^2} \approx 981 \cdot 2^3 \cdot \frac{1}{(200 + 273)} \cdot \frac{180}{0.15^2} = 138{,}500$$

Referring to Table 3.4, this value of the Grashof number is well below the critical value of 10^9. The flow is, hence, laminar. The laminar free convection heat transfer to a vertical wall is given by Table 3.4 as

$$\overline{\text{Nu}}_l^0 = 0.59 (\text{GrPr})^{1/4}$$

Let us assume Prandtl number to be unity. Then $\overline{\text{Nu}}_l^0 = 0.59 \cdot (138{,}500)^{1/4}$. The mass transfer coefficient $\bar{h}^0/C_g$ is then

$$\frac{\bar{h}^0}{C_g} = \frac{\rho_g \alpha_g}{l} \overline{\text{Nu}}_l^0$$

$$= 0.57 \text{ gm/cm}^2/\text{sec}.$$

Diffusion Flames in Liquid Fuel Combustion

The mass transfer driving force $B \equiv C_g(T_\infty - T_W)/Q$, $Q = L + C_l(T_W - T_R)$. Since $T_\infty (= 200\,°C) \gg T_B (= 78.5\,°C)$, let us assume that $T_W \approx T_B$.

$$B = \frac{0.25(200 - 78.5)}{200 + 0.56(78.5 - 20)} = 0.131$$

$$\dot{W}''_W = \frac{\bar{h}^0}{C_g} \ln(B + 1) = 0.57 \ln 1.131 = 0.0696 \text{ gm/cm}^2/\text{sec}.$$

Example 3

A water drop 4 mm in diameter is suspended in dry air (a) at 15 °C and (b) at 1,000 °C. How long does it take to completely vaporize it if the total pressure is 760 mm mercury. Do not make the usual approximations regarding the surface temperature.

Solution

Assume Lewis number one. Note

$$Y_{AW} = 1 \bigg/ \left[1 + \left(\frac{P}{P_{AW}} - 1\right) \frac{M_g}{M_A}\right]$$

$P = 760$ mm of mercury. $M_g \approx 28$, $M_A = 18$. The relation between B_D and T_W is calculated first from the vapor pressure data. A tabular format similar to the following will be found convenient.

$$B_D = \frac{Y_{A\infty} - Y_{AW}}{Y_{AW} - Y_{AR}} = \frac{0 - Y_{AW}}{Y_{AW} - 1} = \frac{Y_{AW}}{1 - Y_{AW}}$$

T_W °C	P_{AW} (mm)	$\frac{P}{P_{AW}}$	$\frac{760}{P_{AW}}$	$\left(\frac{P}{P_{AW}} - 1\right)\frac{M_g}{M_A}$	Y_{AW}	$(1 - Y_{AW})$	B_D
−17.3	1	760	760	1,180.6	$8.5 \cdot 10^{-4}$	0.9991	$8.5 \cdot 10^{-4}$
11.3	10	76	76	116.7	$8.5 \cdot 10^{-3}$	0.9915	$8.6 \cdot 10^{-3}$
34.1	40	19	19	28	0.0345	0.9655	0.0357
51.6	100	7.6	7.6	10.3	0.0888	0.9112	0.0974
83.0	400	1.9	1.9	1.4	0.4167	0.5833	0.7143
100.0	760	1.0	1.0	0.0	1.0000	0.0000	∞

The relation between B_H and T_W is given by

$$B_H \approx \frac{C_g(T_\infty - T_W)}{L} \approx \frac{0.24}{590}(T_\infty - T_W) = 0.00169(T_\infty - T_W)$$

Plotting B_D and B_H as functions of T_W, as shown in Figure 5.14, B and T_W are found.

At $T_\infty = 15\,°C$ $B = 0.0073$
At $T_\infty = 1{,}000\,°C$ $B = 1.5340$

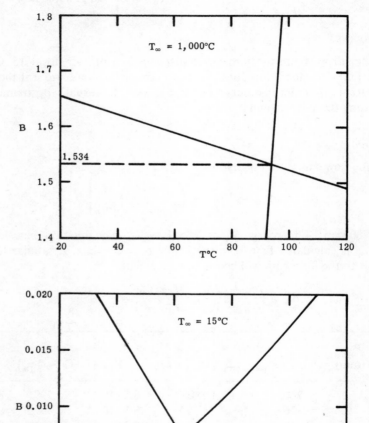

Figure 5.14 B-determination in example 3

Diffusion Flames in Liquid Fuel Combustion

Assuming $\alpha_g \approx 0.243 \text{ cm}^2/\text{sec}$, $\rho_l = 1 \text{ gm/cm}^3$ and $\rho_g \approx 1.22 \cdot 10^{-3} \text{ gm/cm}^3$ (all these values are obtained at $T_\infty = 15 \, °\text{C}$),

$$t_v = d_0^2 \rho_l / [8\rho_g \alpha_g \ln(B + 1)] = 6.02 \text{ sec at } T_\infty = 15 \, °\text{C}$$
$$= 4.26 \text{ sec at } T_\infty = 1{,}000 \, °\text{C}$$

Example 4

An important application of the relationship $B_D = B_H$ in simple vaporization is the working principle of sling psychrometer (which is also known as wet and dry bulb hygrometer). It is used to measure the moisture content in the room air. A thermometer whose mercury bulb is wrapped with a piece of cotton soaked in water reads the so-called "wet-bulb temperature" which is in fact equal to T_W. If Y_{AR} is unity, the equality of B_D and B_H implies the following.

$$\frac{C_g(T_\infty - T_W)}{L + C_l(T_W - T_R)} \equiv \frac{Y_{A\infty} - Y_{AW}}{Y_{AW} - Y_{AR}}$$

Usually, $T_W = T_R$. Hence,

$$Y_{A\infty} = Y_{AW} + (Y_{AW} - 1)C_g(T_\infty - T_W)/L$$

T_∞ is measured by the dry bulb thermometer. T_W is measured by the wet bulb thermometer. Y_{AW} is obtained from saturation tables for the measured value of T_W. Immediately then the moisture content in the ambient air is obtained. A convenient chart is often prepared showing $Y_{A\infty}$ as a function of T_W for various values of T_∞. For a total pressure of 760 mm, it is suggested that the student prepare such a chart.

Since $Y_{AW} - 1$ is always negative, $Y_{A\infty}$ is always lower than Y_{AW}.

Example 5

A turbojet engine has a flight speed of 500 fps. If the combustor length is 5 feet and if the temperature in the combustor is $1{,}000 \, °\text{C}$, what is the maximum allowable size of kerosene droplets if they have to be totally consumed (a) by vaporization alone (b) by vaporization and combustion, before they exit from the combustor?

Solution

(a) *Simple vaporization*

The evaporation time of a droplet of diameter d_0 is given by

$$t_v = \frac{\rho_l d_0^2}{[8\rho_g \alpha_g \ln(B + 1)]} \leq \frac{5 \text{ ft}}{500 \text{ ft/sec}} = 10^{-2} \text{ sec}$$

Hence

$$d_0^2 \leq 8 \cdot 10^{-2} \cdot \frac{\rho_g \alpha_g}{\rho_l} \ln(B + 1)$$

For air, $\rho_g \alpha_g$ at $1{,}000\,°C \approx 0.0002$ gm/cm/sec.
For kerosene, $\rho_l \approx 0.825$ gm/cm³.

$$B \approx \frac{C_g(T_\infty - T_W)}{L} \approx \frac{C_g(T_\infty - T_B)}{L} \approx \frac{0.3(1000 - 250)}{69.5} = 3.24.$$

$$d_0^2 \leq 8 \cdot 10^{-2} \cdot \frac{2 \cdot 10^{-4}}{0.825} \ln(4.24) = 27.9 \cdot 10^{-6} \text{ cm}^2.$$

$$d_0 \leq 5.27 \cdot 10^{-3} \text{ cm}.$$

(b) *Vaporization followed by combustion*

Since combustion enhances the rate of vaporization because of the increased B, the droplet can be much larger than the value calculated above.

$$B \equiv \frac{\Delta H f Y_{O\infty} + C_g(T_\infty - T_W)}{L + C_l(T_W - T_R)}$$

Assuming $T_W \approx T_R$ and $T_W \approx T_B$,

$$B \approx \frac{\Delta H f Y_{O\infty} + C_g(T_\infty - T_B)}{L}$$

The atmosphere around the droplets contains oxygen somewhat lower than 23.2%. It is reasonable to expect it to be about 10%. Using the property values given by Table 5.2,

$$B \approx [0.316 \cdot 10{,}300 \cdot 0.1 + 0.3(1{,}000 - 250)]/69.5$$
$$= 7.93$$

Hence

$$d_0^2 \leq 8 \cdot 10^{-2} \cdot \frac{2 \cdot 10^{-4}}{0.825} \ln(8.93) = 42.5 \cdot 10^{-6} \text{ cm}^2$$

$$d_0 \leq 6.51 \cdot 10^{-3} \text{ cm}$$

Example 6

It is worthwhile to describe here some important features of a turbojet engine shown in Figure 5.15. (Essentially the same description holds for a typical incinerator.) The inlet air is compressed to a high pressure before it is fed into the combustor. Fuel is injected in the form of a spray of fine

Diffusion Flames in Liquid Fuel Combustion

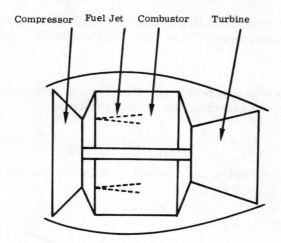

Figure 5.15 A turbojet (for example 6)

droplets into the hot compressed air in the combustor. Spontaneous ignition occurs and the droplets burn as they travel through the chamber. The hot exhaust products are led into the turbine in which they are expanded. A part of the turbine power is used to run the compressor while the rest is used for propulsion. There are several inherent problems associated with the design just described. Firstly, the air flow rate at the usual flight speeds is so high that the possible fuel/air ratio in the combustion chamber will be below the lower ignition limit. Ignition, thereby, will be impossible.

Secondly, due to metallurgical limitations, the maximum temperature of the combustion products is desired to be no more than about 1,000 °C. Relative to the adiabatic flame temperature (see Table 4.3) this temperature is considerably low. The combustion products have to be cooled from the flame temperature ($\approx 2{,}500\,°C$ to $3{,}000\,°C$) to about 1,000 °C before they are allowed to enter into the turbine.

Both these problems are overcome by splitting the air stream from the compressor into the primary and secondary streams. The primary air stream passes through the combustor to yield a fuel/air mixture which is within the ignition limits whereas the secondary air stream bypasses the combustor and mixes with the combustion gases and cools them considerably before they enter into the turbine.

Example 7

A pan 3 inches long and 3 inches wide filled to brim with methanol sits on a table by a window. A gentle breeze of air blows across the window at a speed of 3 feet/second. The air temperature is 28 °C. Find the rate of depletion of methanol when vaporization occurs with and without combustion.

Solution

The situation is as shown in Figure 5.16. The pan may be considered as a flat plate. Assuming representative property values,

$$\mathrm{Re}_l \equiv \frac{u_\infty l}{v_g} = 3 \frac{\mathrm{ft}}{\mathrm{sec}} \cdot \frac{3}{12} \mathrm{ft} \cdot \frac{1}{0.18 \cdot 10^{-3}} \frac{\mathrm{sec}}{\mathrm{ft}^2} = 4{,}170$$

This value is far below the transitional Reynolds number in Table 3.4.

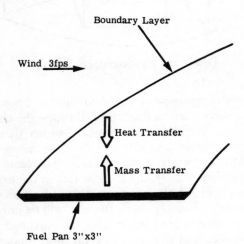

Figure 5.16 A pan of liquid fuel in a cross-wind (for example 7)

Hence the flow is laminar. Laminar forced convection Nusselt number is given by Table 3.4 as

$$\overline{\mathrm{Nu}}_l^0 = 0.664 \, \mathrm{Re}_l^{1/2} \, \mathrm{Pr}^{1/3}$$

For all practical purposes we may assume the Prandtl number to be unity, so that

$$\frac{\bar{h}^0 l}{K_g} \approx 0.664 \sqrt{4{,}170}$$

The average mass transfer coefficient, therefore, is

$$\frac{\bar{h}^0}{C_g} \approx \frac{\rho_g \alpha_g}{l} \cdot 0.664 \sqrt{4{,}170} \approx 1.1 \cdot 10^{-5} \, \mathrm{gm/cm^2/sec}.$$

Diffusion Flames in Liquid Fuel Combustion

For evaporation in the absence of combustion,

$$B \equiv \frac{Y_{F\infty} - Y_{FW}}{Y_{FW} - Y_{FR}}$$

$Y_{F\infty} = 0$, $Y_{FR} = 1$. Y_{FW} is approximately equal to the saturation mass fraction at 28 °C since $T_\infty \ll T_B$.

Hence $Y_{AW} \approx [1 + (760/120 - 1)]^{-1} \approx 0.16$. Therefore, $B \approx 0.1875$ and $\ln(B + 1) \approx 0.174$.

$$\dot{W}''_W = \frac{\bar{h}^0}{C_g} \ln(B + 1) = 1.1 \cdot 10^{-5} \cdot 0.174 = 1.91 \cdot 10^{-6} \text{ gm/cm}^2/\text{sec}.$$

The total rate of evaporation is obtained as $1.11 \cdot 10^{-4}$ gm/sec by simply multiplying $\dot{W}''_W$ by the pan area. For evaporation followed by combustion, with the properties given in Table 5.2,

$$B \equiv \frac{\Delta H f Y_{O\infty} + C_g(T_\infty - T_W)}{L} \approx 3.214$$

$$\ln(B + 1) \approx 1.44$$

Therefore,

$$\dot{W}''_W = 1.1 \cdot 10^{-5} \cdot 1.44 = 1.59 \cdot 10^{-5} \text{ gm/cm}^2/\text{sec}.$$
$$\dot{W}_W = 9.30 \cdot 10^{-4} \text{ gm/sec}.$$

Thus the mass loss rate is nearly ten-fold higher when combustion follows evaporation.

Example 8

Ethanol burns on a wick dipping into a reservoir at 20 °C. If $T_\infty = 20$ °C, find the surface temperature of the wick and the mass transfer driving force.

Solution

Referring to Eqs. 5.56 and 5.57 in the text,

$$B_{OT} \equiv \frac{\Delta H f Y_{O\infty} + C_g(T_\infty - T_W)}{L + C_l(T_W - T_R)}$$

$$B_{FO} \equiv \frac{f Y_{O\infty} + Y_{FW}}{Y_{FR} - Y_{FW}}$$

Assuming Y_{FR} as unity, using Table 5.2,

$$B_{OT} \approx \frac{790 - 0.3 T_W}{188.8 + 0.56 T_W}$$

$$B_{FO} \approx \frac{0.122 + Y_{FW}}{1 - Y_{FW}}$$

From the saturation vapor pressure data, $Y_{FW}(T_W)$ is found by the following tabulation.

T_W °C	P_{FW}	$\dfrac{P}{P_{FW}}$	$\left(\dfrac{P}{P_{FW}}-1\right)\dfrac{M_g}{M_F}$	Y_{FW}	$1-Y_{FW}$	B_{FO}
−31.3	1.0	760	462	0.002	0.998	0.1243
− 2.3	10	76	45.7	0.021	0.979	0.1463
19.0	40	19	11.0	0.084	0.916	0.2250
34.9	100	7.6	4.0	0.199	0.801	0.4000
63.5	400	1.9	0.55	0.646	0.354	2.1700
78.4	760	1.0	0.00	1.000	0.000	

Plotting $B_{OT}(T_W)$ and $B_{FO}(T_W)$, the intersection point gives B as 3.38 and T_W as 70.8 °C.

Example 9

Given $\rho_g \alpha_g = 2 \cdot 10^{-4}$ gm/cm²/sec, $\alpha_g = \nu_g = D = 0.2$ cm²/sec calculate the burning rate for the situation of Example 8 when the wick height is 2 cm.

Solution

We know B from Example 8. If we obtain $\overline{\mathrm{Nu}}_l^0$ from Table 3.4, $\dot{W}_W''$ is calculable. For free convective heat transfer, the Grashof number $gl^3 \beta \Delta T/\nu_g^2$ fixes the mode of the flow. When combustion occurs, it is reasonable to assume that $\beta \Delta T \approx 1$ since $\beta \approx 1/T$ °K for a perfect gas. Hence

$$\mathrm{Gr}_l \approx \frac{gl^3}{\nu_g^2} \approx 196{,}200$$

This is indicative of laminar flow for which,

$$\overline{\mathrm{Nu}}_l^0 = 0.59(\mathrm{Gr}_l \,\mathrm{Pr})^{1/4}$$

$$\frac{\bar{h}^0}{C_g} = \frac{\rho_g \alpha_g}{l} \cdot 0.59(196{,}200)^{1/4}$$

$$= 1.24 \cdot 10^{-3} \text{ gm/cm}^2/\text{sec}.$$

The mass transfer rate is obtained immediately as

$$\dot{W}_W'' = 1.24 \cdot 10^{-3} \ln(B+1)$$
$$= 1.83 \cdot 10^{-4} \text{ gm/cm}^2/\text{sec}.$$

If the surface temperature were assumed equal to the boiling point, B would have been 3.25 and $\dot{W}''_W$ would have been $1.8 \cdot 10^{-3}$ gm/cm²/sec. Thus the approximation of $T_W \approx T_B$ introduces an error of just about 2.0%.

5.12 Concluding Remarks

(a) *Transient Effects*

Usually the droplets are introduced into the combustion chamber at a temperature far below the evaporation temperature. The heat which is received by the droplet, then, heats up more liquid than is transferred. As the evaporation proceeds, the droplet interior temperature gradually increases and the heat lost by conduction into the droplet interior decreases. The heat needed to transfer a unit mass of the reservoir fluid is given by

$$Q = L + C_l(T_W - T_R) + \dot{q}/\dot{W}''_W S$$

where both T_W and T_R are variables with time; $\dot{q}$ is the rate of heat conduction into the interior (cal/sec) and S is the area of mass transfer.* Initially, $\dot{q}$ is quite large and consequently B is rather small. As evaporation continues $\dot{q}$ decreases and B increases. The variation of $\dot{q}$ with time may be examined by considering two extreme cases.

If the liquid conductivity is much smaller than the gas phase conductivity, heat does not penetrate too deep into the droplet interior. The surface temperature quickly rises to its equilibrium value while the interior temperature remains constant at its initial value equal to T_R. Steady state is attained very quickly. In steady state, the heat entering into the droplet is equal (but no more) to the heat needed to transfer $\dot{W}''_W$ gms of the reservoir fluid (i.e., heat entering $= \dot{W}''_W(L + C_l(T_W - T_R))$. When the liquid thermal conductivity is much greater than the gas phase conductivity, the temperature in the droplet may be assumed uniform and slowly rising with time. Consequently, the mass transfer number B gradually increases with time. After a considerable lapse of time, B attains its maximum value corresponding to $Q = L$. Assume that the total time of dissipation of the drop is composed of a preheat time (with negligible evaporation) and a vaporization time (with negligible heating of the drop). The preheat time may be crudely estimated by equating the rate of rise of sensible heat of the droplet to the rate of heat transfer.

$$\rho_l V C_l \frac{dT_l}{dt} = hS(T_g - T_l)$$

---

* In many real life situations, such as in pan burning of liquid fuels, such extraneous heat transfer processes as loss through the rear or gain through the rim of the pan, may become significant. These effects may be accounted for, if $\dot{q}$ is taken as the net heat loss in this equation.

Assuming ρ_l, V, C_l, h, S and T_g constant, integration from $T_l = T_0$ to $T_l = T_W$ gives the preheat time as

$$t_{\text{preheat}} \approx \frac{\rho_l C_l d_0}{6h} \ln\left(\frac{T_g - T_0}{T_g - T_W}\right)$$

With $\text{Nu} = 2$,

$$t_{\text{preheat}} \approx \left[\frac{\rho_l C_l}{12 K_g} \ln\left(\frac{T_g - T_0}{T_g - T_W}\right)\right] d_0^2$$

This time has to be added to the "steady state life-time" of the droplet to obtain the "unsteady state life-time." From Eqs. 5.39 and 5.73

$$t_{v\,\text{unsteady}} \approx \left[1 + \frac{2C_l}{3C_g} \cdot \ln\left(\frac{T_g - T_0}{T_g - T_W}\right) \cdot \ln(B + 1)\right] \frac{\rho_l d_0^2}{8 \rho_g \alpha_g \ln(B + 1)}$$

(b) *Composition of the Burning Liquid*

It is known by experiment that in the evaporation of multicomponent fuel sprays, the concentration of higher boiling components in the remaining drop increases as the spray evaporates; this is particularly true when the air temperature is higher. The consequence of such a "distillation" process is to increase B gradually. Frequently, when high boiling point liquids are burned, the fuel is found to be pyrolyzed partially to carbon whose burning is governed by the laws to be elaborated in the next chapter.

(c) *Suspension of Inert Solids in a Pure Fuel*

The carbonization mentioned above brings two important points into light. The experiments of Ranz and Marshall indicate that when inert solids are suspended in water, the droplet evaporation characteristics differ very little from those of pure water droplet.*

Secondly, under the conditions normally encountered in jet engine combustors, the droplet is known to receive negligible amount of heat by radiation. When solid particles are added to the liquid fuel the increased absorptivity might make radiation nonnegligible.

(d) *Chemical Considerations*

As described in Section 4.1 and Figure 4.2, a diffusion flame can be extinguished by sufficiently increasing the rate of supply of the fuel and oxygen. For

---

* The burning of fuel-soaked wicks is known to be essentially of the same nature as that of pure liquid surfaces. The reason for this is speculated to be same as Ranz and Marshall's finding on water drop evaporation.

example, if the velocity of the air stream is increased, the boundary layer thickness is reduced and consequently the fuel and oxygen supply rates increase due to steeper gradients of composition. In order to accommodate the greater reactant supply in the reaction, the flame zone tends to widen itself. Such an increase in the reaction volume has two competing effects. Firstly, the mean reactant concentration will become larger and as a result, the mean reaction rate tends to increase. Secondly, the mean reaction temperature is lowered. The reduction in the reaction rate due to the temperature decrease overwhelms the increase due to greater volume and reactant concentration. When this critical state is reached, any further decrease in the mean temperature (or increase in the mean reactant concentration) would extinguish the flame.

Keeping the preceding comments in mind, we can summarize this chapter as below. Spalding's theory of mass transfer makes a quantitative study of combustion of liquid droplets (and surfaces of other geometries) possible. The rate of burning, the location and temperature of the flame, and the flame structure may be readily deduced from Schwab–Zeldovich transformation. In particular, the flame approaches the fuel surface as Y_{FW} approaches zero (and therefore B approaches $fY_{O\infty}/Y_{FR}$). The maximum temperature of a diffusion flame is found to be equal to the adiabatic flame temperature of a stoichiometric fuel/oxidant mixture if dissociation is neglected.

References

Blinov, V. I. and Khudyakov, G. N., *Diffusive Burning of Liquids*, Pergamon translation from Russian (1960).

Eckert, E. R. G. and Hartnett, J. P., *Transactions of ASME*, (1957).

Gaydon, A. G. and Wolfhard, H. G., *Flames, Their Structure, Radiation and Temperature*, second ed., Chapman and Hall, London (1960).

Giffin, E. and Muraszew, A., *The Atomization of Liquid Fuels*, John Wiley & Sons, New York (1953).

Godsave, G. A. E., *Fourth Symposium on Combustion*, p. 818, Williams & Wilkins, Baltimore (1953).

Goldsmith, M., *Jet Propulsion*, **26**, p. 172 (1956).

Graves, C. C., *Proceedings of the Third Midwestern Conference on Fluid Mechanics*, p. 759, Minneapolis, Minn. (1953).

Hall, A. R. and Diederichsen, J., *Fourth Symposium on Combustion*, p. 837, Williams and Wilkins, Baltimore (1953).

Kanevsky, J., *Jet Propulsion*, **26**, p. 788 (1956).

Khitrin, L. N., *Physics of Combustion and Explosion*, N.S.F. Israel Program for Scientific Translations (1962).

N.A.C.A. 1300, *Basic Considerations in the Combustion of Hydrocarbon Fuels with Air*. Chapter 1. Atomization and Evaporation of Liquid Fuels by Graves, C. C. and Bahr, D. W., (1957).

Probert, R. P., *Philosophical Magazine*, **37**, p. 94 (1946).

Ranz, W. E. and Marshall, W. R., Jr., Evaporation from Drops, *Part I Chemical Engg. Prog.*, **48**, No. 3, pp. 141–126, (1952). Part II, same Journal, **48**, No. 4, pp. 173–180, (1952).

Rayleigh, On the Instability of Jets, *Proceedings of The London Math. Soc.*, Vol. X, pp. 4–13, (1878).

Rex, J. F., Fuhs, A. E., and Penner, S. S., *Jet Propulsion*, **26**, p. 179 (1956).

Roesch, W. C. and Rose, R. F., A Survey of the Literature on the Subject of Atomization. Progress Report No. 1–46. Jet Propulsion Lab., Caltech., Pasadena (1946).

Spalding, D. B., *Ph.D. Thesis*, Cambridge University, Cambridge (1951).

Spalding, D. B., *4th Symposium on Combustion*, Williams and Wilkins, Baltimore, p. 847 (1953).

Spalding, D. B., *Some Fundamentals of Combustion*, Butterworths, London (1955).

Spalding, D. B., *International Journal of Heat and Mass Transfer*, **2**, No. 4, June 1961.

Wise, H., Lorell, J., and Wood, B. J., *Fifth Symposium on Combustion*, p. 132, Reinhold, New York (1955).

CHAPTER 6

Combustion of Solids

Solid fuels are used in many important applications of combustion since the dawn of man. Logs of wood burning in a fireplace, in a camp-fire, "charcoal" briquettes glowing on a grill, coal burning (often in pulverized form) in locomotive and powerhouse furnaces, metallic compounds burning in aircraft and solid propellant rocket combustion chambers are but a few entries in an almost endless list of examples. The virtue of solids as fuels is enhanced by their relatively high density, flame temperature and heat of combustion. They are easy to handle and store, inexpensive and readily available. In the past, solids are used as fuels mainly in stationary power plants; and it is only very recently, with the advent of gas turbine and rocket, they are considered for aeronautical and space propulsion power plants. The primary problems associated with the use of solid fuels in engines are hinged around the design of fuel delivery systems, residue removal and high temperature materials.

And then there are those examples, too numerous to count, of unwanted fires involving a wide variety of solid fuels. In this chapter, we examine the problem of solid fuel combustion under conditions encountered in combustion chambers and natural fires. Such an examination conveniently yields results which can be expressed in terms of a special type of mass transfer driving force, B.

When subjected to heat, some solids sublime (examples: naphthalene, camphor, etc.) and some melt into a liquid which vaporizes. The flames on such solids stay a short distance, in the gas phase, away from the vaporizing surface. Once established, the enveloping flame provides heat to the evaporating surface by gas phase conduction to result in more mass transfer to the flame zone. Combustion of these solids may be treated by the techniques developed in Chapter 5 for liquid fuel diffusion flames.

Some solids are distilled when subjected to heat. (Examples: coal, wood, paper, cotton, etc.). These solids (sometimes also called charring or pyrolyzing solids) disintegrate irreversibly* in response to heat; products of such disintegration are combustible volatile gases and a carbonaceous solid residue. The volatile gases emitted volumetrically in the interior of the solid seep out of the solid and burn with air in much the same way as liquid fuel vapors. The study of combustion of pyrolyzing solid consequently involves two steps: firstly, to find the rate of evolution of volatiles in the interior of the solid as a function of the surface heat flux condition and secondly, to find the flame stand-off distance and surface heat flux in the gas phase as a function of quantity and quality of the volatiles transpiring from the surface. The latter aspect utilizes the techniques developed in Chapter 5 for liquid fuel diffusion flames with the reservation that the liquid vaporization is a surface process whereas the solid distillation is an internal volumetric process which warrants a detailed understanding of the internal heat and mass transfer phenomena, namely the first aspect. Once all the volatiles of the solid are released, the carbonacious solid residue is attacked by oxygen to yield glowing combustion. Concerning metals as fuels, the boiling points of the metal and its oxide determine the mode of combustion. When the boiling point of the metal oxide is greater than that of the metal itself, then a vapor phase gaseous diffusion flame arises in much the same way as discussed in Chapter 5 on liquid fuels. Nearly all metals pertain to this category except for boron, silicon, titanium and zirconium. These four exceptional metals have boiling points greater than those of their oxides. They burn, as a result, in a manner somewhat similar to carbon.

Those solids whose boiling point approaches the flame temperature (examples: the carbonaceous residue of a pyrolyzing solid, particles of carbon or exceptional metals mentioned above, etc.) constitute an important class of solid fuels. The flame (or the combustion zone, to be precise) approaches the fuel surface as the boiling point, flame temperature and heat of combustion are gradually increased. In some limiting situations the combustion zone in fact coincides with surface. From liquid fuel diffusion flame theory, the flame location on a burning droplet of radius $\mathscr{R}$ is given by Eq. 5.61 as

$$\frac{r_C}{\mathscr{R}} = \frac{\ln(B + 1)}{\ln(1 + f Y_{O\infty}/Y_{FR})} \tag{5.61}$$

As the boiling point of the fuel increases, the transfer number gradually decreases and the flame gradually approaches the fuel surface. Equation 5.61 shows that in the limiting case of flame coinciding with the surface (i.e., $r_C/\mathscr{R} \to 1$), the mass transfer number B is equal to $f Y_{O\infty}/Y_{FR}$; that is, for a

---

* Such disintegration or distillation is named as *pyrolysis*. The word "pyro-lysis" means "death due to heat," or more idiomatically, "thermal decomposition."

Combustion of Solids

pure solid fuel, $B = fY_{0\infty}$. That this inference is indeed true in the combustion of most of the simple solids is shown in this chapter from a consideration of the governing conservation equations.

6.1 Pyrolyzing Solids

(a) Nusselt's Shrinking Drop Theory

Consider, as shown in Figure 6.1, a sphere* (of diameter $d \equiv 2\mathscr{R}$) of a pyrolyzing solid fuel such as wood or coal. Let T_0 be its initial uniform temperature. At time equal to zero, let the surface temperature of the solid be abruptly raised to T_∞ and held there. Heat then flows into the interior of the solid by conduction to raise the temperature of any station continuously with time. The internal temperature history is governed by the following transient conduction equation and conditions.

$$\frac{\partial}{\partial r}\left(K_S r^2 \frac{\partial T}{\partial r}\right) = \rho_S C_S r^2 \frac{\partial T}{\partial t} \qquad (6.1)\dagger$$

$$\left.\begin{array}{lllll} t < 0 & T = T_0 & \text{in} & 0 < r < \mathscr{R} \\ t \geq 0 & T = T_\infty & \text{at} & r = \mathscr{R} \\ & \dfrac{\partial T}{\partial r} = 0 & \text{at} & r = 0 \end{array}\right\} \qquad (6.2)$$

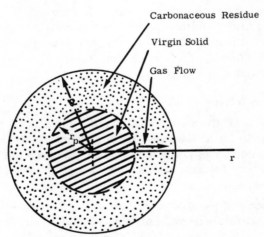

Figure 6.1 Nusselt's shrinking drop model

---

* Sphere is chosen here for explicit illustration. A slab or a cylinder may be handled with equal ease.

† The subscript S stands for the solid and C, for the carbonaceous residue.

Solution $T = T(r, t)$ of Eqs. 6.1 and 6.2 is

$$\frac{T(r, t) - T_0}{T_\infty - T_0} = \frac{2\mathscr{R}}{\pi r} \sum_{n=1}^{\infty} \frac{(-1)^n}{n} \sin \frac{n\pi r}{\mathscr{R}} \exp[-n^2\pi^2\alpha_S t/\mathscr{R}^2] \qquad (6.3)$$

where α_S is the thermal diffusivity of the solid. Equation 6.3 is shown plotted in Figure 6.2. This figure is very convenient to read out the time taken $(\alpha_S t/\mathscr{R}^2)$ by any station $(r/\mathscr{R})$ to attain any temperature $(T - T_0)/(T_\infty - T_0)$. Suppose now that at any station the solid releases its volatiles abruptly when it reaches a temperature equal to a characteristic pyrolysis temperature T_P. The pyrolysis front (i.e., the radius at which $T = T_P$) continuously progresses towards the origin of the sphere as time increases. The rate of volatile production (equal to the rate of weight loss of the sphere) is then given by the speed of inward propagation of the pyrolysis isotherm $T = T_P$.

$$\dot{W}(t) = (\rho_S - \rho_C)4\pi \left(r^2 \frac{\partial r}{\partial t}\right)_{T = T_P} \qquad (6.4)$$

r^2 and $(\partial r/\partial t)_{T_P}$ are obtained from Eq. 6.3 by setting $T = T_P$, a constant. (It is sometimes convenient to graphically differentiate the $r(t)$ relation corresponding to $T = T_P$ from Figure 6.2). Such a calculation first made by Nusselt yields the following relation between the time-averaged pyrolysis rate and sphere diameter.

$$\overline{W} = \mathscr{K}_P W_0 d_0^{-2} \qquad (6.5)*$$

where W_0 is the initial weight of the sphere ($=\rho_S \pi d_0^3/6$) and $d_0 \equiv 2\mathscr{R}$. The constant $\mathscr{K}_P$, whose magnitude for cellulosic fuels lies between 0.005 and 0.01, has units of cm^2/sec and is conceptually similar to the liquid fuel evaporation constant λ. It is a function of the thermal properties of the solid, the pyrolysis temperature T_P, the surface temperature T_∞ (or the heating rate) and the solid initial temperature T_0.

(b) *Comments on Nusselt's Theory*

Obviously, Nusselt's theory is based on the premise of several serious approximations so far as the pyrolysis of charring solids is concerned. In reality various complex physical and chemical processes influence Nusselt's "pure" transient conduction analysis quite strongly. Some of these processes are discussed below.

(i) The process of pyrolysis is not abrupt as Nusselt's theory assumes. Instead, the rate of decrease of the local solid density depends upon the

---

* It is customary to define pyrolysis time t_P as the ratio of the initial weight to the time-averaged pyrolysis rate. Equation 6.5 written for t_P is

$$t_P = d_0^2/\mathscr{K}_P$$

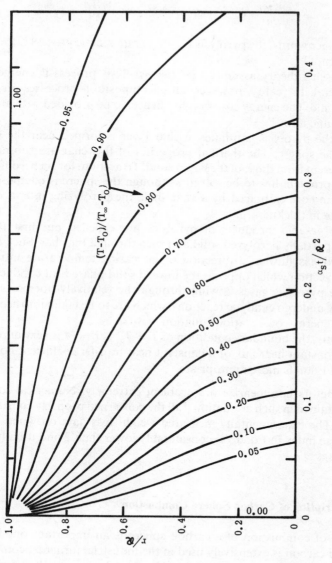

Figure 6.2 Temperature in the sphere as a function of time and location

local temperature and extent of pyrolysis. Such a dependency is usually expressed in some form as following.

$$-\frac{\partial \rho}{\partial t} = k_n(\rho - \rho_C)^n e^{-E/RT} \quad \text{or} \quad k_n(\rho - \rho_C)^n T^m$$

In other words, the pyrolysis is a finite rate kinetic process rather than "abrupt."

(ii) Nusselt's theory assumes that the pyrolysis process is energetically neutral. In reality, however, all decomposition processes are endothermic. The energy equation has hence, to be provided with an endothermic term.

(iii) As the pyrolysis continues, a char layer is formed near the surface of the sphere. The thermal properties of this char are substantially different from those of the virgin solid. To account for such a difference, the problem has to be solved as though the sphere is composed of a virgin core embraced by a char shell, the pyrolyzing interface being finite in thickness.

(iv) The gaseous products of pyrolysis are expelled out first through the partially pyrolyzed solid and then through the char. Such an outward gas flow in a direction of increasing temperature introduces convection which opposes the inward solid phase heat conduction.

(v) The pyrolysis gases flowing through the relatively hot porous char may undergo catalytic reactions, which are to be taken into account in the energy conservation equation.

(vi) Lastly, the boundary condition of $T = T_\infty$ at $r = \mathscr{R}$ is very unrealistic. A constant heat flux or a constant heat transfer coefficient boundary condition is more appropriate.

These factors nonlinearize the problem extremely. Numerical solutions may be obtained in such a situation if all the concerned property constants are specified. The experimentally measured weight loss data on wood suggest that the size index in Eq. 6.5 is considerably lower than 2 and usually about 1.5.

6.2 Description of Carbon Sphere Combustion

The topic of combustion of a carbon sphere is an important one because pulverized carbon is extensively used in the industrial furnaces; combustion of some metals obeys laws similar to carbon combustion; and all pyrolyzing solids, once completely charred, are attacked by oxygen in much the same manner as carbon. In particular, we attempt here to deduce an expression for the burning rate of an individual carbon particle. If such an expression is available, it may be combined with the particle size distribution law to obtain the rate of burning of a cloud of pulverized fuel. Prior to undertaking the task of such a derivation, let us briefly summarize the following basic understanding of the combustion processes of a liquid fuel, a pyrolyzing solid and a simple solid such as carbon.

The rate of burning of any fuel is strongly influenced by the heat flux incident upon the fuel surface. In the case of a simple solid, reaction takes place at its surface whereas in the case of a liquid or a pyrolyzing solid, it takes place in the gas phase. In the case of liquid fuels vaporization takes place at the surface whereas in pyrolyzing solids, the gas production is a subsurface volumetric chemical process. Consequently, the problem of combustion of pyrolyzing solids requires a detailed understanding of both the internal (solid phase) and the external (gas phase) heat and species conservation. Once all the volitiles are expelled the carbonaceous residue of a pyrolyzing solid behaves in the same manner as a simple solid. The combustion of a carbon particle is accompanied by high surface temperatures at which it becomes incandescent. (A homely example of such an incandescence is offered by the glowing charcoals on a grill.) Oxygen diffuses from the free stream to the surface where it directly reacts with the solid to release a great quantity of heat. Much of the heat is lost to the surroundings by radiation. In fact, experiments show that if the carbon particle is larger than $\frac{1}{4}$ mm, radiative heat loss lowers the particle temperature to within 1,100 °K. In order to reduce the radiative loss, the fuel is usually finely pulverized. It is reasonable to expect that whereas the burning rate of liquid fuels and pyrolyzing solid fuels strongly depends upon the rate of heat transfer to the fuel surface, that of simple solids depends strongly upon the rate at which the oxygen diffuses to the fuel surface.

Any heterogeneous reaction involves the following five steps in series.

(i) Oxygen has to diffuse to the fuel surface,
(ii) Diffused oxygen has to be absorbed by the surface,
(iii) Absorbed oxygen has to react with the solid to form absorbed products,
(iv) Absorbed products have to be desorbed from the surface, and
(v) Desorbed products have to diffuse away from the surface.

Since these steps occur in series, the slowest of them determines the burning rate. In the case of carbon combustion steps (ii) and (iv) are known to be extremely fast.

When the particle temperature is low, the particle is small and the flow around it is feeble, step (iii) is known to be much slower than step (i) or step (v). The burning rate then is determined by the chemical kinetics and therefore the process is kinetically controlled. In kinetically controlled regime, the burning rate depends upon temperature exponentially. Since the process of diffusion is a function of particle size and flow, and since it is irrelevant in the kinetic regime, the burning rate is independent of the particle size and flow around it. Furthermore, the concentration of oxygen at the reacting surface is not too different from the free stream concentration. Figure 6.3 indicates this description of the kinetically controlled regime. Figure 6.4 shows the burning rate as a function of the particle temperature.

On the other hand, when the particle and flow velocity are large and temperature is high, step (iii) is known to be much faster than steps (i) and (v). The burning rate is then controlled by the diffusion rate of oxygen to the particle. In the diffusionally controlled regime, the burning rate depends weakly ($\propto T^{0.5-1.0}$) on temperature and strongly on the particle size. The oxygen concentration at the reacting surface is negligibly small. This description is also illustrated qualitatively in Figures 6.3 and 6.4. Carbon and metal combustions are predominantly diffusion controlled. At the low temperatures required to bring about the kinetic control, the particles lose so much heat so rapidly that extinction occurs.

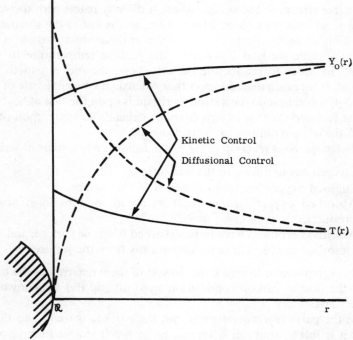

Figure 6.3 Kinetically controlled and diffusionally controlled combustion of a solid particle

Combustion of Solids

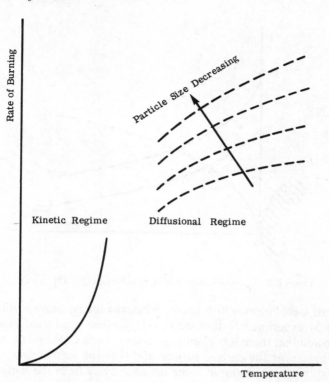

Figure 6.4 Rate of burning versus temperature for the kinetically controlled and diffusionally controlled regimes

Diffusional combustion of a carbon sphere depends upon the extra variable of stoichiometry. Three possibilities arise:

Firstly, at moderately high temperatures, oxygen diffuses to the carbon surface and reaction occurs to form CO_2.

$$C + O_2 \longrightarrow CO_2$$

The fuel/oxygen ratio, f, for this mechanism is 12/32. The product CO_2 diffuses away from the surface. Figure 6.5 shows the expected profiles of Y_O, Y_{CO_2} and T for this scheme.

Secondly, when there is no free oxygen present, the reaction between O_2 and carbon at the surface might yield a richer oxide, viz: CO, instead of the leaner oxide.

$$2C + O_2 \longrightarrow 2CO$$

The stoichiometric fuel-oxygen ratio for this mechanism is naturally higher, $f = 12/16$.

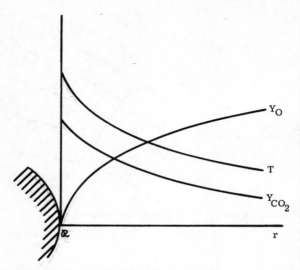

Figure 6.5 Combustion of carbon according to $C + O_2 \rightarrow CO_2$

The third mechanism, which occurs when the temperature is high, is one provoked by experimental observations. Coffin has conducted experiments which showed that there is negligible amount of both CO_2 and O_2 present in the gas phase near the carbon surface and that the temperature and CO_2 profiles exhibit a maximum at some distance away from the surface in the gas phase. Experiments of Hottel and Davis show a definite gas phase re-

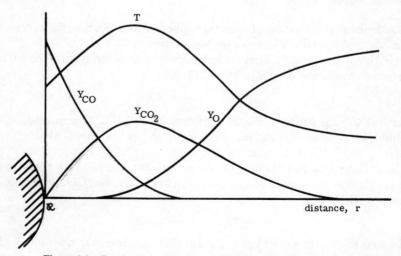

Figure 6.6 Combustion of carbon according to Coffin's hypothesis

Combustion of Solids

action in which carbon monoxide is thought to be burnt by the counter-diffusing oxygen to yield a thin, weakly visible, bluish, envelope flame separated from the carbon surface by a distinct dark zone. A realistic mechanism of carbon burning will therefore be as following. Carbon monoxide diffuses from the carbon surface and burns in gas phase as it meets the oxygen. Part of the carbon dioxide produced in this reaction diffuses back to the carbon surface where it is reduced to CO by the carbon. The distribution patterns of T, Y_O, Y_{CO} and Y_{CO_2} are then similar to those of liquid fuel combustion as shown in Figure 6.6.

6.3 Diffusional Theory of Carbon Combustion

Consider a sphere of carbon, of diameter $2\mathscr{R}$, situated in an oxidizing atmosphere. Let steady state burning take place at the surface of the sphere. In diffusional regime of burning, the rate of weight loss of the sphere is determined by the amount of oxygen diffusing to the surface. Since no chemical reaction occurs in the gas phase, the oxygen conservation equation indicates a balance between diffusion and convection.

$$\frac{d}{dr} r^2 \rho_g D_O \frac{dY_O}{dr} - [\dot{W}'' r^2] \frac{dY_O}{dr} = 0 \tag{6.6}$$

Likewise, the energy equation is

$$\frac{d}{dr} r^2 K_g \frac{dT}{dr} - [\dot{W}'' r^2] C_g \frac{dT}{dr} = 0 \tag{6.7}$$

Continuity of mass is given by $\dot{W}'' 4\pi r^2 = $ constant; i.e.,

$$\dot{W}'' r^2 = \text{constant} \tag{6.8}$$

In order to solve the mass transfer problem, it suffices to obtain the solution of Eq. 6.6 under appropriate boundary conditions. The free stream condition is $Y_O = Y_{O\infty}$. At the surface, the oxygen mass flux is

$$\dot{W}''_{OW} = \rho_g D_O \frac{dY_O}{dr}\bigg|_W - \dot{W}''_W Y_{OW} \quad \text{gm/cm}^2/\text{sec} \tag{6.9}$$

where $\dot{W}''_W$ is the rate of loss of carbon from the particle. If each gram of oxygen consumes f grams of carbon, stoichiometry yields

$$\dot{W}''_{OW} = \dot{W}''_W / f \tag{6.10}$$

The fuel/oxygen ratio depends upon whether the combustion product is CO or CO_2. Equating Eq. 6.9 and 6.10,

$$\left[\rho_g D_O \frac{dY_O}{dr} - \dot{W}'' Y_O \right]_W = \frac{\dot{W}''_W}{f} \quad (6.11)$$

The first integration of Eq. 6.6 with Eq. 6.11 gives,

$$r^2 \rho_g D_O \frac{dY_O}{dr} - [\dot{W}''_W \mathcal{R}^2] Y_O = \dot{W}''_W \mathcal{R}^2 / f$$

Rearranging,

$$r^2 \rho_g D_O \frac{dY_O}{dr} - [\dot{W}''_W \mathcal{R}^2](Y_O + 1/f) = 0$$

Integrating once again,

$$\ln(Y_O + 1/f) = -\frac{[\dot{W}''_W \mathcal{R}^2]}{\rho_g D_O r} + \text{constant}$$

The integration constant is equal to $\ln(Y_{O\infty} + 1/f)$ due to the condition that $Y_O \to Y_{O\infty}$ as $r \to \infty$. Hence,

$$\ln\left(\frac{Y_O + 1/f}{Y_{O\infty} + 1/f}\right) = -\frac{[\dot{W}''_W \mathcal{R}^2]}{\rho_g D_O} \cdot \frac{1}{r} \quad (6.12)$$

Equation 6.12 shows that Y_O increases with increasing r exponentially. Specifically, at $r = \mathcal{R}$, $Y_O = Y_{OW}$. At moderate temperatures when the particle is rather small, Y_{OW} is determined by the excess of diffusion relative to the chemical reaction rate. In diffusional flame regime, however, Y_{OW} will be nearly zero since the reaction rate will be sufficiently rapid to consume all the oxygen reaching the surface. Thus, Eq. 6.12 gives

$$\frac{\dot{W}''_W \mathcal{R}}{\rho_g D_O} = \ln\left(\frac{Y_{O\infty} + 1/f}{1/f}\right) \equiv \ln(B + 1) \quad (6.13)$$

Eliminating $\dot{W}''_W \mathcal{R}/\rho_g D_O$ between Eqs. 6.12 and 6.13, we obtain the following alternative equation for the oxygen mass fraction profile.

$$Y_O = \frac{(B + 1)^{(1 - \mathcal{R}/r)} - 1}{f} \quad (6.12a)$$

B, the mass transfer number, in Eq. 6.13 is defined as

$$B \equiv f Y_{O\infty} \quad (6.14)$$

which is the same as that implied by Eq. 5.61 when the flame coincides with the surface of a pure fuel. When the gas is pure oxygen, $Y_{O\infty} = 1$ so that

Combustion of Solids

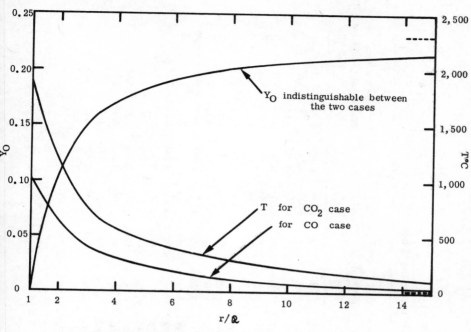

Figure 6.7 Combustion of carbon to CO and CO_2 in air

$B = f$. When the gas is normal air, $Y_{O\infty} = 0.232$ so that $B = 0.232 f$. Figure 6.7 shows Y_O profile around a carbon sphere burning in air. Table 6.1 shows the B-values of various solids along with other pertinent properties.

The time required to completely consume a particle of initial diameter d_0 is deduced from Eq. 6.13 as

$$t_b = \frac{\rho_F d_0^2}{8\rho_g D_O \ln(B+1)} \tag{6.15}$$

Table 6.1 also gives the burning times of 1 micron particles calculated from this equation. Comparison with liquid droplet burning times (i.e., Eq. 5.73 and Table 5.5) indicates that solid particle burning is slower than liquid drop burning by a factor of about thirteen.* The burning time in both the cases is directly proportional to the square of the initial particle diameter. Figure 6.8 shows t_b as a function of the particle size d_0 for carbon and kerosene. The kerosene line may be considered as applicable to all hydrocarbons, in general, since the pertinent physical constants do not vary drastically. Coal

---

* A physical explanation for this can be put in the following words. In liquid fuel combustion, the fuel vapors travel a certain distance from the fuel surface to meet oxygen whose journey is therefore much shorter than in carbon combustion.

TABLE 6.1
Combustion Properties of Carbon and Metals

Fuel	Oxide	ρ_{fuel} gm/cm³	M_{fuel}	M_{oxide}	B.P.$_{fuel}$ °C	B.P.$_{oxide}$ °C	f	B_{oxygen}	B_{air}	$t_b^* \times 10^{-5}$ sec calculated O_2	calculated air	Measured air
Aluminum	Al_2O_3	2.70	27.0	101.9	2,467	2,980	1.120	1.120	0.260	0.90	2.93	—
Beryllium	BeO	1.84	9.0	25.0	2,970	3,900	0.564	0.564	0.131	1.04	3.79	—
Boron	B_2O_3	2.34	10.8	69.6	2,550	1,860	0.451	0.451	0.105	1.57	6.15	2.03
Calcium	CaO	1.55	40.1	56.1	1,487	2,850	2.500	2.500	0.580	0.31	0.85	—
Carbon	CO	1.50ᵃ	12.0	28.0	4,827	−191	0.750	0.750	0.174	0.67ᵇ	2.28ᶜ	1.98
Carbon	CO_2	1.50ᵃ	12.0	44.0	4,827	−79	0.375	0.375	0.087	1.16ᵈ	4.16ᵉ	—
Hafnium	HfO_2	13.30	178.5	210.6	5,400	—	5.590	5.590	1.300	1.76	3.98	—
Lanthanum	La_2O_3	6.16	139.0	325.8	3,469	4,200	5.790	5.790	1.341	0.81	1.82	—
Lithium	Li_2O_3	0.53	6.9	29.9	1,317	1,200	0.865	0.865	0.200	0.21	0.73	—
Magnesium	MgO	1.74	24.3	40.3	1,107	3,600	1.520	1.520	0.353	0.47	1.45	0.225
Plutonium	PuO_2	19.84	242.0	274.0	3,235	—	7.560	7.560	1.755	2.31	4.90	—
Potassium	K_2O	0.86	39.1	94.2	774	—	4.890	4.890	1.134	0.12	0.28	—
Potassium	K_2O_2	0.86	39.1	110.2	774	—	2.440	2.440	0.567	0.17	0.48	—
Silicon	SiO_2	2.35	28.1	60.1	2,355	2,230	0.879	0.879	0.204	0.93	3.23	—
Sodium	Na_2O_2	0.97	23.0	78.0	892	657	1.437	1.437	0.334	0.27	0.85	—
Thorium	ThO_2	11.30	232.0	264.1	4,000	4,400	7.260	7.260	1.685	1.34	2.87	—
Titanium	TiO_2	4.50	48.0	79.9	3,260	2,750	1.497	1.497	0.347	1.23	3.75	—
Uranium	UO_2	19.05	238.0	270.0	3,818	2,500	7.430	7.430	1.726	2.23	4.74	—
Uranium	U_3O_8	19.05	238.0	842.0	3,818	—	5.580	5.580	1.294	2.52	5.73	—
Zirconium	ZrO_2	6.44	91.2	123.2	3,578	5,000	2.850	2.850	0.662	1.20	3.19	—

* For 1 micron ($\equiv 10^{-4}$ cm) particles. $\rho_g D$ is assumed as 5×10^{-4} gm/cm/sec.
ᵃ 1.30–1.70, ᵇ 0.58–0.76, ᶜ 1.97–2.58, ᵈ 1.01–1.32, ᵉ 3.78–4.95.

particle burning time lies between the carbon and liquid lines in Figure 6.8. If the volatile content of the coal considered is low, the coal line approaches the carbon line; if it is high, it approaches the liquid line. It is known that the effect of thermal property variation on burning rate is to increase it two to three fold. Tu's experiments indicate that the influence of movement of the

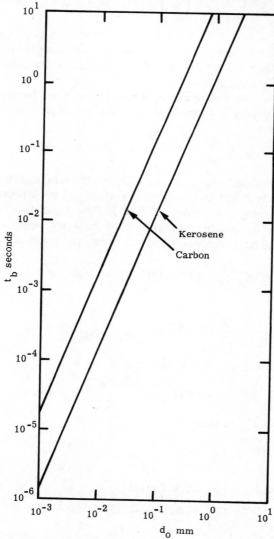

Figure 6.8 Burnout times for carbon and kerosene particles of various diameters (adapted with permission, from: D. B. Spalding, *Fuel XXX*, p. 129, (1951))

oxidizing medium on the burning rate of a carbon sphere is the same as in the case of liquid drop burning; that is, the apparent mass transfer coefficient $(\rho_g D_O/\mathscr{R})$ varies with Reynolds and Schmidt numbers according to the formula

$$\left(\frac{\rho_g D_O}{\mathscr{R}}\right)_{\text{apparent}} \equiv \left(\frac{h^o}{C_g}\right) = \frac{\rho_g D_O}{d}(2 + 0.6\, \text{Re}_d^{1/2}\, \text{Sc}^{1/3})$$

where d is the diameter of the sphere. A velocity of 100 m/sec around a particle of 0.01 mm diameter would increase the heat transfer coefficient from 2 to 10. But relative velocities of this magnitude are almost impossible in real-life combustion systems.

Calculation of the temperature profile around the burning particle is possible if the energy conservation equation is solved. Integrating Eq. 6.7 once,

$$r^2 K_g \frac{dT}{dr} - [\dot{W}''_W \mathscr{R}^2] C_g T = \text{constant}$$

Neglecting radiation,* the left hand side of this equation is equal to the negative of the heat lost by the glowing surface to the surroundings. Under steady state burning conditions, the heat lost by the surface is equal to the heat generated by combustion, the latter quantity being $4\pi \mathscr{R}^2 \dot{W}''_W \Delta H$, where ΔH is the heat of combustion in calories per gram of the fuel.

$$r^2 K_g \frac{dT}{dr} - [\dot{W}''_W \mathscr{R}^2] C_g T = -[\dot{W}''_W \mathscr{R}^2] \Delta H$$

Rearranging,

$$r^2 \rho_g \alpha_g \frac{d(C_g T - \Delta H)}{dr} - [\dot{W}''_W \mathscr{R}^2](C_g T - \Delta H) = 0$$

Integration gives

$$\ln(C_g T - \Delta H) = -\frac{[\dot{W}''_W \mathscr{R}^2]}{\rho_g \alpha_g r} + \text{constant}$$

As $r \to \infty$, $T \to T_\infty$, so that

$$\ln\left(\frac{C_g T - \Delta H}{C_g T_\infty - \Delta H}\right) = -\frac{[\dot{W}''_W \mathscr{R}^2]}{\rho_g \alpha_g r} \tag{6.16}$$

Comparison of Eq. 6.16 with 6.12 gives

$$\left(\frac{C_g T - \Delta H}{C_g T_\infty - \Delta H}\right) = \left(\frac{Y_O + 1/f}{Y_{O\infty} + 1/f}\right)^{D_O/\alpha_g} \tag{6.17}$$

---

* Neglect of radiation obviously is a serious assumption.

Combustion of Solids

This equation may be used to compute the temperature profile from the known mass fraction profile for any given Lewis number. If $D_O = \alpha_g$, Eqs. 6.17, 6.12(a) and 6.14 give the explicit $T(r)$ profile as below.

$$T = (T_\infty - \Delta H/C_g)(B + 1)^{-\mathcal{R}/r} + \Delta H/C_g \qquad (6.18)$$

For any given value of positive B, the temperature falls from T_W at $r = \mathcal{R}$ to T_∞ as $r \to \infty$. Figure 6.7 also shows $T(r/\mathcal{R})$ for carbon burning in air to CO and to CO_2. In particular, the wall temperature is obtained by setting $\mathcal{R}/r$ to unity and using the definition of B.

$$T_W = \frac{C_g T_\infty + f\,\Delta H Y_{O\infty}}{C_g(f Y_{O\infty} + 1)} \qquad (6.19)$$

This equation gives the wall temperature as a function of the combustion parameters. When carbon burns to carbon monoxide in normal air, ($f = 12/16$, $T_\infty \approx 20\,°C$, $Y_{O\infty} \approx 0.232$, $\Delta H \approx 2{,}000$ cal/gm) T_W obtained from Eq. 6.19 will be about $1{,}030\,°C$. If the product were CO_2, ($f = 12/32$, $\Delta H \approx 7{,}300$ cal/gm), $T_W \approx 1{,}950\,°C$. In reality, however, radiative heat loss forces the particle surface temperature to be much lower than that predicted from Eq. 6.19.

It is instructional to consider adiabatic mixing of one gram of the oxygen-bearing fluid at a temperature of T_∞ with $f Y_{O\infty}$ grams of the fuel to allow combustion in which $f Y_{O\infty} \Delta H$ cal. of heat is produced. The resultant total enthalpy $C_g T_\infty + f Y_{O\infty} \Delta H$ cal. is utilized to raise the temperature of $1 + f Y_{O\infty}$ grams of the mixture to T_W. Energy balance of such a mixing, hence, is expressed as

$$C_g(1 + f Y_{O\infty})T_W = C_g T_\infty + f Y_{O\infty} \Delta H$$

which implies the same result as Eq. 6.19. In the process of calculations with Eq. 6.19, one discovers that neglect of $C_g T_\infty$ causes an error of only about 0.5%. Evidently, the surface temperature is rather high when the burning occurs in pure oxygen. When such extreme temperatures are involved, excessive radiant losses, and flash vaporization, usually accompanied by sloughing and sputtering may be expected. If the ratio of mass sputtered (or sloughed off) to the mass vaporized is to be denoted by ϕ, Eq. 6.13 may be modified to the following form.

$$\frac{\dot{W}_T'' \mathcal{R}}{\rho_g D_O} = \ln(f Y_{O\infty}[1 + \phi] + 1) \qquad (6.20)$$

$\dot{W}_T''$ is the sum of rate of vaporization $\dot{W}_W''$ and rate of sloughing $\dot{W}_W'' \phi$.

6.4 Combustion of Carbon With CO Burning in Gas Phase

As mentioned earlier in Section 6.2 (Figure 6.7), experiments of Coffin and Hottel on carbon particle burning indicate that carbon leaves the particle surface in the form of CO which reacts with oxygen in the gas phase to result in a thin envelope flame. The carbon dioxide produced in the envelope flame partly diffuses to the hot particle surface where it is reduced to carbon monoxide by carbon.

Let f_W grams of the carbon react with 1 gram of CO_2 at the surface to form $(1 + f_W)$ grams of CO. As CO diffuses away from the surface, let f_g grams of CO react with 1 gram of oxygen to form $(1 + f_g)$ grams of CO_2. Coffin's experiments show that there is negligible amount of oxygen and carbon dioxide present at the fuel surface. And, naturally, the free stream is pure from CO_2 as well as CO. The species of oxygen and carbon dioxide are conserved in the gas phase according to the following equations.

$$\frac{d}{dr} r^2 \rho_g D_O \frac{dY_O}{dr} - [\dot{W}_W'' \mathcal{R}^2] \frac{dY_O}{dr} + \dot{W}_O''' = 0 \tag{6.21}$$

$$\frac{d}{dr} r^2 \rho_g D_d \frac{dY_d}{dr} - [\dot{W}_W'' \mathcal{R}^2] \frac{dY_d}{dr} + \dot{W}_d''' = 0 \tag{6.22}$$

The subscript d is for the dioxide. The source-sink terms in Eqs. 6.21 and 6.22 are related to one another as $\dot{W}_O''' = -\dot{W}_d'''/(1 + f_g)$. Multiplying Eq. 6.22 by $1/(1 + f_g)$ and adding to Eq. 6.21 with the assumption of equal diffusivities,

$$\frac{d}{dr} r^2 \rho_g D \frac{d\hat{Y}}{dr} - [\dot{W}_W'' \mathcal{R}^2] \frac{d\hat{Y}}{dr} = 0 \tag{6.23}$$

The new conserved variable $\hat{Y}$ is defined by

$$\hat{Y} \equiv Y_O + Y_d/(1 + f_g) \tag{6.24}$$

Integrating Eq. 6.23 once,

$$r^2 \rho_g D \frac{d\hat{Y}}{dr} - [\dot{W}_W'' \mathcal{R}^2] \hat{Y} = \text{constant} \tag{6.25}$$

The constant is evaluated by applying the equation at $r = \mathcal{R}$, the fuel surface. Substituting Eq. 6.24 in Eq. 6.25 and realizing that $Y_{OW} = 0$, the constant is $[\mathcal{R}^2 \rho_g D \, dY_d/dr - (\dot{W}_W'' \mathcal{R}^2) Y_d]_W/(1 + f_g)$. The denominator is $(-1/4\pi)$ times the flux of CO_2 to the surface. If each gram of CO_2 reaching the surface consumes f_W grams of carbon,

$$\dot{W}_W''/f_W = \dot{W}_d''$$

Hence the constant is simply $[\dot{W}_W'' \mathcal{R}^2]/[f_W(1 + f_g)]$. Equation 6.25 consequently simplifies to the following.

$$r^2 \rho_g D \frac{d\hat{Y}}{dr} - [\dot{W}_W'' \mathcal{R}^2] \hat{Y} = \frac{[\dot{W}_W'' \mathcal{R}^2]}{f_W(1 + f_g)}$$

Combustion of Solids

Integration gives

$$\ln\left(\hat{Y} + \frac{1}{f_W(1+f_g)}\right) = -\frac{[\dot{W}''_W \mathscr{R}^2]}{\rho_g Dr} + \text{constant}$$

The constant is evaluated by applying the condition of $Y_O = Y_{O\infty}$ and $Y_d = 0$ at $r \to \infty$. It is, $\ln(Y_{O\infty} + 1/[f_W(1+f_g)])$. Hence,

$$\ln\left(\frac{\hat{Y} + \dfrac{1}{f_W(1+f_g)}}{Y_{O\infty} + \dfrac{1}{f_W(1+f_G)}}\right) = -\frac{[\dot{W}''_W \mathscr{R}^2]}{\rho_g Dr} \qquad (6.26)$$

At $r = \mathscr{R}$, $Y_O = 0$ and $Y_d = 0$ so that $\hat{Y} = 0$. Equation 6.26 then gives the rate of burning of carbon as

$$\frac{\dot{W}''_W \mathscr{R}}{\rho_g D} = \ln(Y_{O\infty} f_W(1+f_g) + 1) \qquad (6.27)$$

The gas phase reaction between CO and O_2 thus modifies the definition of B as $Y_{O\infty} f_W(1+f_g)$. The lifetime of a particle in this situation becomes

$$t_b = \frac{\rho_F d_O^2}{8\rho_g D \ln(Y_{O\infty} f_W(1+f_g) + 1)} \qquad (6.28)$$

If f_g is $2M_{CO}/M_{O_2} = 56/32$ and f_W is $M_C/M_{CO_2} = 12/44$, $Y_{O\infty} f_W(1+f_g) = 0.75 Y_{O\infty}$. The burning time calculated by this scheme for a 1 micron carbon sphere burning in air is the same as that when carbon burns to CO directly. Coffin's experimentally measured time of burning is $1.98 \cdot 10^{-5}$ seconds for a one micron particle, which agrees well with the time calculated by our theory.

6.5 Combustion of Pulverized Coal

Hottel and Stewart published an article in Industrial and Engineering Chemistry in 1940 analyzing the furnace space requirements for the combustion of pulverized coal. Combining the equation for the burning rate of a single coal particle and a Rosin–Rammler type size distribution law, they obtained the relation shown in Figure 6.9 between the fraction of the fixed carbon unburned and time normalized* by the burning time of the "weight mean" particle, $t_{b\,\text{mean}}$. They also showed that their theory agrees satisfactorily with

---

* This is similar to Probert's concept of spray burning treated in Section 5.10.

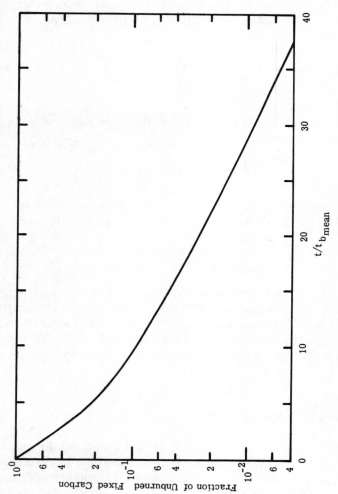

Figure 6.9 Hottel and Stewart's result on burning of pulverized coal (adapted with permission of American Chemical Society, from: H. C. Hottel and I. M. Stewart, *Ind. and Engr. Chemistry*, **32**, p. 719, (1940))

Sherman's experiments in which the unburnt fraction of the fixed carbon is measured at various stations on the centerline of a burning pulverized coal/air jet.

6.6 Concluding Remarks

In this chapter we merely introduced the basic principles of combustion of solid particles. It is clear that simple theories of the present type are usually more than adequate to describe quantitatively the seemingly complex phenomenon. Even though metal combustion is often accompanied by complications such as sputtering, sloughing and formation of oxide insulation shells ("cenospheres") the concept of transfer number developed in this chapter (and listed in Table 6.1) gives an excellent first approximation.

The field of solid fuel combustion is a new territory of research even today. The effect of variable properties is yet virtually unexplored. The papers of Tu, Davis and Hottel, Hottel and Stewart, Hottel and Davis, are highly recommended for work on carbon sphere burning. The work of Coffin is certainly the most definitive one on solid combustion. The paper of Olson and that of Gerstein and Coffin are good summaries of the work on high energy fuels. The book *Heterogeneous Combustion* edited by Wolfhard, Glassman and Green gives an excellent introduction to the various facets of metal combustion. The Combustion Symposia, the journals of Fuel, Combustion and Flame, A.I.A.A. Journal and Industrial and Engineering Chemistry often contain excellent articles on solid combustion. Note here may be made of the new international journal called Combustion Science and Technology.

References

Burke, S. P. and Schumann, T. E. W., *Proc. 3rd. International Conf. on Bituminous Coal*, **2**, p. 485 (1931).

Coffin, K. P. and Brokaw, R. S., A General System for Calculating the Burning Rates of Particles and Drops and Comparison of Calculated Rates for Carbon, Boron, Magnesium and Iso-octane, N.A.C.A. Tech. Note 3929 (1957).

Gerstein, M. and Coffin, K. P., Combustion of Solid Fuels. Chapter in the volume *Combustion Processes*, Princeton High Speed Aerodynamics and Jet Propulsion Series, p. 444 (1956).

Hottel, H. C. and Davis, H., Combustion Rate of Carbon, *Ind. and Eng. Chem.*, **26**, p. 889 (1934).

Hottel, H. C. and Parker, A. S., Combustion Rate of Carbon, *Ind. and Eng. Chem.*, **28**, p. 1334 (1936).

Hottel, H. C. and Stewart, I. M., Space Requirement for the Combustion of Pulverized Coal, *Ind. and Eng. Chem.*, **32**, p. 719 (1940).

Olson, W. and Setze, P. C., High Energy Fuels, p. 883, *7th Symposium* (*International*) *on Combustion*, Butterworths, London (1959).

Spalding D. B., Combustion of Fuel Particles, *Fuel*, **30**, p. 121 (1951).

Sherman, R. A., *Proc. 3rd. International Conf. on Bituminous Coal*, **2**, p. 370 (1932).

Tu, C. M., Davis, H., and Hottel, H. C., Combustion Rate of Carbon Spheres in Flowing Gas Streams, *Ind. and Eng. Chem.* **26**, p. 749 (1934).

Wolfhard, Glassman and Green (editors), *Heterogeneous Combustion*, Academic Press, New York (1964).

CHAPTER 7

Combustion of Gaseous Fuel Jets

In numerous technological applications of combustion such as in metallurgical and boiler furnaces, high temperature ovens, modern propulsion devices or in accidental fuel tank ruptures, a gaseous fuel jet is burned as it issues into air through a narrow injection nozzle at a high velocity. When a spray of fine droplets of a highly volatile liquid fuel is introduced into hot compressed air, it may quickly evaporate to form a combustible vapor jet provided certain conditions are met.

A nonaerated Bunsen flame* offers a simple everyday example of a jet flame. The shape and size of a jet flame and the details of composition and temperature across it are factors of importance in determining the space heating rates, efficiencies of combustion equipments and fire threat to the nearby objects.

To predict these factors under ideal conditions with a simple mathematical model and to inspect the validity of such predictions by comparison with experiments is the main topic of concern in this chapter. The jet is called *plane* (or two-dimensional) if it issues through a narrow slit of infinite length. It is called *cylindrical* (or, more logically, circular) if it issues through a fine circular hole. Most of the real jets are nearly cylindrical. Jets may either be *free* or *confined* depending upon the proximity of the walls bounding the oxidant gas into which the jet issues. The size of the nozzle opening (i.e., slit width or the hole diameter), the velocity of injection and the viscosity of the injected gas determine whether the jet is *laminar* or *turbulent*. The high velocity jet emerging out of the port mixes with the relatively quiescent surrounding air by molecular diffusion (and, if turbulent, by eddy mixing). Naturally, because of the better mixing, turbulent jet flames are considerably shorter than laminar ones. In order to accomplish high space heating rates, flames in most of the indistrial furnaces are rendered turbulent.

---

* A nonaerated Bunsen flame is one to which the fuel gas is supplied through the tube without premixing it with air. This is done by simply shutting off the primary air ports located at the bottom of an ordinary Bunsen burner.

7.1 Plane Free Nonburning Jet

Shown in Figure 7.1 is a unit length of an infinitely long narrow slit in a wall which separates a high pressure hot gas F from a normal pressure cool quiescent gas O of infinite extent in all directions. Both the gases are assumed to be inert and of equal and constant thermo-physical properties. Due to the pressure differential, the compressed gas leaks through the slit into the quiescent gas in the form of a free jet. Let us assume that the slit walls are smooth, frictionless, impervious and adiabatic. Let d_i be the width of the slit. Let u_i, T_i and Y_{Fi} respectively be the velocity, temperature and species F mass fraction with which the jet emerges out of the slit. As shown in the figure let x be the coordinate along the plane of symmetry of the jet with $x = 0$ at the slit. Let y be the lateral coordinate with $y = 0$ at the plane of symmetry. Since the slit walls are assumed frictionless, adiabatic and impervious, u_i, T_i and Y_{Fi} will be uniform across the cross section of the jet at $x = 0$.

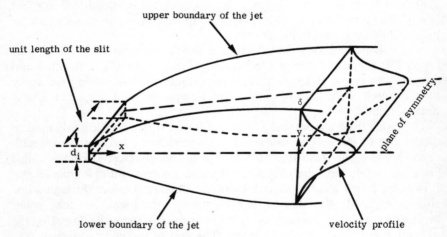

Figure 7.1 Plane free jet

The quiescent room gas O exerts a frictional drag on the high velocity jet of gas F, such a drag having the following implications. The drag is responsible for induction (or entrainment) of the ambient fluid into the moving jet. Due to the mixing, the jet is gradually slowed and cooled down; the average mass fraction of species F in the jet is also gradually lowered. At distances very far away from the slit (i.e., as $x \to \infty$) the influence of injection vanishes. Let the subscript ∞ stand for the ambient conditions. Let u_m, T_m and Y_{Fm} be respectively the maximum x-directional velocity, temperature and species F mass fraction at any cross section. These, of course, occur at the plane of symmetry, $y = 0$. It is then reasonable to expect the profiles

Combustion of Gaseous Fuel Jets

of u/u_m, $(T - T_\infty)/(T_m - T_\infty)$ and Y_F/Y_{Fm} at various cross sections of the jet to be as indicated qualitatively in Figure 7.2.

In real jets, the mixing penetrates to the plane of symmetry within a short distance from the slit. As shown in Figure 7.2, the jet may be divided into two zones—the potential (or developing) zone and the fully developed zone. Shapiro shows that (refer to Section 7.10) the length of the potential zone $\mathscr{L}$ is usually 4 to 6 times the slit width, d_i. We will consider that d_i and $\mathscr{L}$ are very small in our analysis. If we neglect the effects of chemical reactions, thermal radiation and pressure gradients, the profiles of u/u_m, $(T - T_\infty)/(T_m - T_\infty)$ and Y_F/Y_{Fm} at any value of x in the fully developed zone of the jet are known to be "similar" in shape to those at any other value of x. It should be clear to the student by now that the growth of the jet as a result of mixing is similar to the growth of a boundary layer over a solid boundary which confines a flowing fluid. In fact, jets and wakes illustrate a special class of boundary layer type phenomena arising in the absence of solid boundaries. The definition of the jet thickness 2δ, in much the same manner as that of the boundary layer thickness, is arbitrary because the velocity, temperature and composition approach the ambient value asymptotically. There are two customarily used measures of the jet thickness; (a) firstly, by that value of y beyond which the velocity gradients are insignificantly small and (b) secondly, by that value of y at which $(u - u_\infty)$ is half the value of $(u_m - u_\infty)$. In the present work, we follow the first definition.

The jet thickness may also be defined on the basis of either the temperature profile or the composition profile. If Prandtl and Schmidt numbers are unity, the momentum, thermal and composition jet thicknesses are mutually equivalent. This too, we assume here because turbulent jets are of our primary concern.

The temperature, velocity and composition distributions in a jet are obtained by solving the appropriate conservation equations. Just as in the case of a boundary layer over a flat plate, the differences in pressure across a free jet are negligible since the constant pressure of the surroundings impresses itself on the jet. Then, neglecting buoyancy* and balancing the acceleration forces with the viscous forces, the equation of motion for a steady state incompressible constant property† plane jet is obtained as following

$$u\frac{\partial u}{\partial x} + v\frac{\partial u}{\partial y} = (v + \varepsilon)\frac{\partial^2 u}{\partial y^2} \qquad (7.1)$$

$$P = P_\infty; \text{ a constant}$$

---

* See Section 7.8 for a discussion of the assumptions.

† Howarth's transformation may be used to convert the compressible equations to the incompressible form in the case of a two-dimensional flow problem. Here we concern ourselves only with the incompressible problem. A brief description of Howarth's transformation is to be presented in Section 7.8.

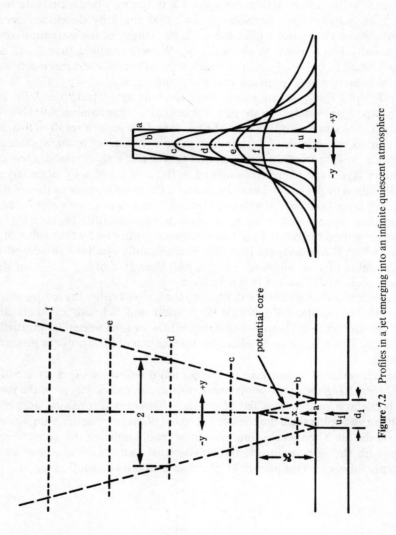

Figure 7.2 Profiles in a jet emerging into an infinite quiescent atmosphere

Combustion of Gaseous Fuel Jets

u and v are, as usual, the velocity components in x and y directions respectively. When the flow is laminar, as discussed in Section 3.7, the turbulent eddy diffusivity ε is zero. When the flow is fully turbulent, the kinematic viscosity v is usually so small compared to ε that it may be neglected. Prandtl's turbulence hypothesis describes ε as a function of the fluid, flow and geometry according to Eq. 3.52 which we rewrite here for the sake of completeness.

$$\varepsilon = C' \delta^2 \left| \frac{\partial u}{\partial y} \right| \tag{3.52}$$

The constant C' is reported by Reichardt as about 0.0185 for a plane jet (and 0.0128 for a cylindrical jet). ε is also given in the following forms sometimes.

$$\varepsilon = C'' \delta u_m$$
$$\varepsilon = C''' v R_{e_i}^m \qquad m \approx 0.85$$

It is customary to assume equal eddy diffusivities for momentum, heat and mass transfer. The conservation of the total mass is stipulated by Eq. 3.59.

$$\frac{\partial u}{\partial x} + \frac{\partial v}{\partial y} = 0 \tag{7.2}$$

The energy and species F conservation equations are written as following

$$u \frac{\partial}{\partial x}(T - T_\infty) + v \frac{\partial}{\partial y}(T - T_\infty) = (\alpha + \varepsilon_t) \frac{\partial^2}{\partial y^2}(T - T_\infty) \tag{7.1a}$$

$$u \frac{\partial Y_F}{\partial x} + v \frac{\partial Y_F}{\partial y} = (D_F + \varepsilon_D) \frac{\partial^2 Y_F}{\partial y^2} \tag{7.1b}$$

The boundary conditions* for Eq. 7.1 are

$$y \geq \delta, \quad 0 \leq x \leq \infty, \quad \left. \begin{array}{c} u \\ (T - T_\infty) \\ Y_F \end{array} \right\} = 0 \qquad \left. \begin{array}{c} \partial u / \partial y \\ \partial(T - T_\infty)/\partial y \\ \partial Y_F / \partial y \end{array} \right\} = 0$$

$$y = 0, \quad 0 \leq x \leq \infty, \quad v = 0 \qquad \left. \begin{array}{c} \partial u / \partial y \\ \partial(T - T_\infty)/\partial y \\ \partial Y_F / \partial y \end{array} \right\} = 0 \tag{7.3}$$

$$x = 0, \quad 0 \leq y \leq d_i/2, \quad u = u_i, \quad (T - T_\infty) = (T_i - T_\infty), \quad Y_F = 1$$

$$x = \infty, \quad 0 \leq y \leq \infty, \quad \left. \begin{array}{c} u \\ (T - T_\infty) \\ Y_F \end{array} \right\} = 0 \qquad \left. \begin{array}{c} \partial u / \partial y \\ \partial(T - T_\infty)/\partial y \\ \partial Y_F / \partial y \end{array} \right\} = 0$$

---

* Due to the inherent symmetry, we need to consider only half of the jet for solution.

Because of our assumptions of $v = \alpha = D_F$ and $\varepsilon = \varepsilon_t = \varepsilon_D$, it is obvious that the conservation equations are identical with identical boundary conditions. Their solutions, hence, will be identical as well. This is precisely the point we exploit in this chapter. We will solve the momentum equation by a simple integral technique and utilize this solution to deduce the temperature and composition profiles.

7.2 Invariance of the Jet Momentum, Enthalpy and Species Contents: Partial Integration of the Equations

Consider integration of Eq. 7.1 with respect to y. Multiplying Eq. 7.2. by u,

$$u\frac{\partial u}{\partial x} = -u\frac{\partial v}{\partial y} \tag{7.4}$$

By the rule of product differentiation, $\partial(vu)/\partial y = u\partial v/\partial y + v\partial u/\partial y$. Substituting this identity into Eq. 7.4, $v\partial u/\partial y = u\partial u/\partial x + \partial(uv)/\partial y$. Equation 7.1, thereby, transforms to

$$\frac{\partial u^2}{\partial x} + \frac{\partial(uv)}{\partial y} = (v + \varepsilon)\frac{\partial^2 u}{\partial y^2} \tag{7.5}$$

Equations 7.2 and 7.5 may now be integrated from $y = 0$ to any value of y in the jet. Performing such an integration once from $y = 0$ to $y = \delta$ and again from $y = 0$ to $y = \delta/2$, we obtain

$$v_\delta = \frac{d}{dx}\int_0^\delta u\,dy \tag{7.6}$$

$$v_{\delta/2} = \frac{d}{dx}\int_0^{\delta/2} u\,dy \tag{7.7}$$

$$\frac{d}{dx}\int_0^\delta u^2\,dy + [uv]_\delta = \left[(v + \varepsilon)\frac{\partial u}{\partial y}\right]_\delta \tag{7.8}$$

$$\frac{d}{dx}\int_0^{\delta/2} u^2\,dy + [uv]_{\delta/2} = \left[(v + \varepsilon)\frac{\partial u}{\partial y}\right]_{\delta/2} \tag{7.9}$$

The normal velocity v is zero at the plane of symmetry and gradually increases with increasing y to reach a constant value for $y \geq \delta$. v_δ is this constant value. It gives the jet entrainment at the particular cross-section under consideration. Let us digress here to deduce an important characteristic of jets from Eq. 7.8. Since by Eq. 7.3, $u = 0$ and $\partial u/\partial y = 0$ at $y = \delta$, Eq. 7.8 reduces to

$$\frac{d}{dx}\int_0^\delta u^2\,dy = 0 \tag{7.10}$$

Combustion of Gaseous Fuel Jets

Upon multiplying this equation by the constant density ρ, it becomes instantly clear that the momentum of the jet is invariant with x and is equal to its value at $x = 0$.

$$\int_0^\delta u^2 \, dy = u_i^2 \, d_i/2, \text{ a constant} \qquad (7.11)$$

Turning now to the energy equation, consider the following equality by product rule.

$$\frac{\partial u(T - T_\infty)}{\partial x} + \frac{\partial v(T - T_\infty)}{\partial y}$$

$$= u \frac{\partial (T - T_\infty)}{\partial x} + v \frac{\partial (T - T_\infty)}{\partial y} + (T - T_\infty)\left(\frac{\partial u}{\partial x} + \frac{\partial v}{\partial y}\right)$$

The last term is zero due to the continuity equation. Equation 7.1(a) becomes

$$\frac{\partial u(T - T_\infty)}{\partial x} + \frac{\partial v(T - T_\infty)}{\partial y} = (\alpha + \varepsilon_t)\frac{\partial^2 (T - T_\infty)}{\partial y^2}$$

integration yields the following equation corresponding to Eq. 7.8.

$$\frac{d}{dx}\int_0^\delta u(T - T_\infty) \, dy + [v(T - T_\infty)]_\delta = \left[(\alpha + \varepsilon_t)\frac{\partial (T - T_\infty)}{\partial y}\right]_\delta \qquad (7.8a)$$

Since both $(T - T_\infty)$ and its derivative are zero at and beyond $y = \delta$, this equation boils down to

$$\frac{d}{dx}\int_0^\delta u(T - T_\infty) \, dy = 0 \qquad (7.10a)$$

This equation (if multiplied by the volumetric specific heat ρC) implies that the enthalpy content of the jet remains independent of x.

$$\int_0^\delta u(T - T_\infty) \, dy = u_i(T_i - T_\infty) \, d_i/2, \text{ a constant} \qquad (7.11a)$$

Similar manipulations on the species F equation give

$$\frac{d}{dx}\int_0^\delta u Y_F \, dy = 0 \qquad (7.10b)$$

Alternately,

$$\int_0^\delta u Y_F \, dy = u_i Y_{Fi} \, d_i/2 = u_i \, d_i/2, \text{ a constant} \qquad (7.11b)$$

Let us now return to the task of integrating Eq. 7.6 to 7.9. Assume that the velocity profile at any section in the fully developed region of the jet may be approximated by straight lines as shown in Figures 7.3 and 7.5. As we shall see later, in spite of its apparent oversimplification, the linear profile yields reasonably accurate results.

$$\frac{u}{u_m} = 1 - \frac{y}{\delta} \qquad 0 \le |y| \le \delta \tag{7.12}$$

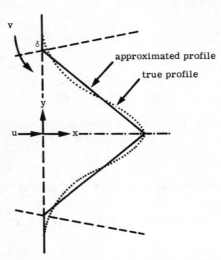

Figure 7.3 Terminology of a jet profile approximation

Remember that u_m, v_δ and δ are as yet unknown undetermined functions of x. Substitution of Eq. 7.12 in Eq. 7.6 to 7.9 simplifies them to the following

$$v_\delta = \frac{d}{dx}(u_m \delta/2) \tag{7.13}$$

$$v_{\delta/2} = \frac{d}{dx}(3u_m \delta/8) \tag{7.14}$$

$$\frac{d}{dx}(u_m^2 \delta) = 0 \tag{7.15}$$

$$\frac{d}{dx}(7u_m^2 \delta/24) + u_m v_{\delta/2}/2 = (v + \varepsilon)u_m/\delta \tag{7.16}$$

Solution of these equations for known conditions of $(v + \varepsilon)$ gives the variation of u_m, v_δ and δ with respect to x and consequently by Eq. 7.12, the velocity profile at any x. Further simplification may be made to arrive at the following set of governing equations.

$$v_\delta = \frac{d}{dx}(u_m \delta/2) \tag{7.17}$$

$$u_m^2 \delta = u_i^2 d_i/2 \tag{7.18}$$

$$\frac{3}{16}\frac{d}{dx}(u_m \delta) = (v + \varepsilon)/\delta \tag{7.19}$$

At $x = 0$, $u_m = u_i$ and $\delta = d_i/2$.

7.3 Solution of Laminar Plane Free Inert Jet

In this case, $\varepsilon = 0$. Eliminating δ between Eqs. 7.18 and 7.19,

$$\frac{d}{dx}(1/u_m) = \frac{64v}{3u_i^4 d_i^2} u_m^2$$

Integrating this equation directly by separating the variables u_m and x,

$$\frac{1}{3u_m^3} = \frac{64v}{3u_i^4 d_i^2} x + \text{constant}$$

The constant of integration is determined by setting $u_m = u_i$ at $x = 0$.

$$\frac{u_m}{u_i} = \left[1 + \frac{64v}{u_i d_i}\frac{x}{d_i}\right]^{-1/3} \tag{7.20}$$

$u_i d_i/v$ is the Reynolds number at which the jet issues out of the slit.

$$\text{Re}_i \equiv u_i d_i/v$$

Combining Eqs. 7.18 and 7.19, the jet thickness is obtained as

$$\frac{2\delta}{d_i} = \left[1 + \frac{64}{\text{Re}_i}\frac{x}{d_i}\right]^{2/3} \tag{7.21}$$

The entrainment velocity is

$$\frac{v_\delta}{u_i} = \frac{16}{3\text{Re}_i}\left[1 + \frac{64}{\text{Re}_i}\frac{x}{d_i}\right]^{-2/3} \tag{7.22}$$

Equation 7.12 takes the following form.

$$\frac{u}{u_i} = \left[1 + \frac{64}{\text{Re}_i}\frac{x}{d_i}\right]^{-1/3}\left[1 - 2\left(\frac{y}{d_i}\right)\left(1 + \frac{64}{\text{Re}_i}\frac{x}{d_i}\right)^{-2/3}\right] \tag{7.23}$$

If the initial mass flux in the jet out of a unit length of the slit is $\rho u_i d_i$ gm/sec, the mass flux at any cross section will be given by the product of the mean velocity through that section, the density and the area of that section.

$$\frac{\dot{W}}{\rho u_i d_i} = 1 + 0.25\left[1 + \frac{64}{\mathrm{Re}_i}\frac{x}{d_i}\right]^{1/3}$$

If the slit under consideration is infinitesimally narrow or if the cross sections under consideration are sufficiently away from the slit, the factor $64vx/u_i d_i^2$ will be much larger than unity. Our results then may be simplified to compare with those given by Schlichting in his treatise on boundary layer theory. In real life, the nozzles are indeed very fine. Hence when the slit is narrow,

$$\frac{u}{u_i} = \left(\frac{\mathrm{Re}_i d_i}{64x}\right)^{1/3}\left(1 - 2\frac{y}{d_i}\left[\frac{\mathrm{Re}_i d_i}{64x}\right]^{2/3}\right) \tag{7.24}$$

$$\frac{v_\delta}{u_i} = \left(\frac{16}{3\,\mathrm{Re}_i}\right)\left(\frac{\mathrm{Re}_i d_i}{64x}\right)^{2/3} \tag{7.25}$$

$$\frac{2\delta}{d_i} = \left(\frac{64x}{\mathrm{Re}_i d_i}\right)^{2/3} \tag{7.26}$$

$$\frac{\dot{W}}{\rho u_i d_i} = 1 + 0.25\left(\frac{64x}{\mathrm{Re}_i d_i}\right)^{1/3} \tag{7.27}$$

The patterns of variation of u_m, v_δ and δ with x suggested by Eqs. 7.20–7.26 are shown in Figure 7.4 for a Reynolds number of 1,000. In particular, it is interesting to note that the entrainment velocity is proportional to the square of the maximum velocity at any cross section. Schlichting's more involved but exact solution is as following

$$\frac{u}{u_i} = 0.4543\left(\frac{\mathrm{Re}_i d_i}{x}\right)^{1/3} \operatorname{sech}^2 \xi \tag{7.24s}$$

$$\frac{v}{u_i} = \frac{0.5503}{\mathrm{Re}_i}\left(\frac{\mathrm{Re}_i d_i}{x}\right)^{2/3}[2\xi \operatorname{sech}^2 \xi - \tanh \xi] \tag{7.25s}$$

$$\xi \equiv 0.2752\left(\frac{y}{d_i}\right)\left(\frac{\mathrm{Re}_i d_i}{x}\right)^{2/3} \left(= 2.2 \text{ at } \frac{y}{\delta} = 1\right)$$

At the edge of the jet, Schlichting's ξ is of the order of 2 so that his solution gives the jet thickness, approximately, as

$$\frac{2\delta}{d_i} \approx \frac{4}{0.2752}\left(\frac{x}{\mathrm{Re}_i d_i}\right)^{2/3} \tag{7.26s}$$

Combustion of Gaseous Fuel Jets

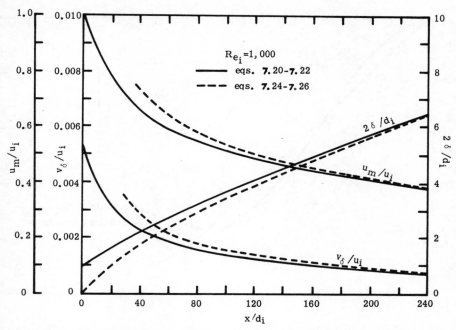

Figure 7.4 Maximum velocity, entrainment velocity and jet thickness as functions of distance from the port for a laminar plane free inert jet

Comparison of Schlichting's results with ours indicates that our linear velocity profile predicts consistent and correct functional forms of the variables even though the numerical coefficients are slightly underestimated. Figure 7.5 depicts a comparison of the exact and linear velocity profiles quantitatively. A point of peripheral interest here concerns with the notation used by Schlichting. Since d_i tends to zero for an infinitesimally narrow slit, the use of d_i as the standard measure of length becomes difficult. Schlichting, therefore, uses the ratio of the jet momentum to the square of the initial velocity as the standard length.

Knowing the solution of the momentum equation, we can now readily deduce the temperature and species F profiles by assuming $\alpha = \nu = D_F$. Thus when $64x/\text{Re}_i d_i \gg 1$,

$$\left(\frac{(T - T_\infty)}{(T_i - T_\infty)}\right) = \frac{Y_F}{Y_{Fi}} = \frac{u}{u_i} = \left(\frac{\text{Re}_i d_i}{64x}\right)^{1/3} \left(1 - 2\frac{y}{d_i}\left[\frac{\text{Re}_i d_i}{64x}\right]^{2/3}\right)$$

$$= 0.4543\left(\frac{\text{Re}_i d_i}{x}\right)^{1/3} \text{sech}^2\left(0.2752\frac{y}{d_i}\left[\frac{\text{Re}_i d_i}{x}\right]^{2/3}\right) \quad (7.28)$$

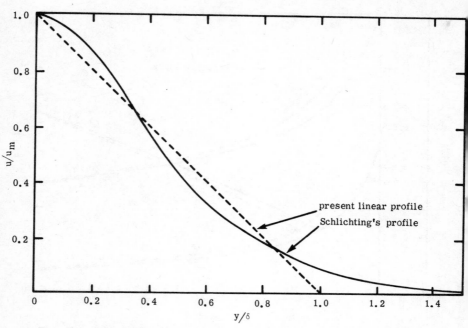

Figure 7.5 Comparison of the exact and linear velocity profiles in a laminar plane jet

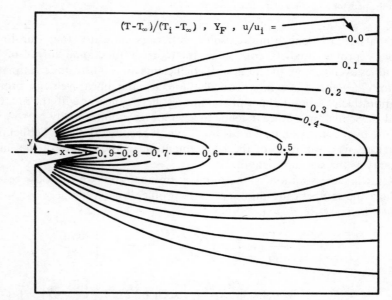

Figure 7.6 Isovelocity (isothermal or isocomposition) contours in a laminar plane jet

Combustion of Gaseous Fuel Jets

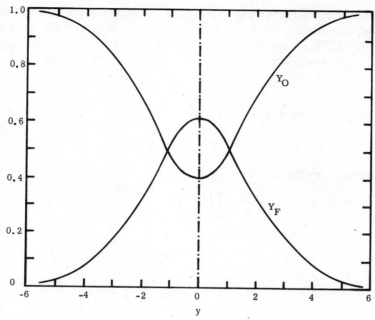

Figure 7.7 Complementary nature of the fuel and oxidant mass fraction profiles in a laminar plane jet

It should be stressed that Eq. 7.28 is invalid for small values of x/d_i. One may now plot iso-velocity (or isothermal or isocomposition) contours in the jet as shown qualitatively in Figure 7.6. Since we assumed that species F and O are inert and since these are the only species present in the system, the profile of $Y_O(x, y)$ is obtained from that of $Y_F(x, y)$ as

$$Y_O(x, y) = 1 - Y_F(x, y)$$

Figure 7.7 shows qualitatively the complementary profiles of Y_O and Y_F at an arbitrary cross section which is sufficiently far from the slit. It can be noted that the location where Y_F is a certain fraction f of Y_O is given by $Y_F/Y_O = f = Y_F/(1 - Y_F)$. That is,

$$Y_F = f/(f + 1)$$

The numbers on the curves of Figure 7.6, thus, also represent $f/(f + 1)$.

7.4 Turbulent Plane Free Inert Jet

Consider the situation in which the slit opening d_i and the velocity of discharge u_i are such that the Reynolds number Re_i is sufficiently large to ensure a fully turbulent jet. The molecular viscosity v is then negligible in comparison with the turbulent diffusivity ε. Assuming ε to be given by Eq. 3.52, the linear profile of velocity condenses it to $\varepsilon = C' u_m \delta$. Equation 7.19 therefore becomes

$$\frac{3}{16} \frac{d}{dx}(u_m \delta) = C' u_m$$

Substituting δ from 7.18,

$$-\frac{1}{u_m^3} du_m^2 = \frac{32}{3} \frac{C'}{u_i^2 d_i} dx$$

Integrating with the condition $u_m = u_i$ at $x = 0$,

$$\frac{u_m}{u_i} = \left[1 + \frac{64}{3} C' \frac{x}{d_i}\right]^{-1/2} \tag{7.29}$$

From Eqs. 7.17, 7.18 and 7.29,

$$\frac{v_\delta}{u_i} = \frac{8C'}{3}\left[1 + \frac{64}{3} C' \frac{x}{d_i}\right]^{-1/2} \tag{7.30}$$

$$\frac{2\delta}{d_i} = \left[1 + \frac{64}{3} C' \frac{x}{d_i}\right] \tag{7.31}$$

Hence,

$$\frac{u}{u_i} = \left[1 + \frac{64}{3} C' \frac{x}{d_i}\right]^{-1/2}\left[1 - 2\frac{y}{d_i}\left(1 + \frac{64}{3} C' \frac{x}{d_i}\right)^{-1}\right] \tag{7.29a}$$

These equations indicate that the turbulent jet maximum velocity u_m is proportional to $x^{-1/2}$. The entrainment velocity is *directly* proportional to the local maximum velocity, the proportionality constant being sometimes called as the *entrainment constant*. Another striking result is that the turbulent jet thickness increases linearly with x. The mass flow rate $\dot{W}$ is proportional to $x^{1/2}$. It is instructional to compare these patterns of dependencies with those of the laminar jet.

Equations 7.29 to 7.31 may be simplified for a fine slit. Schlichting presents the exact solution for this situation as below.

$$\frac{u}{u_i} = \left(\frac{3\sigma}{4} \frac{d_i}{x}\right)^{1/2} \text{sech}^2 \xi \tag{7.29s}$$

$$\frac{v_\delta}{u_i} = \left(\frac{3}{16\sigma} \frac{d_i}{x}\right)^{1/2} [2\xi \text{ sech}^2 \xi - \tanh \xi] \tag{7.30s}$$

$$\frac{2\delta}{d_i} \approx \frac{4}{\sigma} \frac{x}{d_i} \tag{7.31s}$$

Combustion of Gaseous Fuel Jets

where, $\xi \equiv \sigma y/x$; (≈ 2.96 at $y/\delta = 1$). σ is an arbitrary constant obtained from Reichardt's experiments as 7.67. Once again it is encouraging to note that the results of our simple analysis are consistent with those of the exact one.

Because we assume $\varepsilon = \varepsilon_t = \varepsilon_D$, the solution of the momentum equation would also give that of the energy and species equations.

7.5 Cylindrical Free Inert Jet

Jets in industrial applications are mostly cylindrical rather than plane. A cylindrical jet evolves when the jet fluid is discharged through a circular hole in a wall as shown in Figure 7.8.

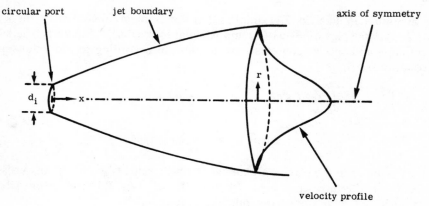

Figure 7.8 Cylindrical free jet

The plane jet analysis presented in Sections 7.1 through 7.4 may easily be modified to describe the behavior of cylindrical jets. The equations of motion and continuity in symmetrical cylindrical coordinates (with the usual assumptions) are as following.

$$ru \frac{\partial u}{\partial x} + rv \frac{\partial u}{\partial r} = (v + \varepsilon) \frac{\partial}{\partial r} r \frac{\partial u}{\partial r}$$

$$\frac{\partial ru}{\partial x} + \frac{\partial rv}{\partial r} = 0 \tag{7.32}$$

x and r respectively are the longitudinal and normal coordinates with the origin at the nozzle. u and v respectively are longitudinal and normal velocity components. Manipulations similar to those made in Section 7.2 transform Eq. 7.32 to the following integral equations.

$$v_\delta = \frac{1}{\delta}\frac{d}{dx}\int_0^\delta ru\, dr \tag{7.33}$$

$$v_{\delta/2} = \frac{2}{\delta}\frac{d}{dx}\int_0^{\delta/2} ru\, dr \tag{7.34}$$

$$\frac{d}{dx}\int_0^\delta ru^2\, dr = 0 \tag{7.35}$$

$$\frac{d}{dx}\int_0^{\delta/2} ru^2\, dr + [ruv]_{\delta/2} = \left[(v + \varepsilon)r\frac{\partial u}{\partial r}\right]_{\delta/2} \tag{7.36}$$

Equation 7.35, like Eq. 7.11, implies that the momentum content of the jet $\int_0^\delta 2\pi r\rho u^2\, dr$ is invariant with x and is equal to $\pi d_i^2 \rho u_i^2/4$. Similarly, the jet enthalpy and species contents remain constant according to the following equations.

$$\int_0^\delta 2\pi r\rho u C(T - T_\infty)\, dr = \pi d_i^2 \rho u_i C(T - T_\infty)/4$$

$$\int_0^\delta 2\pi r\rho u Y_F\, dr = \pi d_i^2 \rho u_i Y_{Fi}/4$$

Assuming, as in the case of a plane jet, a linear velocity profile at any cross section in the fully developed region of the jet, Eqs. 7.33 to 7.36 may be routinely solved to obtain the following results.

Laminar cylindrical Jet:

$$\frac{u}{u_i} = \left[1 + \frac{48}{\text{Re}_i}\frac{x}{d_i}\right]^{-1}\left(1 - 2\frac{r}{d_i}\left[1 + \frac{48}{\text{Re}_i}\frac{x}{d_i}\right]^{-1}\right) \tag{7.37}$$

$$\frac{v_\delta}{u_i} = \frac{4}{\text{Re}_i}\left[1 + \frac{48}{\text{Re}_i}\frac{x}{d_i}\right]^{-1} \tag{7.38}$$

$$\frac{2\delta}{d_i} = \left[1 + \frac{48}{\text{Re}_i}\frac{x}{d_i}\right] \tag{7.39}$$

where,

$$\text{Re}_i \equiv u_i d_i/v$$

Combustion of Gaseous Fuel Jets

For an infinitesimally small hole jet, Schlichting gives the following solution.

$$\frac{u}{u_i} = \frac{3}{32}\left(\frac{\mathrm{Re}_i\, d_i}{x}\right)\frac{1}{(1+\frac{1}{4}\xi^2)^2} \tag{7.37s}$$

$$\frac{v}{u_i} = \frac{\sqrt{3}}{8}\left(\frac{d_i}{x}\right)\frac{(\xi - \frac{1}{4}\xi^3)}{(1+\frac{1}{4}\xi^2)^2} \tag{7.38s}$$

$$\frac{2\delta}{d_i} \approx \frac{32}{\sqrt{3}}\left(\frac{x}{\mathrm{Re}_i\, d_i}\right) \tag{7.39s}$$

where,

$$\xi \equiv \frac{\sqrt{3}}{8}\left(\frac{r}{d_i}\right)\left(\frac{\mathrm{Re}_i\, d_i}{x}\right); \; (\approx 5.2 \text{ at } y/\delta = 1).$$

Turbulent cylindrical Jet:

$$\frac{u}{u_i} = \left[1 + 24C'\frac{x}{d_i}\right]^{-1}\left(1 - \frac{2r}{d_i}\left[1 + 24C'\frac{x}{d_i}\right]^{-1}\right) \tag{7.40}$$

$$\frac{v_\delta}{u_i} = 2C'\left[1 + 24C'\frac{x}{d_i}\right]^{-1} \tag{7.41}$$

$$\frac{2\delta}{d_i} = \left[1 + 24C'\frac{x}{d_i}\right] \tag{7.42}$$

Schlichting's solution is as following.

$$\frac{u}{u_i} = \frac{3}{16\sqrt{\pi}\,\sigma'}\left(\frac{d_i}{x}\right)\frac{1}{(1+\frac{1}{4}\xi^2)^2} \tag{7.40s}$$

$$\frac{v}{u_i} = \frac{\sqrt{3}}{8}\left(\frac{d_i}{x}\right)\frac{(\xi - \frac{1}{4}\xi^3)}{(1+\frac{1}{4}\xi^2)^2} \tag{7.41s}$$

$$\frac{2\delta}{d_i} \approx \frac{16\sqrt{\pi}\,\sigma'}{3}\left(\frac{x}{d_i}\right) \tag{7.42s}$$

where,

$$\xi \equiv \frac{1}{4\sigma'}\sqrt{\frac{3}{\pi}}\left(\frac{r}{d_i}\right)\left(\frac{d_i}{x}\right).$$

σ' is a constant determined by Reichardt's experiments as 0.0161 which renders a value of 0.0128 to the eddy diffusivity constant C'. It is readily obvious that the principal functional variations of Schlichting's theory and the present linear theory are mutually consistent for the cylindrical jet as well.

In particular, note that whether the jet is laminar or turbulent, the jet thickness is a linear function of the distance x. Furthermore, the entrainment is directly proportional to the maximum velocity u_m in both laminar as well as turbulent cases.

Table 7.1 is a comparative summary of our linear solution and Schlichting's solution when the injection port is extremely small. Reichardt's experimental values of turbulent diffusion constants are used in preparing this table. The results summarized in this table are also graphically illustrated in Figures 7.9, 7.10 and 7.11, primarily to demonstrate the different decay patterns with respect to x. The full lines in these figures are from the formulae of Table 7.1. The broken lines are approximated for a finite nozzle port d_i.

It is noticeable in Figure 7.9 that the jet maximum velocity u_m decreases with increasing x much more rapidly when the jet is turbulent than when it is laminar. Cylindrical jets, in general, slow down sooner than plane jets due

TABLE 7.1

Comparison of Schlichting's and the Present Results when $x/d_i \gg 1$.

Variable	Source	Plane Laminar	Plane Turbulent	Cylindrical Laminar	Cylindrical Turbulent
$\dfrac{u_m}{u_i}$	This work	$0.250\left(\dfrac{\mathrm{Re}_i\, d_i}{x}\right)^{1/3}$	$1.59\left(\dfrac{d_i}{x}\right)^{1/2}$	$0.021\left(\dfrac{\mathrm{Re}_i\, d_i}{x}\right)$	$3.26\left(\dfrac{d_i}{x}\right)$
	Schlichting	$0.454\left(\dfrac{\mathrm{Re}_i\, d_i}{x}\right)^{1/3}$	$2.40\left(\dfrac{d_i}{x}\right)^{1/2}$	$0.094\left(\dfrac{\mathrm{Re}_i\, d_i}{x}\right)$	$10.00\left(\dfrac{d_i}{x}\right)$
$\dfrac{2\delta}{d_i}$	This work	$16.0\left(\dfrac{x}{\mathrm{Re}_i\, d_i}\right)^{2/3}$	$0.394\left(\dfrac{x}{d_i}\right)$	$48.0\left(\dfrac{x}{\mathrm{Re}_i\, d_i}\right)$	$0.308\left(\dfrac{x}{d_i}\right)$
	Schlichting	$14.6\left(\dfrac{x}{\mathrm{Re}_i\, d_i}\right)^{2/3}$	$0.520\left(\dfrac{x}{d_i}\right)$	$18.5\left(\dfrac{x}{\mathrm{Re}_i\, d_i}\right)$	$0.100\left(\dfrac{x}{d_i}\right)$
$\dfrac{v_\delta}{u_i}$	This work	$\dfrac{0.333}{\mathrm{Re}_i}\left(\dfrac{\mathrm{Re}_i\, d_i}{x}\right)^{2/3}$	$0.079\left(\dfrac{d_i}{x}\right)^{1/2}$	$0.084\left(\dfrac{d_i}{x}\right)$	$0.084\left(\dfrac{d_i}{x}\right)$
	Schlichting	$\dfrac{0.550}{\mathrm{Re}_i}\left(\dfrac{\mathrm{Re}_i\, d_i}{x}\right)^{2/3}$	$0.156\left(\dfrac{d_i}{x}\right)^{1/2}$	$0.108\left(\dfrac{d_i}{x}\right)$	$0.032\left(\dfrac{d_i}{x}\right)$

Combustion of Gaseous Fuel Jets

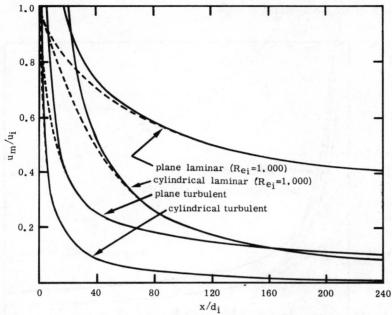

Figure 7.9 Decay of the maximum velocity along the axis of plane and cylindrical laminar and turbulent free inert jets

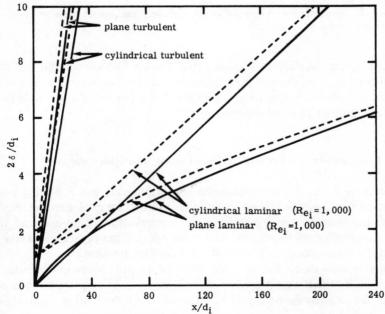

Figure 7.10 Thickness as a function of distance along the axis for plane and cylindrical laminar and turbulent free inert jets

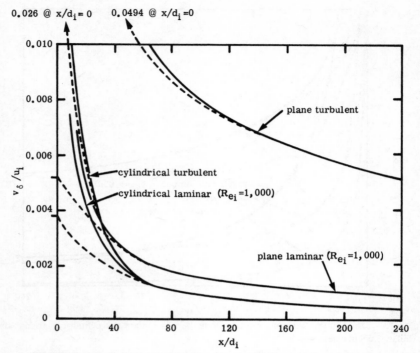

Figure 7.11 Entrainment as a function of distance along the axis for plane and cylindrical laminar and turbulent free inert jets

to their greater surface to volume ratio. Figure 7.10 shows that turbulent jets are fatter and shorter than laminar jets. Consequently, as shown by Figure 7.11, the turbulent jet entrainments are an order of magnitude larger than the laminar jet entrainments.

7.6 Combustion of a Free Jet of Fuel Issuing into Quiescent Air

In Sections 7.1 to 7.5 we considered jets of an inert gas F issuing into a quiescent gas O which extends to infinity. Mainly, we exploited the likeness of the momentum, energy and species conservation equations and their boundary conditions in order to extract the velocity, temperature and composition fields simultaneously when the diffusivities are equal.

Suppose now that a fuel gas (species F) burns as it issues in the form of a jet into air in which the oxygen concentration is $Y_{O\infty}$. The equations governing the jet flame phenomenon may be reduced to a form analogous to that of inert jet equations by defining composite variables according to Schvab–Zeldovich transformation. In this section, for the purpose of illustrating the solution technique, let us consider a cylindrical jet of fuel.

Combustion of a fuel jet is well-known to be a diffusion-limited process. The fuel in the jet and the oxygen in the room are transported towards one another by convective diffusion. At all those stations where the fuel and oxygen are in stoichiometric proportions, combustion takes place rapidly. In an ideal diffusion flame, the reaction zone is so thin that it could be considered as a surface of thickness zero which is impermeable to oxygen as well as the fuel. Consequently, oxygen is absent on the fuel side of the flame and the fuel is absent on the oxygen side.

In a nonaerated Bunsen flame, for example, the reaction zone is visibly blue due to energy emitted by the unstable intermediate chemical species.* For a given set of conditions including the initial chemical and thermodynamic properties, flow situation, and mixing details, the solution of a jet flame problem yields the flame size and shape along with the temperature and composition maps. Such information immediately yields the combustor space heating rate, materials requirements and combustion efficiency.

In our model, we ignore (a) radiative heat loss (b) body forces (i.e., gravitational, electrical, magnetic forces, etc.) and (c) property variation with temperature and composition. Each of these three assumptions is questionable to some degree. However, once the technique is illustrated, they may be relieved. Particularly, the assumption of constant properties deserves an elaboration here.

The occurrence of chemical reactions modifies the flow pattern quite strongly since the chemical heat release raises the reaction zone temperature to about 1,500 °K to 2,000 °K. Such an elevation in temperature (a) increases the mixture viscosity (refer to Chapter 3) and thus slows down the flow and (b) decreases the fluid density. The density decrement requires an increase in the flow velocity so that the prescribed mass flux is accommodated. In a rigorous sense, these two factors couple the equation of motion with the energy and species equations so that the solution is rather complicated. If the problem is two-dimensional, Howarth's transformation with certain restrictions on the transport properties may be used to separate the flow problem from the heat and mass transfer problems. If the problem is cylindrically symmetrical, Mangler's transformation may be used to convert it into two-dimensional form to which Howarth's transform may be applied.

---

* As the carbon formed in the combustion process flows out of the reaction zone, it is usually incandescent. Far out of the flame, the soot cools down to lose its incandescence. One must exercise caution not to mistake the yellowish incandescent zone as the (faintly bluish) "reaction zone."

In this section we ignore the influence of chemical reactions on the flow problem so that the momentum equation may be solved independently of the heat and mass transfer problem. In fact, with this supersimplification, the flow solution does not differ from that of the inert jet flow. Thus, for a cylindrical jet, Eqs. 7.37 and 7.40 give

$$\frac{u}{u_i} = \mathscr{F}(x, r) \tag{7.43}$$

where, if the jet is laminar,

$$\mathscr{F}(x, r) = \left[1 + \frac{48}{\text{Re}_i} \frac{x}{d_i}\right]^{-1} \left(1 - 2\frac{r}{d_i}\left[1 + \frac{48}{\text{Re}_i} \frac{x}{d_i}\right]^{-1}\right)$$

and if the jet is turbulent,

$$\mathscr{F}(x, r) = \left[1 + 24C'\frac{x}{d_i}\right]^{-1} \left(1 - 2\frac{r}{d_i}\left[1 + 24C'\frac{x}{d_i}\right]^{-1}\right)$$

Exact solution for this case when $x/d_i \gg 1$ is given by Schlichting's equation. Let subscripts F and O respectively stand for the fuel and oxygen. Let the subscript P stand for the inert products of reaction including the atmospheric nitrogen. Assuming $\alpha = \nu = D_O = D_F = D_P$ when the flow is laminar and $\varepsilon_t = \varepsilon = \varepsilon_D$ when the flow is turbulent, steady state constant property energy and species equations for the cylindrical jet are as following

$$ru\frac{\partial(T - T_\infty)}{\partial x} + rv\frac{\partial(T - T_\infty)}{\partial r} = (\nu + \varepsilon)\frac{\partial}{\partial r} r \frac{\partial(T - T_\infty)}{\partial r} + \frac{\dot{q}'''r}{\rho C} \tag{7.44}$$

$$ru\frac{\partial Y_j}{\partial x} + rv\frac{\partial Y_j}{\partial r} = (\nu + \varepsilon)\frac{\partial}{\partial r} r \frac{\partial Y_j}{\partial r} + \frac{\dot{W}'''_j r}{\rho} \tag{7.45}$$

Equation 7.45 is valid for all the three species in our system (i.e., j is F for the fuel, is O for oxygen and is P for the products). $\dot{q}'''$ and $\dot{W}'''_j$ are respectively the sources and sinks of energy (cal/cm^3/sec) and species j (gm/cm^3/sec). Let the chemical reaction be described by the following simple, single, overall, stoichiometric equation.

f gms of F + 1 gm of $O \to (1 + f)$ gms of P + $f\Delta H$ cals of heat (7.46)

f and ΔH are respectively the stoichiometric fuel/oxygen ratio and the heat combustion of the fuel (cal/gm). Equation 7.46 suggests that the sources and sinks in Eqs. 7.44 and 7.45 are mutually related as below.

$$-\frac{\dot{W}'''_F}{f} = -\dot{W}'''_O = +\frac{\dot{W}'''_P}{1 + f} = +\frac{\dot{q}'''}{f\Delta H} \tag{7.47}$$

Combustion of Gaseous Fuel Jets

Rearrangement of Eq. 7.47 yields the following five statements of conservation.

$$\frac{\dot{W}_F'''}{f} + \frac{\dot{W}_P'''}{1+f} = 0 \qquad \dot{W}_O''' + \frac{\dot{W}_P'''}{1+f} = 0$$

$$\frac{\dot{W}_F'''}{f} + \frac{\dot{q}'''}{f\Delta H} = 0 \qquad \dot{W}_O''' + \frac{\dot{q}'''}{f\Delta H} = 0$$

$$\frac{\dot{W}_F'''}{f} - \dot{W}_O''' = 0 \qquad (7.48)$$

Using Eq. 7.48, the source-sink terms of Eqs. 7.44 and 7.45 may be eliminated to obtain

$$ru\frac{\partial b}{\partial x} + rv\frac{\partial b}{\partial r} = (v + \varepsilon)\frac{\partial}{\partial r}r\frac{\partial b}{\partial r} \qquad (7.49)$$

where the variable b is defined in any of the following five ways on the basis of the five pairs of the conversation equations suggested by Eq. 7.48.

$$b_{FP} \equiv \left(Y_F + \frac{f}{1+f}Y_P\right) - \frac{f}{1+f}(1 - Y_{O\infty})$$

$$b_{OP} \equiv \left(Y_O + \frac{1}{1+f}Y_P\right) - \left(\frac{1 + fY_{O\infty}}{1+f}\right)$$

$$b_{FT} \equiv \left(\frac{C(T - T_\infty)}{\Delta H} + Y_F\right) \qquad (7.50)$$

$$b_{OT} \equiv \left(\frac{C(T - T_\infty)}{f\Delta H} + Y_O\right) - Y_{O\infty}$$

$$b_{FO} \equiv \left(Y_O - \frac{Y_F}{f}\right) - Y_{O\infty}$$

The boundary and initial conditions of Eq. 7.49 are:

$$\begin{array}{llllr}
x = 0, & 0 \le r \le d_i/2, & b = b_i & & \\
x = \infty, & 0 \le r \le \infty, & b = 0, & \partial b/\partial r = 0 & \\
r = 0, & 0 \le x \le \infty, & v = 0, & \partial b/\partial r = 0 & (7.51) \\
r \ge \delta, & 0 \le x \le \infty, & b = 0, & \partial b/\partial r = 0 &
\end{array}$$

Equation 7.49 is similar to Eq. 7.31. The boundary conditions are also similar. Hence the solution of Eq. 7.49 with Eq. 7.51 is automatically given by Eq. 7.37 (laminar jet) or Eq. 7.40 (turbulent jet).

$$\frac{b}{b_i} = \mathscr{F}(x, r) \tag{7.52}$$

where $\mathscr{F}(x, r)$ is given by Eq. 7.43.

(a) Flame Shape

The infinitely fast combustion reaction occurs at all those stations where the fuel and oxygen meet in stoichiometric proportions to readily consume one another. Hence, at the flame surface both the fuel and oxygen are totally absent.* Using this criterion, b_{FO} at the flame surface is equal to $-Y_{O\infty}$. It is equal to $-(1/f + Y_{O\infty})$ at the nozzle. Substituting these two values in Eq. 7.52, and denoting the flame location by the subscript C,

$$\frac{Y_{O\infty}}{Y_{O\infty} + \dfrac{1}{f}} = \mathscr{F}(x_C, r_C) \tag{7.53}$$

Equation 7.53 implicitly gives the flame shape. The iso-velocity contours (qualitatively similar to those shown in Figure 7.6) indeed represent the shape of the flame provided the parameter on the contours is $fY_{O\infty}/(fY_{O\infty} + 1)$ ($= b_C/b_i$). For example, if carbon monoxide ($f = 28/16$) were to burn in a room of pure oxygen ($Y_{O\infty} = 1$), the parameter is $1/(1 + 16/28) = 0.635$; the flame then is expected to fall on the contour of 0.635 iso-velocity. If the CO were to burn in normal air, $Y_{O\infty} \approx 0.232$, the flame lies on the contour of 0.289 iso-velocity. It is clear from Fig. 7.6 that the flame in air will be much fatter and longer than in pure oxygen.

For a given flow through a given nozzle, the farthest distance a flame would stretch $\bar{x}_C$ is known as the flame height. It is obtained by setting $r_C = 0$ in Eq. 7.53. For a laminar cylindrical jet flame, $\bar{x}_C$ is given by

$$\frac{fY_{O\infty}}{fY_{O\infty} + 1} = \left(1 + \frac{48}{\mathrm{Re}_i} \frac{\bar{x}_C}{d_i}\right)^{-1} \tag{7.54}$$

For a turbulent cylindrical jet flame,

$$\frac{fY_{O\infty}}{fY_{O\infty} + 1} = \left(1 + 24C' \frac{\bar{x}_C}{d_i}\right)^{-1} \tag{7.55}$$

---

* The same conclusion may be drawn by the following alternative logic. The fuel is absent on one side of the reaction zone and the oxygen is absent on the other side. Therefore at the flame, both the species must be absent.

Combustion of Gaseous Fuel Jets

Resolving these two equations for the flame height, for a laminar flame,

$$\bar{x}_C = \frac{u_i d_i^2}{48 v f Y_{O\infty}} \quad (7.56)$$

and for a turbulent flame,

$$\bar{x}_C = \frac{d_i}{24 C' f Y_{O\infty}} \quad (7.57)$$

These two equations indicate several important characteristics of jet flames.

(i) The height of a laminar flame is directly proportional to the volumetric fuel feed rate $\pi d_i^2 u_i/4$, inversely proportional to the diffusion coefficient D, stoichiometric factor f and air purity factor $Y_{O\infty}$. Figure 7.12 shows $\bar{x}_C$ as a function of the fuel feed rate for various values of $vfY_{O\infty}$. For most hydrocarbon fuels burning in air the parameter $vfY_{O\infty}$ is approximately 0.015 cm²/sec. If the burning were in pure oxygen, $vfY_{O\infty} \approx 0.06$ cm²/sec.

(ii) The height of a turbulent flame is independent of the volumetric fuel feed rate but is linearly proportional to the nozzle diameter and

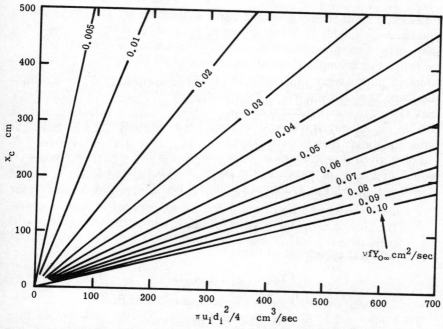

Figure 7.12 Height of a laminar cylindrical jet flame as a function of the fuel feed rate and $vfY_{O\infty}$.

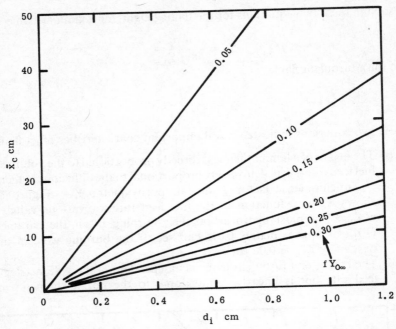

Figure 7.13 Height of a turbulent cylindrical jet flame as a function of the port diameter and $fY_{O\infty}$

inversely proportional to the eddy diffusion constant C', the stoichiometric factor f and the air purity factor $Y_{O\infty}$. Figure 7.13 shows $\bar{x}_c$ as a function of d_i for various values of $fY_{O\infty}$ with a (constant) mixing constant C' of 0.0128. It is clear from Figs. 7.12 and 7.13 that the height of a turbulent flame is much smaller than that of a laminar one.

(iii) Jost's derivation of Eq. 7.56 explicitly brings into light the fundamental characteristics of the laminar flame. Suppose that a fuel is issuing through a nozzle of diameter d_i at a velocity u_i. The time it takes for the fuel to reach the tip of the flame is proportional to $\bar{x}_c/u_i$ seconds. The time it takes for oxygen to diffuse across the jet to the axis is proportional to d_i^2/D. Equating these two times and resolving for the flame height,

$$\bar{x}_c \propto u_i d_i^2/D$$

This relation is in excellent agreement with our Eq. 7.56.

(iv) Equation 7.57 may be deduced from Eq. 7.56 by the following procedure which shows certain inherent synonymity between the laminar and turbulent flames. The laminar and turbulent jet equations of conservation are analogous except for the replacement of v by ε. Hence Eq. 7.56 must be valid for a turbulent flame if ε is substituted in place of v. It is known in the literature that the jet turbulent viscosity ε is proportional to the product of the jet width and the maximum x-component of velocity. This product itself is invariant with distance for cylindrical turbulent jets as indicated by the following empirical expression

$$\varepsilon/v \approx 0.01\,\mathrm{Re}_i$$

Hence, $\varepsilon \approx 0.01 u_i d_i$. Substituting this value of ε in place of v in Eq. 7.56 yields the turbulent flame height as

$$\bar{x}_C \approx \frac{d_i}{0.48 f Y_{O\infty}}$$

The excellent agreement between this result and Eq. 7.57 is an evidence of the consistency in the underlying physical arguments.

(v) Lastly, a synthesis of the preceding observations on the flame heights yields a relation between $\bar{x}_C/d_i$ and Re_i as qualitatively shown in Figure 7.14. For a given $f Y_{O\infty}$, the $\bar{x}_C/d_i$ increases linearly with Re_i in the laminar regime and remains independent of Re_i in the turbulent regime. Increased $f Y_{O\infty}$ has a definite tendency to shorten the flame for all Re_i.

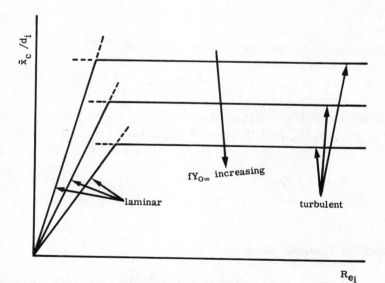

Figure 7.14 Synthesis of laminar and turbulent cylindrical jet flame heights (theory)

(b) Composition Profiles Across the Flame

In cross sections beyond $\bar{x}_c$, the composition is described by the mixing of the inert products with air. The profiles in this region may be determined by the inert jet equations of Section 7.5 with the initial conditions prescribed at $x = \bar{x}_c$.

In cross sections within $x < \bar{x}_c$, the presence of the flame slightly complicates the profiles. The technique of determining these profiles is discussed below.

Consider the variation of b_{FO} (defined by Eq. 7.50 as $(Y_O - Y_F/f) - Y_{O\infty}$). At the port, since $Y_O = 0$ and $Y_F = 1$, $b_{FOi} = -(Y_{O\infty} + 1/f)$. Because within the flame at any cross section (i.e., in $0 \leq |r| \leq r_C$ at any x) no oxygen is present, Eq. 7.52 gives

$$\frac{Y_F + fY_{O\infty}}{1 + fY_{O\infty}} = \mathscr{F}(x, r) \tag{7.58}$$

where $\mathscr{F}(x, r)$ is defined by Eq. 7.43 or its equivalent. Solving Eq. 7.58 for Y_F,

$$Y_F = (1 + fY_{O\infty})\mathscr{F}(x, r) - fY_{O\infty}; \qquad 0 \leq |r| \leq r_C \tag{7.59}$$

This equation gives the fuel mass fraction profile. Since outside the flame (i.e., $r_C \leq |r| \leq \delta$) no fuel is present, Eq. 7.52 becomes

$$\frac{Y_O - Y_{O\infty}}{-(1 + fY_{O\infty})/f} = \mathscr{F}(x, r)$$

Rearranging, the oxygen mass fraction profile is given by

$$Y_O = Y_{O\infty} - \left(\frac{1 + fY_{O\infty}}{f}\right)\mathscr{F}(x, r); \qquad r_C \leq |r| \leq \delta \tag{7.60}$$

Consider now the profile of b_{FP} (defined as $(Y_F + fY_P/(1 + f)) - (1 - Y_{O\infty})f/(1 + f)$). b_{FP} at the port is equal to $(1 + fY_{O\infty})/(1 + f)$. Hence Eq. 7.52 gives

$$\frac{Y_F + \dfrac{fY_P}{1 + f} - \dfrac{f(1 - Y_{O\infty})}{1 + f}}{\dfrac{(1 + fY_{O\infty})}{1 + f}} = \mathscr{F}(x, r)$$

Solving for Y_P outside the flame ($Y_F = 0$)

$$Y_P = \left(\frac{1 + fY_{O\infty}}{f}\right)\mathscr{F}(x, r) + (1 - Y_{O\infty}); \qquad r_C \leq |r| \leq \delta \tag{7.61}$$

Combustion of Gaseous Fuel Jets

Similarly considering b_{OP} whose value at the port is $-(1 + fY_{O\infty})/(1 + f)$,

$$\frac{Y_O + \dfrac{Y_P}{1+f} - \dfrac{(1+fY_{O\infty})}{1+f}}{-\dfrac{(1+fY_{O\infty})}{1+f}} = \mathscr{F}(x, r)$$

Within the flame, $Y_O = 0$ so that

$$Y_P = (1 + fY_{O\infty})(1 - \mathscr{F}(x, r)); \qquad 0 \leq |r| \leq r_C \quad (7.62)$$

Thus, Eqs. 7.59, 7.60, 7.61 and 7.62 give the profiles of the three species at any cross section. Figure 7.15 shows the underlying concept of the preceding profile derivation. Note from Eqs. 7.59 and 7.62 that the sum $Y_F + Y_P$ is equal to unity within the flame and from Eqs. 7.60 and 7.61 that the sum $Y_O + Y_P$ is equal to unity outside the flame. Furthermore, at the flame, $Y_{PC} = 1$.

(c) Temperature Profile Across the Flame

In cross sections $x > \bar{x}_C$ the temperature profile may be determined by the inert thermal jet equations to follow the gradual decay due to mixing. In cross sections within $x < \bar{x}_C$, the profile may be obtained in a manner similar to the product profile.

Within $0 \leq |r| \leq r_C$ at any section consider the profile of b_{OT} whose value at the port is $C(T_i - T_\infty)/f\Delta H - Y_{O\infty}$. According to Eq. 7.52,

$$\frac{\dfrac{C(T - T_\infty)}{f\Delta H} + Y_O - Y_{O\infty}}{\dfrac{C(T_i - T_\infty)}{f\Delta H} - Y_{O\infty}} = \mathscr{F}(x, r)$$

But within the flame, $Y_O = 0$. Hence, resolving for $(T - T_\infty)$,

$$(T - T_\infty) = \left[(T_i - T_\infty) - \frac{Y_{O\infty} f \Delta H}{C}\right]\mathscr{F}(x, r) + \frac{Y_{O\infty} f \Delta H}{C}; \qquad 0 \leq |r| \leq r_C \quad (7.63)$$

Outside the flame (i.e., $r_C \leq |r| \leq \delta$), $Y_F = 0$. Therefore, from the definition of b_{FT},

$$\frac{\dfrac{C(T - T_\infty)}{\Delta H} + Y_F}{\dfrac{C(T_i - T_\infty)}{\Delta H} + 1} = \mathscr{F}(x, r)$$

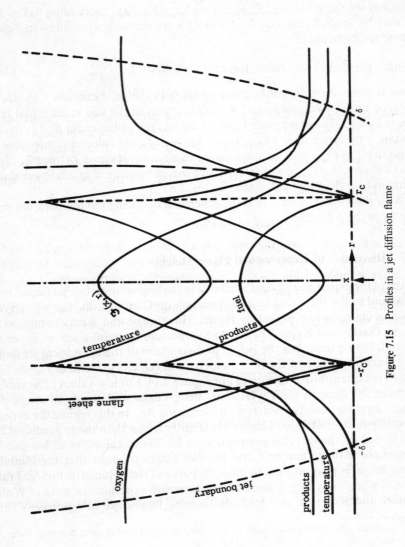

Figure 7.15 Profiles in a jet diffusion flame

Combustion of Gaseous Fuel Jets

Rearranging for $(T - T_\infty)$,

$$(T - T_\infty) = [(T_i - T_\infty) + \Delta H/C]\mathscr{F}(x, r); \qquad r_c \leq |r| \leq \delta \tag{7.64}$$

Since at the flame surface, $\mathscr{F} = \mathscr{F}_c = fY_{0\infty}/(1 + fY_{0\infty})$ according to Eq. 7.53 it may be verified that Eqs. 7.64 and 7.63 are mutually consistent to yield a flame temperature of

$$(T_c - T_\infty) = [(T_i - T_\infty) + \Delta H/C]\frac{fY_{0\infty}}{1 + fY_{0\infty}} \tag{7.65}$$

This is the maximum possible temperature in the jet. Equation 7.65 has an interesting physical meaning.* When $fY_{0\infty}$ gms of the fuel at an initial temperature of T_i is burned with 1 gm. of the oxygen in the room at T_∞, the resultant enthalpy release $fY_{0\infty} C(T_i - T_\infty) + fY_{0\infty}\Delta H$ calories is adequate to raise the $1 + fY_{0\infty}$ gms of the mixture to a temperature of T_c from T_∞. In a real diffusion flame, however, radiative losses result in a much lower flame temperature than that calculated by Eq. 7.65.

Also shown in Figure 7.15 is the scheme of deducing the temperature profile from the b-profiles.

7.7 Discussion of Experimental Flame Heights

Most of the experimental data available in the literature on free jet flames may be found in the third and fourth Symposia on Combustion. The third Symposium contains two papers by Hottel, Hawthorne and Weddel and one by Wohl, Gazley and Kapp. The fourth Symposium contains a survey paper by Hottel. Hottel describes the height and structure of the cylindrical jet flame as a function of the fuel, the port diameter and velocity in terms of Figure 7.16 which is a quantitative version of our Figure 7.14. For low values of Reynolds number the flame is clearly laminar, sharp-edged and steady in shape. Its height increases nearly linearly with increasing Re_i. In this regime the experimentally measured flame heights are slightly larger than those predicted by the forced jet theory; this is surmised to be due to the effect of buoyancy. Hottel correlated Rembert's and Gaunce's data to show that the laminar flame height is proportional to the *square root* of the volumetric fuel feed rate $\pi d_i^2 u_i/4$ when buoyancy is present. The extensive experimental data of Wohl, Gazley and Kapp on city gas† and butane flames are in agreement with

---

* A similar observation was made in the study of liquid fuel droplet burning. Refer to Eq. 5.66.

† Approximate composition of city gas by volume is given by Wohl et al. as following.

H_2	35.3%	$C_{2.5}H_{4.2}$	5.5%	CO_2	4.9%
CH_4	15.0%	CO	11.9%	N_2	21.2%
C_2H_6	5.4%	O_2			

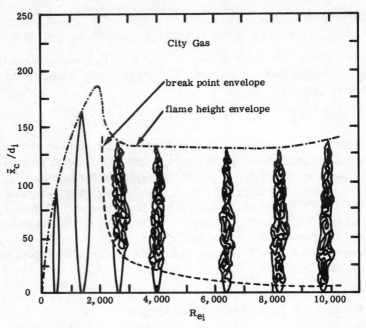

Figure 7.16 Synthesis of laminar and turbulent cylindrical jet flame heights (experimental) (adapted with permission of The Combustion Institute, from: H. C. Hottel and W. R. Hawthorne, *3rd International Symposium on Combustion*, p. 255, (1949))

Hottel's correlation. Wohl's data on butane flames fit the following formula in cm-sec units

$$\bar{x}_C \approx 8.55(u_i d_i^2)^{1/2} \tag{7.66}$$

When the nozzle Reynolds number is about 2,000, the tip of the flame becomes wrinkled due to the onset of turbulence. With further increase in Re_i, the flame height goes through a maximum while the turbulence propagates rapidly down from the tip towards the nozzle.

The transition from laminar to turbulent flame occurs at some critical Reynolds number in the range of 2,000 to 10,000. The reason for such a wide range of transition Reynolds number is the strong dependency of gas viscosity on temperature. Flames whose adiabatic temperatures are relatively high are expected to become turbulent at relatively higher nozzle Reynolds numbers. Conversely, flames whose adiabatic temperatures are relatively low are susceptible to become turbulent relatively sooner. Table 7.2 shows the critical Re_i for various fuels burning in air as reported by Hottel and Hawthorne.

For higher Reynolds numbers, nearly the entire flame is broken into tur-

TABLE 7.2
Critical Port Reynolds Numbers for
Various Flames in Air

Fuel	$Re_{critical}$
Hydrogen	2,000
City gas	3,300– 3,800
Carbon monoxide	4,800– 5,000
Propane	8,800–11,000
Acetylene	8,800–11,000
Hydrogen (with primary air)	5,500– 8,500
City gas (with primary air)	6,400– 9,200

With the permission of the Combustion Institute, from: H. C. Hottel, p. 254, Third International Symposium on Combustion (1949).

bulence, is less luminous and is constant in height. At Re_i about 10,000, the flame starts to lift away from the nozzle. Similar behavior has been noted by other investigators, notably by Wohl et al. and Yagi.

Turbulent flame height is observed to be independent of the fuel volumetric feed rate but linearly dependent upon the nozzle diameter. This is consistent with our theory (Eq. 7.57). Neglecting buoyancy (i.e., for a high velocity fuel jet issuing through a fine nozzle) Hawthorne derives the following formula which takes density differences into account for the turbulent flame height.

$$\frac{\bar{x}_C}{d_i} = 5.3\left[\frac{T_f}{(1+f)T_i}\left(\frac{M_P}{M_F} + \frac{M_P}{fM_O}\right)\right]^{1/2} \tag{7.67}$$

T_f is the adiabatic flame temperature and M is the molecular weight. This equation is proved to be valid for a wide variety of fuels (burning in air) including propane (pure as well as mixed with hydrogen), hydrogen, acetylene, carbon monoxide and city gas (pure as well as mixed with CO_2). For city gas, Wohl determined $\bar{x}_C/d_i$ to be about 110.

Yagi and Saji, in an excellent paper published in the fourth Symposium on Combustion, reported experiments which substantiated their contention that a fully turbulent jet of pulverized coal may be treated as though it were a gaseous fuel jet.

7.8 Some General Comments

Let us pause here briefly to review the understanding of the jet flames we acquired thus far.

(a) On the Technique of Solution

When the pressure gradients do not complicate the flow problem we have discovered that the governing equations and their boundary conditions for flow, heat and mass transfer (with as well as without combustion) may be reduced to a single form so that when the diffusivities are equal and when the reactions (if any) are infinitely fast, the profiles of all the following variables overlap.

$$\frac{u}{u_i}, \frac{(T-T_\infty)}{(T_i-T_\infty)}, Y_F, (1-Y_O) \quad \text{when combustion is absent;}$$

$$\frac{u}{u_i}, \frac{b_{FO}}{b_{FOi}}, \frac{b_{FP}}{b_{FPi}}, \frac{b_{OP}}{b_{OPi}}, \frac{b_{FT}}{b_{FTi}}, \frac{b_{OT}}{b_{OTi}} \quad \text{when combustion is present.}$$

Furthermore, if the influence of the reaction on the flow is neglected, u/u_i profiles remain unaltered with or without combustion. Thus the solution of the incompressible flow is adequate to deduce the temperature and composition profiles in any jet whether burning is present or not. The flame shape is then imminent. In fact, measurement of the distribution of any single conserved variable in a jet is adequate to fully describe the combustion of a similar jet. This realization indeed simplifies study of jet flames in complex flow situations. For instance, consider a jet of fuel burning in a feeble flow of air shown in Figure 7.17. This situation differs from that dealt with in the present chapter so far by the fact that the jet is no longer issuing into a quiescent air. If one produces a geometrically similar isothermal jet of air issuing into an air flow and measures the velocity field in the jet, such measurements immediately yield the combustion characteristics of a similar jet of fuel. Alternatively, one might choose to map out the temperature field in a similar hot jet and use the isothermal contours to locate the flame in a jet of fuel.

(b) Real Jet Flames

To summarize the assumptions made in our study of gaseous fuel jet diffusion flames, the following are the crucial ones.

(i) Steady state exists.
(ii) The jet issues into an infinite, boundless medium. (i.e., it is a free jet.)
(iii) Radiative heat transfer is negligible.
(iv) Properties remain independent of temperature and composition. Flow is incompressible.
(v) $v = \alpha = D_j$ for laminar flames.
(vi) $\varepsilon = \varepsilon_t = \varepsilon_D \gg v$ for turbulent flames.

Combustion of Gaseous Fuel Jets

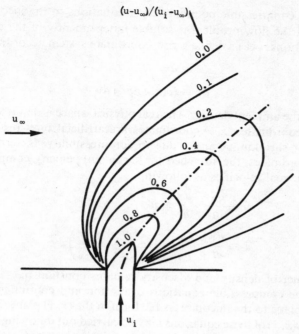

Figure 7.17 Free transverse jet

(vii) The combustion reaction is simple, extremely fast and may be expressed by a single gross stoichiometric relation.

(viii) Buoyancy forces are negligible in comparison with the initial inertial forces of the jet.

Let us discuss these assumptions briefly. The steady state assumption needs no comment. Jets in real life applications in furnaces and engines are confined in a container which may either be open (such as a turbojet engine) or closed (such as a power plant furnace). The problem of confined jets is the topic proper for Section 7.9.

Neglect of the radiative heat losses results in a higher flame temperature. However, inclusion of the radiative term in the energy conservation equation complicates it to such an extent that our spirit of simplicity is marred.

The constant property incompressible assumption is easily relaxed for the case of a two-dimensional jet by Howarth's transformation. When the flow is compressible, since the fluid density is a function of temperature and composition, the equation of motion is coupled with the energy and species equations. The solution of these coupled, simultaneous, nonlinear, partial differential equations becomes a formidable task. Howarth's transformation converts the compressible boundary layer equations to the incompressible form so that the flow problem is soluble independently of the energy and species equations. Let us define a new coordinate system according to

$$d\tilde{x} \equiv dx$$
$$d\tilde{y} \equiv (\rho(x, y)/\rho_\infty)\, dy \tag{7.68}$$

where ρ_∞ is the ambient density. The real physical space is described in terms of the x, y coordinates. $\tilde{x}, \tilde{y}$ coordinates are artificial ones resulting from stretching or shrinking x and y due to compressibility. In terms of these artificial coordinates, the stretched (or shrunken) velocity components are related to the real velocities as following

$$\tilde{u} \equiv \frac{d\tilde{x}}{dt} = \frac{dx}{dt} \equiv u$$
$$\tilde{v} \equiv \frac{d\tilde{y}}{dt} = \frac{\rho}{\rho_\infty} v + u \int_0^y \frac{\partial}{\partial x}\left(\frac{\rho}{\rho_\infty}\right) dy \tag{7.69}$$

If the product of density and viscosity remains constant (i.e., $\rho\mu \equiv \rho^2 v =$ constant) the compressible equations of motion and continuity transform from (x, y) plane to the imcompressible form in the $(\tilde{x}, \tilde{y})$ plane. Solution for $\tilde{u}(\tilde{x}, \tilde{y})$ and $\tilde{v}(\tilde{x}, \tilde{y})$ of these equations may be carried out following our already familiar methods. The resultant $\tilde{u}$ and $\tilde{v}$ are then inverted to the u and v components in the (x, y) plane by Eqs. 7.68 and 7.69.

The requirement of $\rho\mu =$ constant needs a further comment here. According to Chapter 3, (Eq. 3.18) the viscosity depends upon temperature and molecular weight according to

$$\mu \propto (TM)^{1/2}$$

Experimental measurements show that $\mu \propto T^m$ where m is somewhere between 0.7 and 0.85. Since density is proportional to (M/T), if the molecular weights of various species in the problem do not differ drastically or if the composition gradients are minor, the product $\rho\mu =$ constant is a tolerable approximation to reality.

The energy and species equations may similarly be reduced to the incompressible form by assuming that the specific heat is constant (and equal to some mean value) and that $\rho^2 \alpha$ is equal to $\rho_\infty^2 \alpha_\infty$, a constant and $\rho^2 D_j$ is equal to $\rho_\infty^2 D_{j\infty}^2$, another constant.

Assumption (v) is often an unavoidable one since the accuracy of the $\Pr \neq \mathrm{Sc} \neq \mathrm{Le}$ solution is not usually worth the extra effort. Chapter 3 enables us to estimate the orders of magnitude of these dimensionless transport properties. Some explicit values are presented in Tables 7.3 and 7.4. $\Pr = \mathrm{Sc} = \mathrm{Le}$ is certainly not the most serious assumption of our analyses.

TABLE 7.3

Pr, Sc and Le for Various Gases at $T = 0\,°C$ and $P = 1$ atm

Gas	Sc	Pr	Le
Argon	0.75	0.67	1.12
Nitrogen	0.74	0.71	1.04
Methane	0.70	0.74	0.95
Oxygen	0.74	0.72	1.03
Carbon Dioxide	0.71	0.75	0.95
Hydrogen	0.73	0.71	1.03
Simple kinetic theory	0.83	0.67	1.25

TABLE 7.4

Temperature Dependence of Pr, Le and Sc of Air and O_2 at $P = 1$ atm

$T\,°K$	Sc		Pr		Le	
	air	O_2	air	O_2	air	O_2
300	0.810	0.740	0.710	0.720	1.140	1.030
350	0.841	0.810	0.700	0.695	1.200	1.160
400	0.870	0.840	0.692	0.680	1.260	1.240
450	0.894	0.860	0.680	0.680	1.310	1.270
500	0.910	0.880	0.680	0.682	1.340	1.290
550	0.925	0.890	0.677	0.691	1.370	1.290
600	0.945	0.900	0.678	0.710	1.400	1.270
700	0.960	—	0.682	—	1.410	—
800	0.980	—	0.690	—	1.420	—
900	0.985	—	0.695	—	1.420	—
1,000	0.990	—	0.698	—	1.420	—

Assumption (vi) is found by experience to be quite acceptable.

As to assumption (vii), jet flames are known to be diffusionally controlled to the extent that chemical reactions are usually much faster than diffusion and conduction. However, no flames are "ideally" diffusional since no chemical reactions are, in reality, "infinitely" fast.* Finite rate chemical kinetics produce departures in the behavior of a diffusion flame from that predicted by "ideal" theories. These departures include the following aspects.

Real diffusion flames are not infinitesimally thin surfaces since finite rate reactions need finite volumes to occur in. Such a finite space requirement

---

* For this reason, "ideal" diffusion flames are only theoretically possible. The ideal analyses have hence to be tempered ultimately for the non-infinite reaction rates.

causes the simultaneous presence of oxygen and fuel in the reaction space as shown in Figure 7.18. The temperature profiles consequently are flattened whereas the maximum temperature is lowered.

The isotherms in an ideal and a real flame of a hypothetical fuel jet burning in a hypothetical oxidant are shown in Figure 7.19. This figure clearly demonstrates the lowering of the flame temperature and broadening of its thickness due to finite rate kinetics.

For a precise prediction of the shape of a real flame one needs to take buoyancy into account. This is particularly important in upward free jet flames. In industrial applications, however, the initial momenta are so large that neglect of buoyancy is not so serious.

Two auxiliary comments are warranted here. Firstly, our theory of turbulent jet diffusion flames is based upon time-averaged profiles of velocity, temperature and composition. In reality, the random movement of eddy pockets imposes random fluctuations on these time-averaged profiles. Accordingly, the gas mixture at the boundaries of the flame contains pockets (or islets) of fuel which jump back and forth creating a poorly defined brush of a reaction zone. That is, the boundaries of a turbulent flame are not fixed in space and time (as are the laminar flame boundaries) but are strained, folded, wrinkled, convoluted and broken. These convolutions occur so randomly and

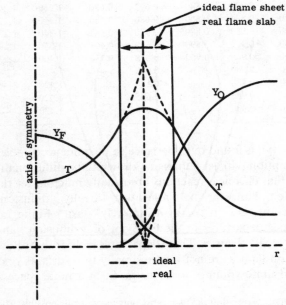

Figure 7.18 Ideal and real temperature and composition profiles in a diffusion flame

Combustion of Gaseous Fuel Jets

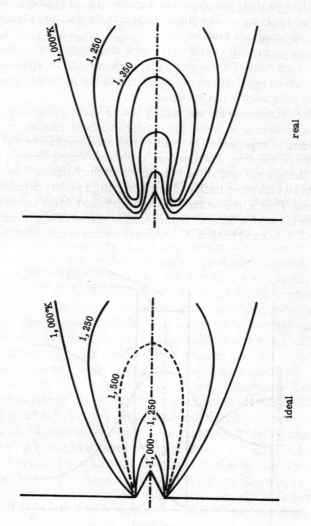

Figure 7.19 Ideal and real isothermal contours in a free jet flame

rapidly that the flame seems bulkier and longer to an observers' eye. Due partly to this reason, Hottel and Hawthorne's visually measured turbulent flame heights are larger than those predicted by the time-mean stoichiometric contour.

Lastly, Howarth's transformation works only for a plane flow problem. At present, no transform is available to reduce compressible cylindrical jet problem to incompressible form.* Our incompressible treatment, nevertheless, gives reasonable trends.

7.9 Confined Flames

Jet flames in gas turbines, ram jets and industrial furnaces are confined within a finite volume combustion chamber, the air in which is neither pure not quiescent. The facts that the extent of the oxidizer is finite and bound by solid surfaces and that the chemical heat release introduces pressure gradients, exceedingly complicate the solution of the flow problem. Some simple geometries are considered theoretically by Burke and Schumann and Landis and Shapiro. We will discuss these theories in this section. Due to the complexity of the problem of combustion of jets in a container, much of our present understanding comes from empirical studies such as those of Thring and Newby (flames in a furnace), Forstall and Shapiro (combustion in a duct), Hawthorne, Weddel and Hottel (turbulent jet in a blast furnace), Sawai and his associates (duct combustion), etc.

Confined jet flames may be broadly classified into two groups—the transverse (or normal) group in which the jet of fuel is injected in a direction perpendicular to the air flow and the longitudinal group in which the jet is injected in a direction parallel to the air flow. Both the longitudinal and normal jets are important in the interest of a combustor designer.

Studies available in the literature on normal jets are mostly experimental. Hottel, Hawthorne and Weddel found that the pattern of a flame of a horizontal hydrogen jet into a blast furnace is greatly influenced by the confining walls. It is conceivable that the single most important factor governing the behavior of a normal jet is the magnitude of the main flow velocity u_∞ relative to the jet port velocity v_i. If $\dot{W}_F''$ is the mass flux $(\rho_F v_i)$ of the fuel gas entering perpendicularly upwards into an air flow of $\dot{W}_a''(= \rho_\infty u_\infty)$ through an orifice of diameter d_p in the wall of a horizontal duct, shown in Figure 7.20,

---

* As hinted earlier, one may find a tricky way out of this dilemma by transforming the cylindrically symmetric flow problem to a plane problem by Mangler's transformation and then accounting for compressibility by Howarth's transformation. Space does not permit further discussion here. Interested reader is referred to Schlichting.

Combustion of Gaseous Fuel Jets

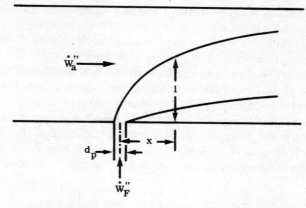

Figure 7.20 Confined jet flame in a transverse flow

the upper edge of the jet, l, as a function of x, the downstream distance, is given by the following equation.

$$\left(\frac{l}{d_p}\right)^{1.65} = 2.91\left(\frac{\dot{W}_F''}{\dot{W}_a''}\right)\left(\frac{x}{d_p}\right)^{0.5} \tag{7.74}$$

7.10 Longitudinal Confined Jet Flames

Much of the research on confined flames happens to be concerned with longitudinal ones. The reasons for this are multifarious. Primarily, the mode of fuel injection into gas turbines and ramjets (see Bevan's work) results in a longitudinally confined jet flame.

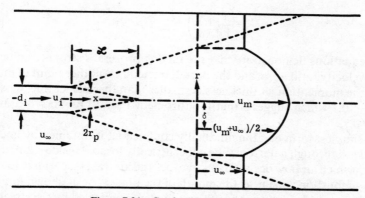

Figure 7.21 Confined longitudinal jet

As we saw earlier, study of mixing, heat and momentum transfer in inert jets yields quite a bit of information pertinent to the combustion behavior of a similar jet. Landis and Shapiro (and Forstall and Shapiro) investigated the turbulent mixing of a small high velocity heated jet issuing into a relatively slower cooler parallel flow in a cylindrical duct. Their model is schematically shown in Figure 7.21 in order to establish the nomenclature. Let the subscripts i, ∞ and m respectively stand for the conditions at the port, in the main flow and on the axis of symmetry. Let $\mathscr{L}$ be the length of the potential core and r_p be the radius of the potential core at any x less than $\mathscr{L}$. Let δ (and δ_t) be the radius at which the decay of velocity (and temperature) is half complete. Then Shapiro's observations may be summarized by the following equations.

$$\left(\frac{u - u_\infty}{u_m - u_\infty}\right) = \frac{1}{2}\left(1 + \cos\frac{\pi r}{2\delta}\right) \tag{7.75}$$

$$\left(\frac{u_m - u_\infty}{u_i - u_\infty}\right) = \left(\frac{\mathscr{L}}{x}\right) \tag{7.76}$$

$$\left(\frac{T - T_\infty}{T_m - T_\infty}\right) = \frac{1}{2}\left(1 + \cos\frac{\pi r}{2\delta_t}\right) \tag{7.77}$$

$$\frac{r_p}{d_i} = \frac{1}{2}\left(1 + 3\frac{u_\infty}{u_i} - \frac{1}{4}\frac{x}{d_i}\right); \quad 0 \leq x \leq \mathscr{L} \tag{7.78}$$

$$\frac{\mathscr{L}}{d_i} = 4\left(1 + 3\frac{u_\infty}{u_i}\right) \tag{7.79}$$

$$\frac{\delta}{d_i} = \frac{1}{2}\left(\frac{\mathscr{L}}{x}\right)^{(u_\infty - u_i)/u_\infty}; \quad x \geq \mathscr{L} \tag{7.80}$$

$$\frac{\delta_t}{d_i} = \frac{2}{1.4}\left(\frac{\delta}{d_i}\right); \quad x \geq \mathscr{L} \tag{7.81}$$

These equations demonstrate that the mixing process is strongly influenced by the velocity ratio u_∞/u_i and the port diameter d_i. Another point of interest is that the momentum jet thickness is smaller than thermal jet thickness by a factor of 0.7 which suggests that the turbulent Prandtl number is approximately 0.7.

The simplest form of a longitudinally confined jet is obtained by injecting the fuel gas through a narrow tube concentrically located inside a larger tube through which air flow occurs. Thus, the fuel and air travel separately to arrive at the combustion zone near the mouth of the inner tube as shown in Figure 7.22. The sort of flow expected in this geometry is indicated by the streamline

Combustion of Gaseous Fuel Jets

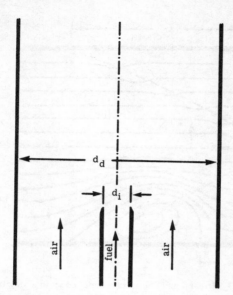

Figure 7.22 Apparatus to produce a confined jet flame in a longitudinal flow

pattern qualitatively sketched in Figure 7.23. The flow is, naturally, symmetrical around the axis of the tubes. For the sake of brevity, two possible streamline patterns are shown in the two halves of Figure 7.23. When the fuel jet issues at a low velocity relative to the air flow, the streamlines are expected to be deflected only slightly as indicated in the upper half of the figure. Thring and Newby give the following approximate relation between x_2 (the distance downstream from the nozzle at which the jet hits the wall) and the duct diameter d_d.

$$x_2 \approx 2.25 \, d_d$$

When the fuel jet issues at a relatively high velocity, the entrainment will be so intense that once all the air flow ($\dot{W}''_a$) is entrained, a recirculation vortex is generated in the downstream at a distance x_1 from the fuel nozzle as shown in the lower half of the figure. Thring and Newby provide an approximate relationship between x_1, nozzle size and the fuel and air mass flow rates as below.

$$x_1 \approx 2.5 \, d_i(1 + \dot{W}''_a / \dot{W}''_F)$$

Burke and Schumann considered the noncirculating type longitudinal jet flame with the following assumptions which are already familiar to us.

(i) Reaction takes place in a sheet of space whose thickness is zero. If air is available in excess of the stoichiometric requirement, the flame sheet

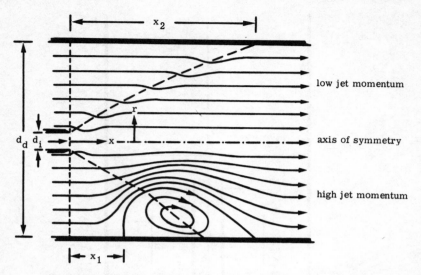

Figure 7.23 Streamline patterns of a confined jet in a longitudinal flow

forms (roughly) a cone which covers the fuel supply. If the air supply is deficient of the stoichiometric requirement the flame sheet forms an inverted truncated cone which covers the air supply.
(ii) The reaction zone is defined as that location where the air and the fuel are in stoichiometric proportions.
(iii) Diffusion alone (and not the details of the chemistry) is important in determining the shape and size of the flame. Diffusion coefficient is independent of temperature and composition.
(iv) Expansion of gases due to heating and the product gas behavior are disregardable.
(v) Velocities of the fuel and air streams are equal. (This is experimentally accomplished by using fuel and air mass flows proportional to d_i^2 and $(d_a^2 - d_i^2)$ respectively. Different sets of tubes are necessary to test mixtures of different strengths.)

With these assumptions, Burke and Schumann choose the task of finding the flame shape as a function of such variables as the diameters of the tubes, flow rate, velocity, stoichiometric ratio, diffusion coefficient, pressure etc. Let x be the axial coordinate with $x = 0$ at the mouth of the fuel tube. Let r be the radial coordinate with $r = 0$ at the axis. Let d_i be the diameter of the inner tube and d_a be that of the outer tube as shown in Figure 7.22. Subscript i stands for the conditions at $x = 0$. Let subscripts F and O respectively stand

Combustion of Gaseous Fuel Jets

for the species of fuel and oxygen. Steady state cylindrically symmetric constant property equations of species conservation then take the following form.

$$\frac{\partial}{\partial r} r\rho_g D_O \frac{\partial Y_O}{\partial r} - \rho_g u r \frac{\partial Y_O}{\partial x} + r\dot{W}_O''' = 0 \quad (7.82)$$

$$\frac{\partial}{\partial r} r\rho_g D_F \frac{\partial Y_F}{\partial r} - \rho_g u r \frac{\partial Y_F}{\partial x} + r\dot{W}_F''' = 0 \quad (7.82a)$$

u is the velocity of the fuel and the oxidant, assumed constant. According to simple, single, fast, reaction stoichiometry, 1 gram of oxygen completely consumes f grams of fuel so that the sink terms are related as $\dot{W}_F''' = f\dot{W}_O'''$. Using this relation Eq. 7.82 reduces to the following form when the diffusivities are equal.

$$\frac{\partial}{\partial r} r\rho_g D \frac{\partial (Y_F - Y_O f)}{\partial r} - \rho_g u r \frac{\partial (Y_F - Y_O f)}{\partial x} = 0 \quad (7.83)$$

The boundary conditions are as following.

$$x = 0, \quad 0 \leq r \leq d_i/2, \quad Y_F - Y_O f = Y_{Fi}$$

$$d_i/2 \leq r \leq d_d/2, \quad Y_F - Y_O f = -Y_{Oi} f$$

$$x \geq 0, \quad r = 0 \text{ and } r = d_d/2, \quad \partial(Y_F - Y_O f)/\partial r = 0$$

With these boundary conditions Eq. 7.83 is soluble to obtain $(Y_F - Y_O f)$ as a function of x and r. Since Y_F is zero outside the flame and Y_O is zero within it, the flame location is given by $Y_F - Y_O f = 0$. With this argument, Burke and Schumann obtain the following implicit formula for the flame shape (x_C, r_C).

$$\sum_{n=1}^{\infty} \frac{1}{\kappa_n} \left[\frac{(J_1(\kappa_n d_i/2) \cdot J_0(\kappa_n r_C))}{J_0^2(\kappa_n d_d/2)} \right] e^{-\kappa_n^2 D x_C/u} = E \quad (7.84)$$

where E is a constant defined by

$$E \equiv \left(\frac{d_d^2 Y_{Oi} f}{4 d_i (Y_{Oi} f + Y_{Fi})} - \frac{d_i}{4} \right) \quad (7.84a)$$

J_0 and J_1 are Bessel functions which may be found tabulated in mathematical handbooks.* $\kappa_1, \kappa_2, \kappa_3, \ldots$ etc. are the positive roots† of the equation $J_1(\kappa d_d/2) = 0$. Figure 7.24 shows the flame shapes deduced from Burke and Schumann's calculations for methane burning in air of various oxygen concentrations and the following conditions. $d_d = 2$ inches, $d_i = 1.0$ inch, and $u = 0.61$ inch/second (this velocity corresponds to 1 cubic foot/hour of

---

* For example, C.R.C. Handbook or Jahnke and Emde's *Tables of Functions*.
† From Jahnke and Emde's tables, $\kappa_1 d_d/2 = 3.8317$, $\kappa_2 d_d/2 = 7.0156$, $\kappa_3 d_d/2 = 10.1736$, $\kappa_4 d_d/2 = 13.3232$, $\kappa_5 d_d/2 = 16.4706$, etc.

volumetric flow). The diffusion coefficient is assumed to be 0.0763 inch²/second. In the case of combustion in oxygen-rich air, the flame is based upon the fuel tube. This is named as *overventilated* flame. Its height is determined by setting $r_c = 0$ in Eq. 7.84. On the other hand, if combustion were to take place in air of poorer oxygen content, the flame is based upon the air flow annulus. Burke and Schumann call this an *underventilated* flame. Setting $r_c = d_d/2$ in Eq. 7.84 gives its height. It is logical from Figure 7.24 to expect that an overventilated flame results when Y_{Oi} is greater than 0.5.

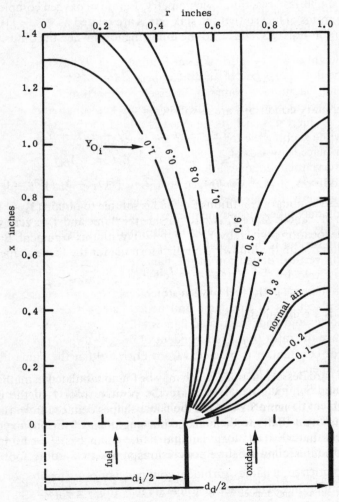

Figure 7.24 Burke-Schumann flame shapes

Combustion of Gaseous Fuel Jets

Since the flame heights, in general, are large enough and since κ_n increases rapidly with n, the exponential term in Eq. 7.84 decays quite rapidly. In fact, in order to visualize its implications, Eq. 7.84 may be rewritten crudely by retaining only the first term in the summation series.

$$\frac{J_1(3.8317\, d_i/d_d) \cdot J_0(7.6634\, r_C/d_d)}{3.8317\, J_0^2(3.8317)} e^{-58.6 D x_C / u d_d^2} = \frac{1}{2}\left[\frac{d_d}{d_i}\left(\frac{Y_{Oi}f}{Y_{Oi}f + Y_{Fi}}\right) - \frac{d_i}{d_d}\right] \tag{7.85}$$

This equation shows that the flame height is proportional to $u d_d^2/4D$, a finding which is consistent with our free jet analyses. The following points drawn from the preceding equation form the gist of the work of Burke and Schumann,

  (i) Height of any cylindrical flame is unaffected by the size of the tubes if the flow of gas and of air are kept constant.
  (ii) The height of a flame is inversely proportional to the diffusion coefficient provided all other factors are kept unaltered.
  (iii) If an inert gas such as nitrogen is added to the fuel to reduce Y_{Fi}, (while keeping the total flow in the inner tube unaltered), the flame is displaced inwards so that taller underventilated flames and shorter overventilated flames are resulted.
  (iv) Addition of primary oxygen to the fuel gas similarly displaces the flame inwards. However, if the oxygen is added excessively combustion may take place in the premixed stream in which case the present theory is invalid. (The basic assumption that combustion occurs at a thin surface is then violated).
  (v) The vertical dimension of the flame is directly proportional to the flow velocity.
  (vi) An increase in the fuel/oxygen stoichiometric ratio f causes an inward displacement of the flame and hence makes overventilated flames shorter and underventilated flames taller.
  (vii) Since the ratio D/u is independent of pressure when the total flow is kept constant, it is predicted from Eq. 7.84 that the flame height is independent of the pressure.

Burke and Schumann analysis should be equally valid for the time-averaged turbulent flames provided the molecular diffusivity D is replaced by the eddy diffusivity ε. (Refer to the fourth comment following Eq. 7.56 and 7.57).

The facts that $x_C \propto u d_d^2$ for a laminar flame and $x_C \propto d_d$ for a turbulent flame lead to the following three conclusions of practical utility.

  (i) As with free jet flames, turbulent confined flames are much shorter than laminar ones. The space heating rate may thus be enhanced greatly by rendering the flames turbulent.

(ii) The space heating rate may further be increased by feeding the fuel and air through a large number of adjacent small orifices rather than through a fewer large orifices.

(iii) It is clear from Figure 7.24 that as the nitrogen of the air is gradually replaced by oxygen, the underventilated flame height gradually increases to a large value when Y_{Oi} is between 0.5 and 0.7. Any further enrichment of the air with oxygen would convert the flame into an overventilated type whose height decreases with increasing Y_{Oi}.

The flame height dependency on the flow rate of air (of a fixed oxygen content) relative to the flow rate of the fuel is similar to that described above. For example, if the air supply rate is smaller than the stoichiometrically required amount, a truncated (relatively short) underventilated flame results. Combustion then will be incomplete and part of the fuel escapes into the exhaust. If, on the other hand, the air supply rate is stoichiometrically optimum, a flame which extends too far downstream may occur—thus causing a very low space heating rate. If the air supply rate is greater than the stoichiometric requirement, the flame becomes overventilated and part of the air leaves with the exhaust carrying away a great deal of heat. Such heat loss in the exhaust seriously impairs the combustor efficiency. This discussion points out to a dilemma related to the fuel/air feed rate ratio most desirable in a given type of a practical duct type combustor. A widely used solution for this dilemma is to supply air at a *slightly* greater rate than required stoichiometrically so that the flame is short enough, combustion is complete and yet the air content in the flue gases is low. We conclude the discussion on Burke and Schumann's longitudinally confined flames at this junction.

Barr and Mullins studied Burke and Schumann's problem (theoretical as well as experimental) to delineate the influence of dilution of air* with inert gases such as CO_2, H_2O, extra N_2, etc. and of introduction air and fuel at different velocities.

Rummel investigated turbulent flames confined in an open hearth steel furnace. Mainly his discoveries include the following two.

(i) Mixing takes longer time and consequently the diffusion flame height is greater if the wall thickness of the inner tube of Figure 7.22 is increased.

(ii) If the air and fuel jets impinge on one another at an angle of 90° between them, the mixing is most efficient and the flame is extremely short.

Sawai, Kunugi and Jinno published a paper (in the fourth Symposium on Combustion) on their work which is similar to that of Rummel.

---

* Air diluted by inert gases is termed by Barr as *vitiated air*.

Combustion of Gaseous Fuel Jets

Thring and Newby considered turbulent jet flames associated with open hearth steel furnaces. Ignoring free convection they have investigated small scale cold mixing models experimentally and the influence of density difference between the fuel and air theoretically. Their main finding is that the density differences are compensated if the diameter of the port is modified as following for use in the usual jet equations.

$$d'_i = d_i(\rho_F/\rho_a)^n \tag{7.86}$$

where,

$n = 0.5$ for a cylindrical jet and
$ = 1.0$ for a plane jet.

7.11 Comparative Importance of Droplet Vaporization and Jet Mixing in Combustors

As described in Chapter 5, in a gas turbine a liquid fuel is injected into a hot air stream in the form of a spray of fine droplets. Upon contact with the hot air, the drops are vaporized. If the fuel is of relatively low B and high boiling point, the droplet vaporization and combustion characteristics determine the minimum length of the combustion chamber required for complete combustion. If on the other hand, the fuel is highly volatile, the spray of droplets soon evaporates to form a vapor jet the mixing and combustion characteristics of which determine the minimum combustor length.

Spalding has proposed a technique to evaluate the relative importance of droplet vaporization and jet mixing. Consider a cylindrical turbulent spray of a liquid fuel. The gaseous jet flame height is given by Eq. 7.57 which when $24C'x_C/d'_i \gg 1$ takes the following form.

$$\bar{x}_C \approx \frac{d'_i}{24C'}\left(\frac{fY_{O\infty} + 1}{fY_{O\infty}}\right) \tag{7.57}$$

d'_i is the nozzle diameter modified according to Thring's equation. If we consider, in the spray, a droplet which travels along the axis of the jet, its velocity diminishes with increasing x according to Eq. 7.43. When $24C'x/d'_i \gg 1$,

$$\frac{u_m}{u_i} \approx \frac{d'_i}{24C'x}$$

u_i is the injection velocity. All those droplets not traveling on the axis travel at a velocity lower than u_m. They therefore have longer time available for evaporative combustion. Consequently, if we focus our attention on the droplets traveling on the axis, we can arrive at the lower limit of the evaporative burning. Now setting $u_m = (\partial x/\partial t)_{r=0}$, Eq. 7.43 may be intergrated to obtain the distance traversed by the drop in any given lapse of time.

$$\frac{x}{d_i'} = \left(\frac{u_i t}{12C' d_i'}\right)^{1/2} \tag{7.87}$$

Substituting the burning time of a droplet in place of t, Eq. 7.87 gives the distance a droplet of initial diameter d_0 requires to completely burn. Denoting this distance as $\bar{x}_b$,

$$\frac{\bar{x}_b}{d_i'} = \left(\frac{u_i \rho_F d_0^2}{12C' d_i' 8\rho_g D \ln(B+1)}\right)^{1/2} \tag{7.88}$$

A comparison of $\bar{x}_c$ and $\bar{x}_b$ may be made by dividing Eq. 7.88 by Eq. 7.57.

$$\frac{\bar{x}_b}{\bar{x}_c} = \left(\frac{fY_{O\infty}}{fY_{O\infty}+1}\right)\left[\left(\frac{48u_i C'}{d_i'}\right)\left(\frac{\rho_F d_0^2}{8\rho_g D \ln(B+1)}\right)\right]^{1/2} \tag{7.89}$$

If $\bar{x}_b \gg \bar{x}_c$, the droplet evaporative combustion determines the minimum required duct length. This is true when the injection velocity is large, initial droplet size is large, diffusivity is small and B is small. The droplets then burn with isolated envelope flames. If, on the other hand, the injection velocity is small, nozzle diameter is large, the fuel is extremely volatile (i.e., B is large), diffusivity is large and the droplet size is small, $\bar{x}_c \gg \bar{x}_b$ so that the droplets quickly evaporate to disappear. The vaporous jet burns as it mixes with air in the fashion described in this chapter. The duct length then is determined by Eq. 7.57.

In gas turbines $\bar{x}_c \approx \bar{x}_b$ so that both droplet vaporization and jet mixing play equally important roles in determining the minimum duct length. Thring and Newby show that in industrial furnaces $\bar{x}_c \gg \bar{x}_b$ and hence the jet mixing plays the governing role. Burgoyne's work indicates that if the maximum droplet size in a spray is less than 10 microns the droplets evaporate quickly and jet mixing gives the flame height.

7.12 Concluding Remarks

The jet flame is a very interesting topic for theoretical as well as experimental study. Most of its properties are predictable with sufficient accuracy by radically simplified theories.

The solutions for the free boundary problems indicated in Figures 7.25 to 7.27 may now be easily obtained from the concepts developed in this chapter.

Combustion of Gaseous Fuel Jets

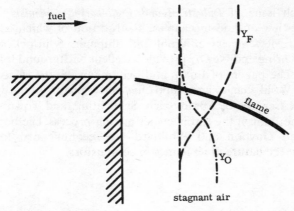

Figure 7.25 Flow of a fuel over a step in a quiescent room of air

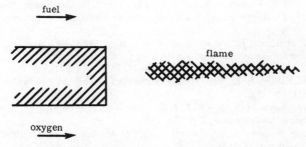

Figure 7.26 Modified Marble–Adamson problem

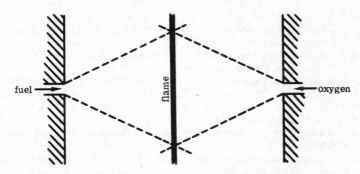

Figure 7.27 Opposed jet diffusion flame

The March issue of *Industrial and Engineering Chemistry* every year contains a review of fluid mechanics, a subsection of which is on jets and wakes. The review papers of Wohl and Shipman, Scurlock and Grover, Hottel, and Thring and Newby provide excellent background for a beginner in the field. The papers of Burke and Schumann, Hottel, Hawthorne and Weddel, and Wohl, Gazley and Kapp offer a clear picture of the development of ideas. The treatises of Abramovich, Schlichting, and Townsend present a thorough account of the turbulent jet mixing process. The books by Jost, Spalding, and Gaydon and Wolfhard are recommended for extensive discussion on the nature of jet flames in combustors.

References

Abramovich, G. N., *Theory of Turbulent Jets*, M.I.T. Press, Cambridge (1963).
Barr, J., *Fuel*, London, **28**, p. 181, 200, 205, 225 (1949).
Bevans, R. S., *High-Output Combustion of Volatile Liquid Fuels*, Sc.D. Thesis in Chemical Engg., M.I.T., Cambridge (1946).
Bevans, R. S., M.I.T., Report on Navy Contract No. a(s) 5152, Cambridge (1946).
Burgoyne, J. H., Discussion in *Selected Combustion Problems*, p. 392, Butterworths, London (1954).
Burke, S. and Schumann, T., *Ind. and Engg. Chem.*, **20**, p. 998 (1928). Also p. 2 in the volume on *First and Second International Symposia on Combustion* (1965).
Corrsin, S., *N.A.C.A. Wartime Rept.*, A.C.R. No. 3L-23-W-94 (1943).
Corrsin, S. and Kistler, A. L., Free-Stream Boundaries of Turbulent Flows, N.A.C.A. Rept. 1244 (1955)
Corrsin, S. and Uberoi, M. S., *N.A.C.A. Tech.*, Note No. 1865 (1949).
Curtet, R., *Combustion and Flame*, **2**, p. 383 (1958).
Forstall, W. and Shapiro, A. H., *Journal of Applied Mechanics*, p. 381 (1950), also M.I.T. Meteor Report No. 39 (1949).
Gaunce, H. Unpublished Research at M.I.T., Cambridge (1939).
Gaydon, A. G. and Wolfhard, H. G., *Flames, Their Structure, Radiation and Temperature*, Chapman and Hall, London (1953).
Hawthorne, W. R., Weddel, D. S., and Hottel, H. C., Mixing and Combustion in Turbulent Gas Jets, p. 266, *Third International Symposium on Combustion* (1949).
Hottel, H. C., Burning in Laminar and Turbulent Fuel Jets, p. 97, *Fourth International Symposium on Combustion* (1953).
Hottel, H. C. and Hawthorne, W. R., Diffusion in Laminar Flame Jets, p. 254, *Third International Symposium on Combustion* (1949).
Jost, W., *Explosion and Combustion Processes in Gases*, McGraw-Hill, New York (1946).
Landis, F. and Shapiro, A. H., *Proc. Heat Transfer and Fluid Mech. Inst.*, Stanford Univ. Press (1951).
Mullins, B. P., *Fuel*, London, **32**, p. 214 (1953).
Mullins, B. P., Combustion in Vitiated Air, p. 447, *Selected Combustion Problems*, Butterworths, London (1954).
Reichardt, H., Gesetzmassigkeiten der freien Turbulenz, V.D.I. Forschungsheft, 414 (1942); Also *Z. Angew. Math. Mech.*, **21** (1941), translation published in *Journal of Royal Aero. Soc.*, **47**, p. 167 (1943).

Rembert, E. W. and Haslam, R. T., *Ind. and Engg. Chem.*, **17**, p. 1236 (1925).
Richardson, J. M., Howard, H. G., and Smith, R. W., Concentration Fluctuations in a Turbulent Gas Jet, p. 814, *Fourth International Symposium on Combustion* (1953).
Rummel, K., Der Einfluss des Mischvorganges auf die Verbrennung von Gas and Luft in Feuerungen, Dusseldorf; Also *Arch. Eisenhuttenw*, **10**, pp. 505, 541 (1937).
Rummel, K., *Arch. Eseṅhuttenw.*, **11**, pp. 17, 67, 113, 163, 215 (1938).
Sawai, I., Kunugi, M., and Jinno, H., Turbulent Diffusion Flames, p. 805, *Fourth International Symposium on Combustion* (1953).
Schlichting, H., *Boundary Layer Theory*, McGraw-Hill, New York (1968).
Scurlock, A. C. and Grover, J. H., Experimental Studies on Turbulent Flames, p. 215, *Selected Problems in Combustion*, p. 215, Butterworths, London (1954).
Spalding, D. B. *Some Fundamentals of Combustion* Butterworths, London (1955).
Thring, M. W. and Newby, M. P., Combustion Length of Enclosed Turbulent Jet Flames, p. 789, *Fourth International Symposium on Combustion* (1953).
Townsend, A. A., *Structure of Turbulent Shear Flow*, Cambridge University Press (1956).
Wohl, K., Gazley, C., and Kapp, N., Diffusion Flames, p. 288, *Third International Symposium on Combustion* (1949).
Wohl, K. and Shipman, C. W., Diffusion Flames, p. 365 in *Combustion Processes*, Princeton volume in High Speed Aerodynamics and Jet Propulsion Series (1956).
Yagi, S. and Saji, K., Problems of Turbulent Diffusion and Flame Jet, p. 771, *Fourth International Symposium on Combustion* (1953).

CHAPTER 8

Flames in Premixed Gases

8.1 Scope of This Chapter

A combustible mixture (also synonymously termed as an "explosive," "reactant" or "reactive" mixture) may be formed by mixing a fuel with an oxidant (both being considered here as gases) in proper proportions prior to ignition. We have dealt with such mixtures, rather briefly, in Chapter 4 in connection with the topic of ignition and extinction. It is our understanding from Chapter 4 that chemical kinetics play the controlling role in combustion phenomena such as ignition and extinction which occur at relatively low temperatures. In high temperature phenomena such as diffusion flames, on the other hand, the chemical reaction details are not crucial since the overall burning process is in the main governed by the relatively slow processes of heat and mass transfer.

We will return, in this chapter, to the flames in gaseous mixtures which are prepared prior to combustion. In such flames, we will note later, chemical kinetics and flow (of heat and mass) play more or less equally important roles. It is possible to form a combustible gaseous mixture by injecting liquid fuel into an engine. Under certain conditions discussed in Chapter 5, the droplets quickly vaporize and the resultant vapor mixes with air to form the combustible mixture.

Before proceeding further, let us briefly pause here to review our definition of a *flame*. A flame represents a spatial domain in which rapid exothermic reactions take place often (but not necessarily always) emitting light. Fristrom and Westenberg suggest some nontrivial reactions which may be considered as "flames" according to this definition. These reactions are listed in Table 8.1. Since for a premixed flame the reactants are mixed prior to their arrival to the reaction zone, one may visualize it to be composed of an infinite number of adjacently located infinitesimally small Burke-Schumann type diffusion flames.*

---

* The reaction zone is also often called as the "flame zone," "flame front," "reaction wave," etc.

Flames in Premixed Gases

TABLE 8.1
Some Chemical Substances Capable of Yielding Flames*

Reactants	Products
O_3	O_2, (O)
N_2H_4	NH_3, N_2, H_2
$H_2 + Br_2$	HBr, (Br), (H)
$H_2 + O_2$	H_2O, (OH), (H), (O)
$C_xH_y + O_2$	H_2O, CO, CO_2, (OH), (O), (H), H_2
$C_2N_2 + O_2$	CO, N_2
$B_2H_6 + O_2$	HBO, BO, B_2O_2, H_2O, H_2, B_2O_3
$B_2H_6 + N_2H_4$	H_2, $(BN)_x$
$Na + CH_3Br$	NaBr, C_2H_6
$H + CCl_4$	HCl, C_2Cl_6
$N + O_2$	NO, N_2

* From Fristrom and Westenberg's *Flame Structure*, McGraw-Hill (1964). With the permission of McGraw-Hill.

As shown in Figure 4.19, the flame front separates the hot combustion products from the cold reactant mixture. In the flame front itself temperature, composition, density and velocity vary spatially from the reactant values to the product values. Depending upon the kinetic mechanism by which the reactants transform to the products, several intermediate (unstable, short-lived) species might occur in the flame front. The patterns of distribution of all these variables across the front describe what is known as the *structure* of the flame.

Depending upon the conditions of the reactant mixture supply to the reaction zone, a premixed flame may either be stationary or propagating. The problem of flame propagation is of paramount interest in both scientific as well as practical view points. In order to control and possibly prevent propagation of explosions in mine shafts and roadways filled with combustible mixtures, in order to achieve a high but controlled rate of combustion in such closed vessels as the cylinder of a gasoline engine, in order to stabilize a flame in a fast moving reactant stream such as in ramjets and turbojets and in other instances too numerous to enumerate here, one needs a thorough understanding of the flame propagation.

Inasmuch as a propagating flame may be considered as a wave of chemical reaction sweeping across a flowing gas, it offers an excellent proving ground for the analytical skills of a fluid dynamicist, a heat and mass transfer specialist and a physical chemist, all put together into a well-rounded applied mathematician.

In this chapter we will briefly introduce the student to the principles of detonations and deflagrations emphasizing upon the latter. The experimental methods and results of flame propagation will be discussed before going into the presentation of some simple theories. The topic of turbulent flames is then considered along with the concept of flame stability and stabilization.

8.2 Detonations and Deflagrations

Consider a long horizontal tube of constant cross section filled with a combustible mixture as shown in Figure 8.1. Suppose a flame is initiated at the left hand end of the tube by some means as suggested in Chapter 4.

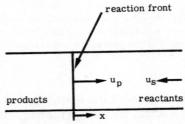

Figure 8.1 Propagation of a reaction front in a combustible mixture confined in a long horizontal tube

Assume that the reaction wave propagates with a steady speed along the tube to the right. Fixing $x = 0$ on the moving wave, the situation is transformed into one in which the wave is at rest and the combustible mixture is fed into the wave at a constant rate. The mixture of products behind the flame front and that of reactants ahead may be considered as uniform, nonviscous and nonconducting. With subscripts s and f respectively standing for the reactant and product mixture variables, one-dimensional equations of mass, momentum and energy conservation may be integrated from s-state to f-state* to obtain

$$\rho_s u_s = \rho_f u_f \quad \text{(Continuity)} \tag{8.1}$$

$$\rho_s u_s^2 + P_s = \rho_f u_f^2 + P_f \quad \text{(Bernoulli's Equation of Momentum)} \tag{8.2}$$

$$\frac{u_s^2}{2} + h_s = \frac{u_f^2}{2} + h_f \quad \text{(Energy)} \tag{8.3}$$

---

* In both the s- and f-states, the gradients of velocity, temperature and composition are assumed to be zero.

Flames in Premixed Gases

The enthalpy h is so defined that it includes the chemical potential energy as well. Equations 8.1 and 8.2 may readily be combined to obtain *Raleigh equation*.

$$\rho_s^2 u_s^2 = \rho_f^2 u_f^2 = \frac{P_s - P_f}{\frac{1}{\rho_f} - \frac{1}{\rho_s}} \tag{8.4}$$

Specifically,

$$u_f = \frac{1}{\rho_f}\left(\frac{P_s - P_f}{\frac{1}{\rho_f} - \frac{1}{\rho_s}}\right)^{1/2}$$

$$u_s \equiv u_0 = \frac{1}{\rho_s}\left(\frac{P_s - P_f}{\frac{1}{\rho_f} - \frac{1}{\rho_s}}\right)^{1/2} \tag{8.4a}$$

Rearranging Eq. 8.3,

$$h_f - h_s = \frac{\rho_s^2 u_s^2}{2\rho_s^2} - \frac{\rho_f^2 u_f^2}{2\rho_f^2}$$

Noting the algebraic identity $(\alpha^2 - \beta^2) \equiv (\alpha + \beta)(\alpha - \beta)$ and utilizing Eq. 8.4, we obtain *Hugoniot equation*.

$$h_f - h_s = \frac{P_f - P_s}{2}\left(\frac{1}{\rho_f} + \frac{1}{\rho_s}\right) \tag{8.5}$$

Separating the chemical energy from sensible enthalpy which may be written in terms of the sensible internal energy, pressure and specific volume, the following alternative form of Hugoniot equation may be obtained.

$$\bar{C}_V(T_f - T_s) = \frac{P_f + P_s}{2}\left(\frac{1}{\rho_s} - \frac{1}{\rho_f}\right) + \Delta H \tag{8.5a}$$

where $\bar{C}_V$ is the constant volume specific heat averaged from T_s to T_f and ΔH is the heat of the combustion reaction.

For a given set of initial conditions (P_s, T_s, ρ_s) of a given mixture $(\Delta H, \bar{C}_V)$, Eq. 8.5 stipulates a series of final states (P_f, T_f, P_f) possible from the energy point of view. For a given set of initial dynamic conditions (P_s, u_s, ρ_s), Eq. 8.4 stipulates a series of final states (P_f, u_f, ρ_f) possible from the flow point of view. In a typical problem the final state(s) satisfying both the energy and flow points of view is (are) given by the point(s) of intersection of the Raleigh and Hugoniot curves in $P - 1/\rho$ plane.

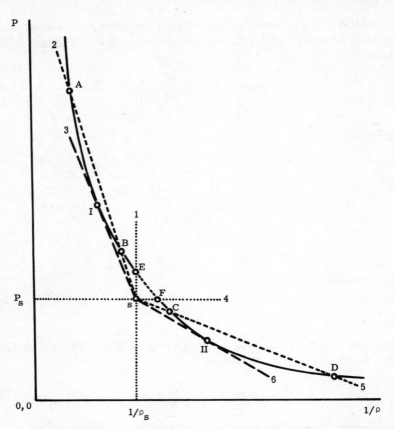

Figure 8.2 Hugoniot curve on a P vs $1/\rho$ plane to distinguish between detonations and deflagrations

Figure 8.2 shows a Hugoniot curve for a fixed value of ΔH and $\bar{C}_V$. Through any given initial state (P_s, ρ_s) falling under this curve, there exist two ranges of the mass flux* in which reaction wave propagation is physically possible—the upper range corresponding to *detonation waves* and the lower range corresponding to *deflagration waves*. The upper range is limited by Rankine lines labelled 1 (for which $\rho_s u_s = \infty$) and 3. The lower range is limited by lines labelled 4 (for which $\rho_s u_s = 0$) and 6. Mass fluxes "greater" than ∞ and less than zero are incompatible with our physical sense. Hence the part EF of Hugoniot curve which lies in the quadrant bounded by the lines 1 and 4 is physically invalidated. This invalidity may be seen from Eq. 8.4 by noting

---

* $(\rho_s u_s)^2$ is the negative of the slope of Rankine lines. See Eq. 8.4.

Flames in Premixed Gases

that along the arc EF of Hugoniot curve, $(P_s - P_f)/(\rho_f^{-1} - \rho_s^{-1})$ will be negative and therefore suggests imaginary values of u_s and u_f.

Consider the intermediate Rankine line labelled 2 in Figure 8.2. It cuts Hugoniot curve at two points, A and B, suggesting two possible detonation states. Point B denotes a *weaker* detonation whereas A denotes a *stronger* detonation. As the slope of line 2 is reduced, A and B approach one another to finally coincide at I as the mass flux approaches that corresponding to line 3. The point I is known as *Chapman-Jouget detonation* state.

This description reveals that the density and pressure behind a detonation wave are larger than those ahead of the wave. As a result, the burnt mixture flows in the same direction as the detonation wave. Furthermore, according to Bernoulli equation, the velocity of the burned mixture will be lower than that of the fresh mixture. If the mass flux is smaller than that corresponding to Rankine line 3, detonation is not possible for the particular Hugoniot curve under consideration (i.e. for the particular ΔH and $\bar{C}_V$).

Similarly, when the mass flux is low, for any Rankine line falling between 4 and 6 there occur two possible (C and D) deflagration states. The lower state is known as *weaker* deflagration and the upper one, *stronger*. As the mass flux is gradually increased, these two states finally approach the *Chapman-Jouget deflagration* point II. It is clear that for deflagrations the values of density and pressure behind the wave are smaller than those ahead of it. Thus (a) the burned mixture flows away from the flame front and (b) the velocity of the burned gases is larger than that of the fresh gas.

A detonation wave is physically a combustion process initiated by a shock wave. It travels at a supersonic velocity and consequently involves extremely high pressure differentials. A normal (deflagration) flame, on the other hand, propagates at a subsonic velocity, does not involve shock compression and hence pressure differentials involved are negligible.

It is possible to show from entropy considerations and second law of thermodynamics that strong deflagrations are physically unreal. Similarly, considering transport phenomena one can show that weak detonations fail to occur.

The velocity of a detonation wave (relative to the fresh mixture and normal to itself) is expected to be of the same order of magnitude as the velocity of thermal motion of the molecules in the burnt mixture. Typically, for a stoichiometric hydrogen/oxygen mixture at $T_s = 291\ °K$, $P_s = 1$ atm. the detonation velocity, final temperature and pressure respectively may be calculated as 2,806 meters/second, 3,583 °K and 18.05 atm. Measured velocities agree with this computed value well within 1%. Addition of nitrogen or extra oxygen lowers the detonation velocity whereas addition of extra hydrogen raises it. Equilibrium dissociation effects have to be taken into account to satisfactorily predict detonation velocities. Equation 8.4(a)

TABLE 8.2

Detonation Velocities at $T_s = 298\ °K$, $P_s = 1$ atm.*

Mixture	u_0 meter/sec
$2H_2 + O_2$	2,821
$2CO + O_2$	1,264
$CH_4 + 2O_2$	2,146
$CH_4 + 1.5O_2 + 2.5N_2$	1,880
$C_2H_6 + 3.5O_2$	2,363
$C_2H_4 + 3O_2$	2,209
$C_2H_4 + 2O_2 + 8N_2$	1,734
$C_2H_2 + 1.5O_2$	2,716
$C_2H_2 + 1.5O_2 + N_2$	2,414
$C_3H_8 + 3O_2$	2,600
$C_3H_8 + 6O_2$	2,280
i-$C_4H_{10} + 4O_2$	2,613
i-$C_4H_{10} + 8O_2$	2,270
$C_5H_{12} + 8O_2$	2,371
$C_5H_{12} + 8O_2 + 24N_2$	1,680
$C_6H_6 + 7.5O_2$	2,206
$C_6H_6 + 22.5O_2$	1,658
$C_2H_5OH + 3O_2$	2,356
$C_2H_5OH + 3O_2 + 12N_2$	1,690

* From Laffitte as quoted by Jost, in *Explosion and Combustion Processes in Gases*, McGraw-Hill (1946).

indicates that low densities give higher detonation velocities. That this is a valid expectation is proved by Bernard Lewis' experiments in which either a lighter gas (i.e., helium) or a heavier gas (i.e., argon) was added to the H_2/O_2 mixtures. Table 8.2, excerpted from Jost's book, provides some data on detonation velocities. For further study of detonations the student is referred to the treatises of Courant and Friedrichs, Lewis and von Elbe, Jost, Zeldovich, etc.

In the rest of this chapter we focus our attention on deflagrations.

8.3 Deflagrations—Some Basic Characteristics

As mentioned in Section 8.2, deflagrations are considerably slower flames with speeds ranging from a few centimeters to a few hundred centimeters per second. A deflagration wave is, in many ways, much more complex than a detonation wave. Jost points out, for example, that the detonation velocity is quite independent of the external disturbances, if only because it is much greater than the velocity of any disturbance process.

Flames in Premixed Gases

The gas mixture is heated and expanded across the flame front owing to the heat release due to combustion. In order to accommodate the prescribed total mass flux the gases have to accelerate in the flame front; that is, the burnt mixture flows much faster than the fresh mixture. Figure 8.3 schematically shows the velocity profile (among other details) across the flame. The pressure difference across the flame front is given by Eq. 8.4(a).

$$P_s - P_f = \rho_s^2 u_s^2 \left(\frac{1}{\rho_f} - \frac{1}{\rho_s} \right) \tag{8.4a}$$

Typically, for a hydrocarbon/oxygen flame, $u_s \approx 300$ cm/sec, $\rho_s \approx 0.0015$ gm/cm^3, ($P_s \approx 15$ psia) and $T_f/T_s \approx 10$ so that $P_s - P_f \approx 0.2$ psi. This pressure differential is about 1% of the supply pressure itself. We therefore may neglect this pressure drop and assume that a deflagration flame is a constant pressure process.

The speed of flame propagation, tube diameter, mixture viscosity and the wall roughness jointly determine whether the flame is laminar or turbulent. In the main of this chapter, we will consider laminar flame propagation. The concepts will then be extended to turbulent flame propagation which, of course, is a topic of most practical interest.

In order to construct some important definitions let us consider explicitly the propagation of an ideal one-dimensional laminar flame. The profiles of velocity, temperature, composition and pressure are qualitatively shown in

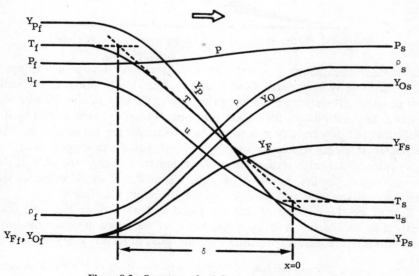

Figure 8.3 Structure of a deflagration flame front

Figure 8.3. As demonstrated later in this chapter, it is quite possible that some intermediate unstable species exist *in* the flame.

The *flame front thickness* δ may be conveniently defined as the ratio of the maximum temperature difference $(T_f - T_s)$ to the maximum temperature gradient dT/dx which occurs at the point of inflexion of the profile.

$$\delta \equiv \frac{(T_f - T_s)}{(dT/dx)_{max}} \qquad (8.6)$$

Most of the flames we are interested in here are at most only a few millimeters in thickness. Measurement of the structure profiles in such a thin spatial domain requires extreme care and skill.

In reality, flame fronts are far from being ideally one-dimensional (i.e., flat and disc-shaped). Due to friction at the tube walls the flame proceeds faster near the axis of the tube than near the walls. Thus viscosity renders a somewhat parabolic shape to the flame front. And then there is the inevitable buoyancy which distorts the paraboloid into an assymetric shape illustrated in Figure 8.4. Furthermore, because of a small amount of heat loss to the walls,

Figure 8.4 The realistic shape of a flame front propagating in a tube

the flame will be quenched near the walls. (In fact, as we have seen in Chapter 4, if the tube is very narrow the excessive heat losses prevent the flame propagation altogether). Since the propagating flame is in contact with the walls at any station only for a short duration of time, the heat gained by the wall does not penetrate deep into it.* As a consequence, the minimum tube diameter through which a flame would marginally propagate (i.e., the quenching distance) is independent of the material of the tube wall. If the flame were stationary, however, the tube wall material would influence the minimum diameter which could contain and preserve the flame.

Let u_P be the speed at which the flame traverses along the tube with respect to an observer fixed in space. Let u_s be the speed at which the fresh reactant mixture moves with respect to the fixed observer. *The fundamental*

---

* Alternatively, it may be said that the thermal mass and conductivity of the walls are much larger than those of the gases confined in them.

Flames in Premixed Gases

flame speed u_0 is defined as the velocity of a *laminar flame* front in a direction *normal to itself* and *with respect to the fresh reactant mixture*. It is clear with this definition that the propagation speed u_P, the fresh gas speed u_s and the fundamental flame speed (which shall hereafter be referred to as simply "flame speed") are mutually related according to the following equation.*

$$u_0 = u_P \pm u_s \qquad (8.7)$$

If the flame propagates in the same direction as the flow of the fresh mixture, the minus sign is valid; if the flame propagates against the fresh mixture flow, the plus sign is valid. The fundamental flame speed is a characteristic kinetic constant of the mixture. It strongly depends upon the combustion reaction kinetics, presence of diluents, temperature and pressure. Figures 8.10 and 8.11 show the variation of u_0 with the fuel/oxidant ratio for various mixtures. Focusing our attention on any of these curves, we note that a maxima is exhibited by a mixture *near* stoichiometric composition. This pattern of variation with composition is similar to that of equilibrium adiabatic flame temperature† T_f; (Refer to Figure C.10, Appendix C).

Equation 8.7 is useful to deduce several important characteristics of a premixed flame. If the fresh mixture velocity is so adjusted that the flame is fixed at a station, $u_P = 0$; the fundamental flame speed is then equal in magnitude to the supply velocity. In practice, however, it is extremely difficult to adjust u_s so precisely as to keep $u_P = 0$ since even the slightest disturbances are adequate to cause the flame nonstationary. If u_s is smaller than u_0, (refer to Figure 8.1) the flame would propagate into the fresh mixture to bring about what is known as *flash-back*. If u_s were higher than u_0, on the other hand, the oncoming supply pushes the flame far into the burnt mixture to result in what is known as *blow-off*.

The concepts of "blow-off" and "flash-back" may be better explained by considering the Bunsen flame as following.

Consider the Bunsen flame shown in Figure 8.5. The high velocity fuel gas issuing through the orifice into the barrel induces air through the ports of primary air. If the control ring is positioned in such a way that the ports are completely blinded, just the pure fuel gas flows up through the barrel to emit at its mouth in the form of a jet which, when mixed with the room air according to the laws derived in Chapter 7, yields a bright diffusion flame.

---

* An everyday example analogous to the present situation is that of a person walking up on an escalator which itself is rolling up. If u_s is the speed of the escalator and u_P is the speed with which the person seems to move to a fixed observer, then the person's speed relative to the escalator is given by Eq. 8.7 (i.e., $u_0 = u_P - u$).

† u_0 is a basic *thermokinetic* property of a given mixture in much the same manner as are the minimum ignition energy, quenching distance, flash-back and blow-off limits, etc. The adiabatic flame temperature is, on the other hand, a basic *thermodynamic* property of the mixture.

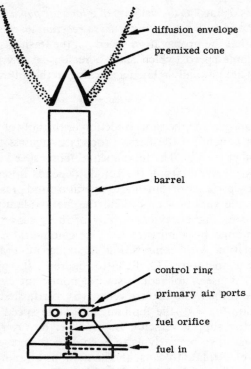

Figure 8.5 A Bunsen burner

If the primary air ports now are gradually opened to allow induction of air into the barrel, a mixture of air and fuel flows through the barrel; a bluish cone of a premixed flame appears within the diffusion flame envelope. With further increase of the primary air feed, the premixed blue cone gradually reduces in size while the diffusion envelope slowly fades away. If the primary air is stoichiometrically adequate, the premixed flame cone is all that remains. Unlike the diffusion flame, a premixed flame is less luminous and much shorter.

Figure 8.6 indicates—qualitatively—the detailed shape of a Bunsen premixed flame cone. It can be seen that it is conical only in an approximate sense. The apex of the flame is rounded off and at its base the flame does not coincide with the barrel mouth; instead, a dead space and a slight but noticeable overhang exist.

Two primary agents responsible for flame propagation in gas mixtures are conduction of heat and diffusion of some unstable intermediate active atoms and radicals (which promote chain reactions) from the flame zone into

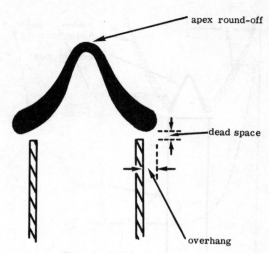

Figure 8.6 Bunsen flame cone

the fresh mixture. As we shall see in Sections 8.5 and 8.6, the relative importance of these two agents differs for different mixtures under different circumstances. As the radius of curvature of the flame front becomes comparable in magnitude with the flame thickness, heat and active particle diffusion becomes more intense. This will increase the fundamental flame speed. At the apex of the cone, therefore, the flame tends to push harder into the fresh mixture; and, hence the observed apex rounding off. At very low pressures, the flame front will be thicker and the rounding of the apex becomes particularly noticeable.

The dead space near the rim of the barrel is a consequence of heat loss (and possibly loss of active chain carriers) from the flame to the metallic wall. Such a loss quenches the reaction to cause the dead space. Because according to Eq. 8.4(a) the pressure inside the flame cone is slightly larger than that outside, it is possible for the fresh mixture to escape unreacted through the dead space into the room air; in so doing, the escaping mixture introduces the "overhang" phenomenon.

Based on these concepts, the variation of the flame speed with radial position in a Bunsen cone is expected to be as qualitatively shown in Figure 8.7. If one wants to obtain reliable flame speeds by the Bunsen burner method, it is clear that one has to exclude those regions of the cone which are influenced by the apex round off and base quenching/overhang phenomena.

Figure 8.8 describes the manner in which an imbalance between the flame speed profile $u_0(r)$ and component, normal to the flame front, of the supply velocity profile $u_{sn}(r)$ distorts an ideal flame cone into the shape indicated by

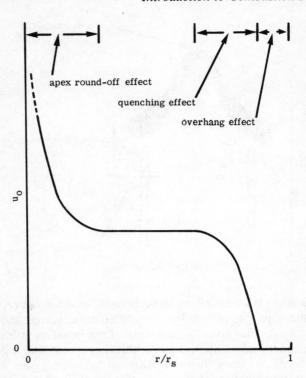

Figure 8.7 Fundamental flame speed at various radial locations in the Bunsen flame cone

Figure 8.6. At all those radial locations where $u_s > u_0$, the flame is thrusted outwards while at locations where $u_0 > u_s$, it is pushed inwards.

Another point of paranthetical importance in this discussion arises due to the unequal diffusion of various species across the flame front. When some species prefer to diffuse faster than others (i.e., preferential diffusion), the flame structure is considerably altered. The sooting of a Bunsen flame at the apex is, for instance, a direct consequence of preferential diffusion.

Suppose that we have a steady burning Bunsen premixed flame cone. If the velocity of the fresh mixture supply is now increased, the flame "lifts off" and "blows away" from the burner. On the other hand, if the supply rate of the fresh mixture is gradually decreased, the flame becomes "flatter," "button-shaped," moves closer to the barrel mouth, changes its shape from convex to concave and finally, "flashes back" into the barrel to extinguish itself usually with an audible "pop."

The experience of blow-off and flash-back suggests that while the flame constantly strives to propagate itself into the fresh reactant mixture, the supply velocity of the latter pushes it back to make it stationary (i.e.,

Flames in Premixed Gases

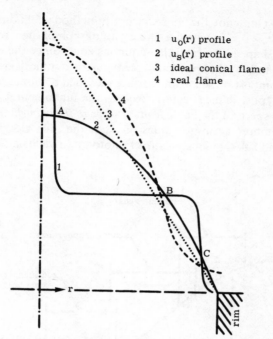

Figure 8.8 Distortion of the flame cone due to the imbalance between the flame speed profile $u_0(r)$ and the supply speed profile $u_s(r)$

$u_P = 0$). If the supply velocity is greater than the fundamental flame speed, the flame is blown away. If u_s is lower than u_0, the flame flashes back. When $u_0 = -u_s$, a stable flame results.

At this point we will terminate this discussion. We will return to the concept of blow-off later in this chapter in connection with the topic of flame stabilization.

8.4 Some Experimental Details

There are three basic types of experiments frequently used to measure fundamental flame speeds. These are (1) burner methods, (2) tube methods and (3) soap bubble and bomb methods. In burner methods the flame remains stationary. In burner and tube methods it is invariably necessary to know the area of the flame front in order to compute the flame speed. Hence, we shall first describe some techniques of accurately determining this area. These could be put into two categories—photographic and particle track techniques

The simplest but not the most accurate method of finding flame front area, of course, is one in which a direct photograph of the flame is obtained to record the shape and size of the luminous zone. Since the luminous zone is produced mainly by the hot incandescent (often, to some degree, particulate) products of combustion, the flame surface recorded by direct photography is closer to the "post flame" region. Because the fundamental flame speed is defined as the speed of the flame front *relative to the unburnt mixture* what is required, to obtain accurate values of u_0, is the area closest to $x = 0$ in Figures 8.3 and 8.9. Obviously a direct photograph yields a surface far from this.

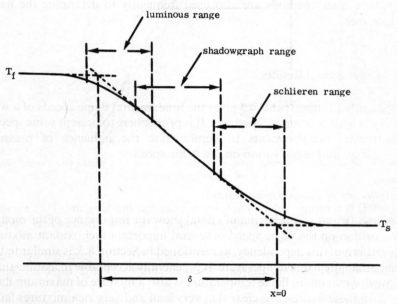

Figure 8.9 Regions recorded by direct, shadowgraph and schlieren photography

Shadowgraph, schlieren and interferograph are three special flame photographic techniques. These techniques exploit two basic physical phenomena. Firstly, the presence of a flame introduces density nonuniformities in space (due to temperature and composition nonuniformities). Secondly, when a beam of light passes across the interface of two media of different refractive indices, it is deflected. Since the path length with deflection is greater than without, the time taken by the beam to travel from the source to the screen is longer with the flame than without it and gives an indication of the density field traversed. While the interferograph gives an indication

Flames in Premixed Gases

of the density field itself, schlieren indicates the field of density gradient and shadowgraph indicates the field of second derivative of density.

Assuming for the moment that the density field in a flame is a consequence of the temperature field alone ($\rho \propto 1/T$), the shadowgraph of a flame shows the region of maximum dT/dx (i.e., the region near the point of inflexion of the $T(x)$ profile). The schlieren, on the other hand, indicates the region close to where d^2T/dx^2 is maximum. This occurs as shown in Figure 8.9 at that face of the flame which is closest to the fresh mixture. For this reason, the area obtained from a schlieren photograph is the most suitable flame area to determine u_0. Shadow and direct photographs yield larger and erroneous flame areas but they have the advantage of simpler optical set up.

Particle track methods are also used frequently to determine the flame surface area.

8.5 Experimental Results

Among other things, Table 4.3 gives the fundamental flame speeds of a wide variety of hydrocarbon/air mixtures. It is proper here to present some specific experimental measurements to demonstrate the influence of pressure, temperature and composition on the flame speed.

(a) *Influence of Fuel/Oxidant Ratio*

Figures 8.10 and 8.11 (Hartmann's data) show the importance of the mixture composition on the flame speed of several important fuel/oxidant mixtures. The pattern of this dependency, as mentioned in Section 8.3, is similar to that of the adiabatic flame temperature. It is generally acceptable to assume that a mixture of maximum flame temperature is also a mixture of maximum flame speed. In these figures, it is clear that very lean and very rich mixtures fail to support a propagable flame. This means, if the mixture contains too little fuel or too little oxidant it fails to keep a propagating flame. It is this observation which is responsible for the existence of the upper and lower flammability limits. (Defining ignition as the *initiation* of a *propagable* flame in a reactant mixture, we have dealt with flammability and ignitability limits synonymously in Chapter 4.) It is important to note from Figures 8.10 and 8.11 that the flame speeds at the flammability limits fall to zero quite steeply.

The maximum flame speed occurs for most mixtures at the stoichiometric composition. For flames with air as the oxidant, it occurs slightly on the richer side of stoichiometric point. For H_2/air and CO/air flames this shift is quite apparent.

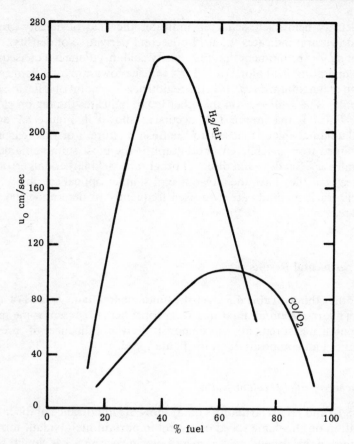

Figure 8.10 Effect of mixture composition on the fundamental flame speed

(b) *Influence of the Fuel Structure*

Gerstein, Levine and Wong measured the flame speeds of various families of hydrocarbons mixed with air in order to understand the influence of the fuel molecular structure. Their findings are briefly presented below.

As the fuel molecular weight increases, the range of flammability (and hence the flame speed curve of Figures 8.10 and 8.11) becomes narrower. Figure 8.12 shows the maximum flame speed as a function of the number of carbon atoms in the fuel molecule for three families. For saturated hydrocarbons (i.e., alkanes such as ethane, propane, butane, pentane, hexane, etc.), the flame speed remains nearly independent (≈ 70 cm/sec) of the number of

Flames in Premixed Gases 287

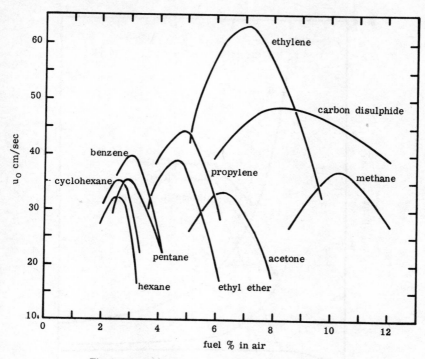

Figure 8.11 Effect of percent fuel on the flame speed

C-atoms in the molecule, n_C. For unsaturated hydrocarbons,* u_0 is high for small n_C, falls steeply as n_C increases to 4 and then falls slowly with further increase in n_C to approach the saturated value of u_0 when $n_C \gtrsim 8$.

(c) *Influence of Pressure*

Lewis studied, by constant volume bomb method, the effect of pressure on the flame speed of various hydrocarbon/O_2/N_2/Ar/He mixtures. Assuming that a proportionality such as $u_0 \propto P^n$ exists, he determined the index n for various mixtures. Figure 8.13 (in which the experimental points are not included for want of clarity), shows Lewis' pressure index n for various hydrocarbon flames. In essence, when low (i.e., < 50 cm/sec), the flame speed

---

* i.e., olefins and acetylenes. Olefins (alkene class) are unsaturated compounds with a double bonded pair of C-atoms. Examples are ethylene, propylene, 1-butene, 2-pentene, 1-hexene, etc. Acetylenes (alkyne class) are those with a triple bonded pair of C-atoms. Examples are propyne, 1-butyne, 1-pentyne, 1-hexyne, etc.

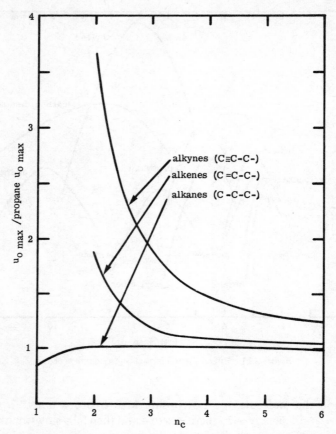

Figure 8.12 The maximum flame speed as a function of the number of carbon atoms in a fuel molecule for saturated and unsaturated hydrocarbons (with the permission of the Combustion Institute, from: T. W. Reynolds and M. Gerstein, p. 194, *3rd International Symposium on Combustion*, (1949))

increases with decreasing pressure; in the range 50–100 cm/sec, it is independent of pressure; and when high (i.e., >100 cm/sec), it decreases with decreasing pressure.

In view of the flame theories to be discussed in Section 8.6, the preceding $u_0(P)$ dependency indicates that the overall order of the combustion reaction is less than 2 for flames with $u_0 < 50$ cm/sec, equal to 2 for those with $50 < u_0 < 100$ cm/sec, and greater than 2 for those with $u_0 > 100$ cm/sec. For example, Spalding concludes, from Lewis' results, that mixtures with $u_0 = 25$ cm/sec show a pressure dependence implying an overall reaction order of 1.4 while mixtures with $u_0 = 800$ cm/sec imply an order of 2.5.

Flames in Premixed Gases 289

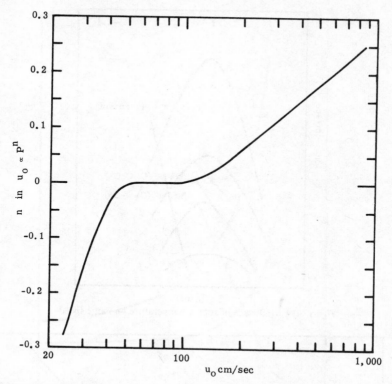

Figure 8.13 Influence of pressure on flame speed (adapted from B. Lewis, *Selected Combustion Problems* (AGARD), p. 177, Butterworths, (1954))

(d) *Influence of Initial Mixture Temperature*

Dugger and Heimel performed experiments which show the dependency of u_0 on T_s. Their experiments confirmed the suspicion of Mallard and Le Chatelier that the flame speed increases with preheating. In general, these experimental results may be expressed as qualitatively shown in Figure 8.14. Dugger's measured variation of u_0 with T_s is shown in Figure 8.15 for three mixtures. An acceptable relation such as $u_0 \propto T_s^m$ may be deduced from these data with m somewhere between 1.5 and 2.

Strong as this dependence may be, it is too weak to be explained on the basis of an Arrhenius type temperature effect on the reaction rate.

(e) *Influence of Flame Temperature*

Figure 8.16 shows the maximum flame speed for several mixtures as a function of the final flame temperature T_f. The data for this figure are compiled by Bartholome and Sachsse. The dependency clearly is very strong; and

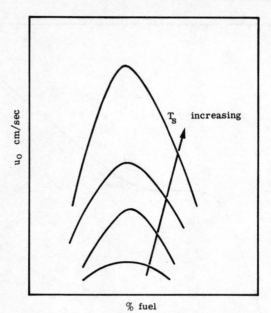

Figure 8.14 Influence of supply temperature on flame speed

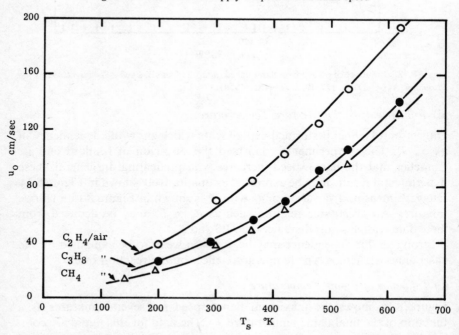

Figure 8.15 Influence of supply temperature on flame speed (adapted with permission of American Chemical Society, from: G. L. Dugger, et al., *Ind. and Engg. Chem.* **47**, p. 114, (1955))

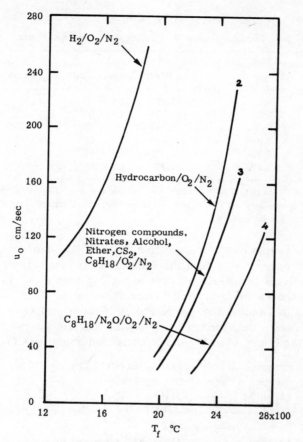

Figure 8.16 Effect of the flame temperature on flame speed (adapted from E. Bartholome, *Naturwiss.*, **36**, p. 171, (1949) and E. Bartholome and H. Sachsse, *Zeit. Electrochem.*, **53**, p. 183, (1949))

stronger if the flame temperature is higher. From these data we may logically conclude that so far as the influence of temperature is concerned, it is the flame temperature that essentially determines u_0. However, for most of the considered mixtures, the flame speed rises much faster than the flame temperature. Dissociation reactions which are favored at high temperatures are responsible for this manifestation.

Dissociation reactions also introduce free radicals into the flame. These free radicals act as chain carriers to promote the reaction and hence the propagation. Generated at or near the flame temperature, they diffuse to the preflame zones. The lighter these free radicals and atoms are, the easier they can diffuse. Thus, atoms of H (and to some degree, those of O and radicals of OH) are notorious to enhance flame speeds. Table 8.3 shows that the

TABLE 8.3

H-atom Concentrations in the Combustion Zone of CO Flames*

Distance from the flame cm	$T\,°K$	C_H equilibrium	C_H actual
0	2,390	2.4×10^{-3}	2.4×10^{-3}
0.002	2,630	8.5×10^{-4}	2.37×10^{-3}
0.004	2,330	2.3×10^{-4}	2.34×10^{-3}
0.006	2,030	4.0×10^{-5}	2.32×10^{-3}
0.008	1,740	4.4×10^{-6}	2.29×10^{-3}
0.010	1,460	2.2×10^{-7}	2.26×10^{-3}

* From Tanford, C. The Role of Free Atoms and Radicals in Burner Flames. *Third Symposium on Combustion*, p. 140 (1949). With permission of the Combustion Institute.

actual local H-atom concentrations in a (wet) CO/O_2 flame are at least an order of magnitude larger than the equilibrium concentrations corresponding to the local temperatures, thus indicating that these atoms indeed diffuse from the reaction zone to the prereaction zones.

According to Tanford, the role played by H-atoms in CO/O_2 flames and in H_2/O_2 flames is described by the following reaction scheme in which the symbol $\mathscr{R}$ stands for CO or H_2 and $\mathscr{P}$, correspondingly, for CO_2 or H_2O.

$H_2 \longrightarrow 2H$ initiation in the burnt gas.

$H \longrightarrow$ diffuses to the unburnt gas.

$H + O_2 + M \longrightarrow HO_2 + M$
$HO_2 + \mathscr{R} \longrightarrow \mathscr{P} + OH$ $\Big\}$ chain continuation
$OH + \mathscr{R} \longrightarrow \mathscr{P} + H$

$H + H + M \longrightarrow H_2 + M$ termination

The termination reaction may occur either at the vessel walls (at low pressures) or in the bulk gas phase (at high pressures).

Hydrogen has similar catalytic effects on the combustion of coal and of some metallic chlorides. Replacement of nitrogen by helium in many combustible mixtures reduces the specific heat to consequently raise the flame temperature which results in an increase in active H-atom concentration and hence the flame speed.

Based upon these and similar observations, Tanford contends that the diffusion of free radicals is of profound importance in flame propagation in certain mixtures. Figure 8.17 indicates the influence of C_H on u_0 as measured by Gaydon, by Bartholome and by Jahn. Simon tested Tanford's hypothesis for combustion in air of 35 hydrocarbons which include alkanes and alkenes (both normal and branched), alkynes, benzene and cyclohexane. She plotted

Flames in Premixed Gases

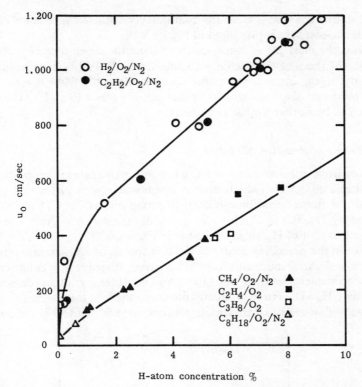

Figure 8.17 Influence of the H-atom concentration of the flame speed of various hydrocarbons burning in oxygen and air (from E. Bartholome, *Zeit. Electrochem.*, **54**, p.169, (1950))

u_0 as a function of the sum of H-atom, O-atom and OH-radical concentrations each multiplied by the respective relative diffusion coefficient. It is found that a single correlation between u_0 and $(6.5\ C_H + C_O + C_{OH})$ describes all these flames, thus implying a single combustion mechanism for all of them.

(f) *Influence of Inert Additives*

Additives such as CO_2, N_2, He and Ar are chemically inert. They do influence, however, the physical properties (such as conductivity, specific heat, etc.) of the mixture. Many investigators studied the effects of these additives. Added to H_2/O_2, CO/O_2 and CH_4/O_2 mixtures, carbon dioxide and nitrogen produce similar effects—they (a) reduce the flame speed, (b) narrow the flammability range and (c) shift the maximum u_0 towards smaller

% fuel content. These effects are qualitatively depicted in Figure 8.18. A quantitative measure is presented in Figure 8.19.

Mainly, the inert gas addition seems to effect the flame speed by effecting the ratio of thermal conductivity to specific heat. Addition of helium (or especially, argon) increases this ratio and, therefore, the flame speed.

The influence of excess oxidant or fuel (see Figures 8.10 and 8.11) may be expected to be similar to that of an inert.

(g) *Influence of Reactive Additives*

It is clear from our discussion of CO/air flame propagation that addition of a small amount of H_2 (which itself is combustible with air) would greatly increase the flame speed due to chain reaction effects. As CO is gradually replaced by H_2, the u_0-% fuel plot gradually moves away from the curve of CO/air to that of H_2/air as indicated in Figure 8.20.

Unlike in the preceding example, if flame speeds of the separate mixtures differ only slightly, the transition of the curves displays a peculiarity. For example, consider a mixture of CO/air. And consider a gradual replacement of CO by CH_4. The curve gradually moves to the left; but in the process of doing so, it first rises to a maximum when a mere 5% of CO is replaced by

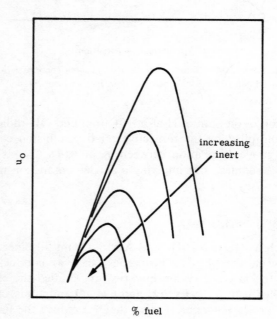

Figure 8.18 Influence of inert species on flame speed

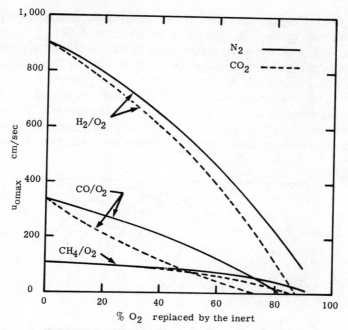

Figure 8.19 Influence of inert species on flame speed

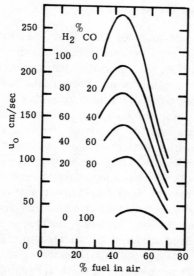

Figure 8.20 Flame speed of H_2 + CO mixtures burning in air

CH_4. Figure 8.21 shows this. The obvious reason is the enhancement of CO/air reaction rate by the hydrogen atoms generated from CH_4.

In connection with flame propagation in mixtures of two or more combustible mixtures (which are assumed not to interact to alter the reaction mechanisms), Mallard and Le Chatelier law has to be stated here.

If two or three mixtures with equal flame speeds are mixed in any proportions, the flame speed of the resultant mixture remains unaltered. Therefore, *mixtures of limit mixtures are themselves limit mixtures.* Likewise, mixtures of maximum flame speed are themselves maximum flame speed mixtures.

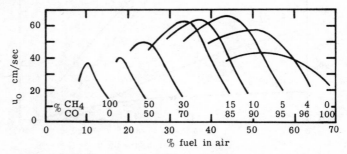

Figure 8.21 Flame speed of CH_4 + CO mixtures burning in air

8.6 Flame Propagation Theory

Flame propagation theory involves solution of the equations of fluid flow, heat transfer and diffusion of various species in order to obtain a detailed picture of the flame structure (i.e., the spatial distribution of pressure, velocity, temperature and concentration of all species) and the fundamental flame speed. The input and boundary conditions for such a solution include (a) the initial mixture conditions, (b) transport properties, (c) thermodynamic properties of all species and (d) details of chemical reaction mechanism and kinetics. Theories which attempt such a complete solution are known in the literature as *comprehensive*. These theories yield a basic understanding of the flame behavior with which one may develop new ways of improving (or impeding, as the case may be) the combustion performance in a given situation. All too often, however, an engineer who either may be designing combustors and afterburners or may be rating the fire hazard of a fuel/oxidant containment does not need the detailed flame structure. Instead, a formula for the flame speed is all that he requires. Theories which provide just this information are naturally simpler and involve approximations suggested by experiments.

Flames in Premixed Gases

Some of these simple theories, known as thermal theories, assume that the principal process governing the flame propagation is conduction of heat from the reaction zone to the prereaction zone. Other simple theories, known as diffusional theories, advocate that the controlling process is the diffusion of chain carrying active radicals and atoms from the reaction zone to the prereaction zone. In reality, however, very few flames are governed either only by thermal or only by diffusional mechanism. The oxidation of hydrogen and that of CO in the presence of moisture or hydrogen are two exceptions in which, as discussed in the preceding section, diffusional mechanism exerts a dominating role. Similarly, in the decomposition of hydrazine, of azomethane, of methyl nitrate and of N_2O, as discussed in Section 4.7, thermal mechanism is predominant. Hydrocarbons burning with oxygen or air exhibit both thermal as well as diffusional features. A detailed survey of all the flame propagation theories (up to 1952) is presented by Evans. In this section we will briefly describe the simple thermal and diffusional theories.

(a) Zeldovich—Frank-Kamenetski Thermal Theory

The principal proponents of thermal theories come from Semenov's school in Russia. Consider steady propagation of a plane deflagration wave. Assume that the reaction rate follows a strong temperature dependency (i.e., an Arrhenius type law). The differential equation describing conservation of energy indicates a balance between conduction, convection and chemical reaction. If K is the mean conductivity of the mixture and C is its specific heat,

$$K\frac{d^2T}{dx^2} - [\rho u]C\frac{dT}{dx} - \dot{W}_F''' \Delta H = 0 \qquad (8.8)$$

If the flame propagates from left to right and if heat loss from the burnt gases to the wall is neglected, the boundary conditions for Eq. 8.8 will be as following.

$$\begin{array}{lll} x = -\infty & T = T_f & dT/dx = 0 \\ x = +\infty & T = T_s & dT/dx = 0 \end{array} \qquad (8.9)$$

The mass flux $[\rho u]$ is independent of x according to the continuity equation. $\dot{W}_F'''$ is the rate of the combustion reaction in grams of fuel disappeared per cubic centimeter per second. It is given by some rate equation such as

$$\dot{W}_F''' = -k_n C_O^{n-j} C_F^j e^{-E/RT} \qquad (8.10)$$

where n is the overall order of the reaction and j is the order with respect to the fuel. C_O and C_F are the concentrations respectively of the oxidant and fuel. ΔH is the heat of combustion in calories per gram fuel.

The reaction rate, as shown in Figure 8.22, is a function of x since the composition and temperature are.

Equation 8.10 indicates that the reaction rate in the oncoming supply mixture is small but not zero. Since T_s is low compared to T_f and since the activation energy is usually about 40,000 to 60,000 cal/mole, the exponential term and hence the reaction rate will indeed be close to zero; and it is convenient to assume it as zero.*

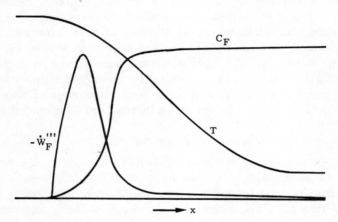

Figure 8.22 Thermal theory of flame propagation

Zeldovich simplifies the problem by assigning an "ignition temperature" T_i. He modifies the spatial distribution of reaction rate by noting that (a) the ignition temperature is close to the final temperature and (b) the reaction proceeds to a negligible extent when temperature is lower than T_i. He divides the flame front into two regions—the preheat zone and the reaction zone—as shown in Figure 8.23. Since reactions are negligible in the preheat zone, Eq. 8.8 simplifies to

$$K \frac{d^2 T}{dx^2} - [\rho u] C \frac{dT}{dx} = 0 \qquad (8.11)$$

with the boundary conditions,

$$\begin{array}{lll} x = -x_i & T = T_i & \\ x = +\infty & T = T_s & dT/dx = 0 \end{array} \qquad (8.12)$$

---

* If this reaction rate is non-negligible all the reactants would be completely consumed by the time they approach the flame front. This matter may be found discussed in the combustion literature under the title *Cold Boundary Difficulty*.

Flames in Premixed Gases

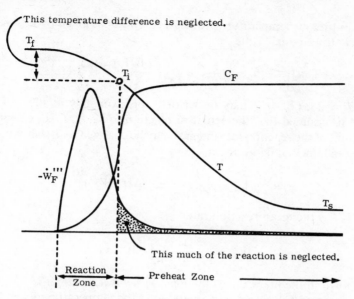

Figure 8.23 Zeldovich–Frank-Kamenetski approximation

Integrating Eq. 8.11 once and evaluating the temperature gradient at the ignition point we obtain

$$\left.\frac{dT}{dx}\right|_i = \frac{[\rho u]C}{K}(T_i - T_s) \tag{8.13}$$

In the reaction zone, since $T_i \approx T_f$, the convection term may be neglected to simplify Eq. 8.8 to

$$K\frac{d^2T}{dx^2} - \dot{W}_F''' \Delta H = 0$$

Rewriting,

$$\frac{dT}{dx} \cdot d\left(\frac{dT}{dx}\right) = \frac{\Delta H}{K} \dot{W}_F''' \, dT$$

Note $(dT/dx)_f = 0$ and integrate from T_i to T_f to obtain

$$\left.\frac{dT}{dx}\right|_i = \left(\frac{2\,\Delta H}{K}\int_{T_i}^{T_f} \dot{W}_F''' \, dT\right)^{1/2} \tag{8.14}$$

For a continuous temperature profile, the two slopes given by Eqs. 8.13 and 8.14 have to be equal. Thus,

$$[\rho u] \equiv \rho_s u_0 = \frac{K}{C(T_i - T_s)} \cdot \left(\frac{2 \Delta H}{K} \int_{T_i}^{T_f} \dot{W}_F''' \, dT\right)^{1/2}$$

Since $T_i \approx T_f$, $(T_i - T_s)$ may be written approximately as $(T_f - T_s)$. Also since it is assumed that the extent of reaction for $T < T_i$ is negligible, the lower limit of the reaction rate integral may be changed to T_s. Thus resolving for the fundamental flame speed,

$$u_0 = \left(\frac{K}{\Lambda \rho_s^2 C^2 (T_f - T_s)} \cdot \Delta H \cdot \overline{\dot{W}_F'''}\right)^{1/2} \qquad (8.15)$$

where $\overline{\dot{W}_F'''}$ is the mean reaction rate.

$$\overline{\dot{W}_F'''} = \frac{1}{(T_f - T_s)} \int_{T_s}^{T_f} \dot{W}_F''' \, dT$$

Λ is known in the literature as the *flame speed eigenvalue*; it is equal to $\frac{1}{2}$ in the present result. It depends strongly upon the temperature dependency of reaction rate and to some degree, on Lewis number. Depending upon the assumptions made in any particular theory, the value of Λ will be approximately equal to, but slightly less than, $\frac{1}{2}$.

Spalding provides charts of Λ as a function of the Lewis number and the shape of the rate-temperature plot. He claims that these values of Λ are suitable for a variety* of reactions whose rates are expressible explicitly in terms of temperature.

The influence, indicated by Eq. 8.15, of pressure, temperature and composition on the flame speed, in a thermally controlled situation, is consistent with the experimental observations discussed in Section 8.5.

(b) Diffusion Theory of Tanford and Pease

We have seen in Section 8.5 that in most of the hydrocarbon/oxygen (air) flames, both thermal and active particle diffusion effects come into picture. A purely diffusional explanation is possible for flame propagation in moist CO/O_2 and H_2/O_2 mixtures.

Tanford and Pease consider a system in which several active chain carrying species exist. The gist of their theory is as following.

The conservation equations for all the stable (i.e., reactants and products) and unstable species express a balance between the local diffusive and

---

* Spalding considers (a) a simple single step reaction, (b) a chain reaction which satisfies a steady state chain carrier concentration condition and (c) a branched chain reaction.

Flames in Premixed Gases

convective fluxes and production and/or consumption. Local chain carrier concentrations are increased by production due to dissociation and are decreased by consumption due to their chain carrying activity. Boundary condition for these species is provided by equilibrium calculation at the flame temperature. The combustion rate (i.e., disappearance of the fuel or appearance of the product) is assumed to be given by

$$\dot{W}_F''' = -\sum_{i=1}^{N} k_i C_i C_F$$

where C_i is the local concentration of the chain carrier $i = 1, 2, \ldots N$ and C_F is the local fuel concentration. k_i is the specific reaction rate constant associated with the unstable species i. Invoking then the total mass conservation (viz: the reactant flow through the flame is equal to the product flow out of the flame), the final expression for the fundamental flame speed is deduced as

$$u_0 = \left(\sum_{i=1}^{N} \frac{k_i C_F P_{if} D_i}{X_\mathscr{P} B_i} \right)^{1/2} \qquad (8.16)$$

where P_{if} is the equilibrium partial pressure of the unstable species i in the final flame gases, D_i is the diffusion coefficient of i through the reactant mixture, $X_\mathscr{P}$ is the mole fraction of the products and B_i is a correction factor to account for the loss of i due to collisions with walls.

As mentioned in Section 8.5, Simon verified Tanford's theory for a variety of hydrocarbon/air flames. She also shows that $u_{0\max}$ versus T_f yields a single correlation for many hydrocarbons and suggests that this apparent validity of thermal theory is perhaps an outcome of the similarity between heat conduction and species diffusion equations.

We will close this section with a comment on the property values. Since the considered processes occur at high temperatures, one has to be rather judicious in selecting the necessary transport properties. Hilsenrath's property tables and Maxwell's data book are recommended in this connection.

8.7 Propagation of Turbulent Premixed Flames

Turbulent flames are of more practical importance than laminar flames. Premixed turbulent flames propagate faster than laminar ones. Considering Eq. 8.15, the space heating rate in a system involving premixed flames is

$$\Delta H \cdot \overline{\dot{W}}_F''' = \frac{\Lambda \rho_s^2 C^2 u_0^2 (T_f - T_s)}{K} \quad \text{cal/cm}^3/\text{sec} \qquad (8.17)$$

This equation shows that an increase in the flame speed would result in a higher space heating rate. Rendering the flames turbulent, the flame speed may be increased far beyond the fundamental (i.e., *laminar*) flame speed. Added to this, the flame surface area is greatly increased by turbulence as it multilates, folds and often breaks up the flame front. The result is a greatly exaggerated space heating rate.

The effect of turbulence on flame propagation was discovered quite accidentally. Experiments using tube method indicated that the "fundamental" flame speed measured in tubes of larger diameters is larger. This was later realized to be a consequence of turbulence which increases with an increase in tube size. Note here must be made, however, that once turbulence is triggered on, what indeed we measure is no longer the "fundamental flame speed" whose definition requires the flame to be laminar. While the fundamental flame speed is a characteristic property of the reacting mixture alone, the turbulent flame speed (hereafter called burning speed, in order to distinguish it) depends upon the dimensions of the tube as well. Transport of heat and mass in laminar flames occurs by molecular diffusion phenomena. In turbulent flames, however, eddy mixing which is a function of geometry contributes to this transport.

Referring to Chapter 3, turbulence is characterized by the eddy size— large or small scale—and the mixing length. The eddy diffusivity ε is a measure of the mutual interaction among the eddies in the same manner as the kinematic viscosity v is a measure of the molecular interactions. ε is known to be approximately proportional to the Reynolds number in tube flow. (Refer to Chapter 7.) Transition from laminar to turbulence is expected in tube flow if Reynolds number ($\equiv \rho u d / \mu$) is greater than 2,300. Thus, flow in a tube becomes turbulent if the tube size or the flow rate is increased.

Figure 8.24 shows Damkohler's (Bunsen burner) measurements of flame speed at various Reynolds numbers. He found that the flame speed is (a) independent of Reynolds number when $Re < 2,300$, (b) proportional to the square root of Reynolds number in the range $2,300 \lesssim Re \lesssim 6,000$ and (c) proportional to the Reynolds number if $Re \gtrsim 6,000$. Obviously, only item (a) above obeys our definition of the *fundamental* flame speed; items (b) and (c) are influenced by turbulence and hence the measured flame speeds depend on geometry and flow. Denoting the turbulent flame speed by the symbol S_T, Damkohler explains his measurements as following.

(a) Small Scale Turbulence

In the range $2,300 \lesssim Re \lesssim 6,000$, turbulence is of fine scale; that is, the eddy size and mixing length are much smaller than the flame front thickness. The

Flames in Premixed Gases

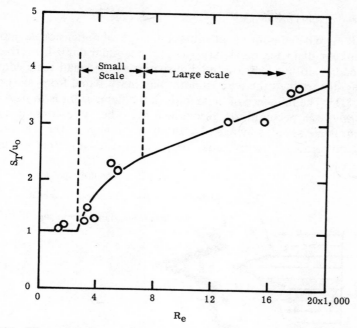

Figure 8.24 Effects of Reynolds number on the flame speed (from G. Damkohler, *Zeit. Electrochem.*, **46**, p. 601, (1940))

effect of these fine scale eddies is to enhance the intensity of transport processes within the combustion wave. Under these circumstances, transport of heat and species is proportional to the eddy diffusivity ε rather than the molecular diffusivity D_i (or $K/\rho_s C$). Equations 8.15 and 8.16 indicate that the flame speed is directly proportional to $\sqrt{K/\rho_s C}$ or $\sqrt{D_i}$. Hence it is logical to expect the burning speed, of a small scale turbulent flame, to be proportional to $\sqrt{\varepsilon}$. Thus

$$\frac{S_T}{u_0} = \left(\frac{\varepsilon}{K/\rho_s C}\right)^{1/2} \approx \left(\frac{\varepsilon}{D_i}\right)^{1/2} \approx \left(\frac{\varepsilon}{\nu}\right)^{1/2}$$

From Section 7.6, $\varepsilon/\nu \approx 0.01 \text{ Re}$ for flow in a tube. Therefore,

$$\frac{S_T}{u_0} \approx 0.1 \text{ Re}^{1/2} \tag{8.18}$$

This equation indeed predicts the trend of Damkohler's small scale burning speed measurements.

(b) *Large Scale Turbulence*

When Re ≳ 6,000 the turbulent eddies are large, of dimensions comparable with the tube diameter, much larger than the laminar flame front thickness. These eddies do not increase the diffusivities as the small scale eddies do, but they distort the otherwise smooth "laminar" flame front as shown in Figure 8.25. The influence of these folds in the flame front is to *increase the flame front area* per unit cross section of the tube. As a consequence, the apparent flame speed is increased without any change in the instantaneous local flame structure itself. Damkohler estimated to show that the increase

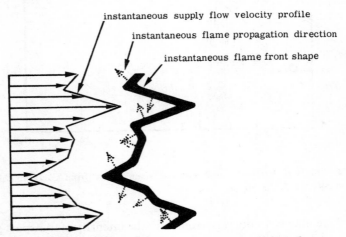

Figure 8.25 An exaggerated view of the turbulent flame front

in surface area is proportional to the characteristic wrinkle (fold) size which is proportional to the magnitude of velocity fluctuation (i.e., of turbulence intensity). Since ε is proportional to the product of intensity and mixing length and since $\varepsilon/v \approx 0.01$ Re,

$$\frac{S_T}{u_0} \propto \text{area} \propto \text{fluctuation} \propto \varepsilon \propto \text{Re} \tag{8.19}$$

This explanation describes Damkohler's large scale burning speeds quite satisfactorily.

It is possible to conceive that at very high Reynolds numbers the turbulence may get so intense that the wrinkles and folds in the flame front ultimately end up breaking the front. The resultant "pieces" (islets or lumps) of flame may now jump ahead of the mean flame location into the fresh gas and conversely, lumps of the fresh gas may jump into the flame. This situation is extremely complex and at present our understanding of it is too sketchy

Flames in Premixed Gases

8.8 Flame Stabilization

In order to accomplish large thrusts in a turbojet or a ramjet engine, the supply velocity of the reactant mixture is desired to be extremely high; it is not unusual for this velocity to be as high as ten times the maximum possible turbulent flame speed of a given mixture. Experience shows that the flame is blown away when the supply velocity exceeds the flame speed. The maximum supply velocity with which fresh mixture may be brought to the flame front without blowing it away is known as *blow-off velocity*. This important limiting velocity depends upon a host of factors which includes the nature of the fuel and oxidant, their ratio, mixture temperature, combustion chamber pressure, turbulence in the approach stream, burner geometry, burner wall roughness temperature, etc.

(a) Stability of a Bunsen Flame

Figure 8.8 indicated the mechanism which determines the shape of a Bunsen flame. Based upon this simple mechanism we can explain the phenomena of blow-off and flash-back. Figure 8.26 shows flame speed and normal component of the supply velocity u_{sn} in four different situations. Only the region

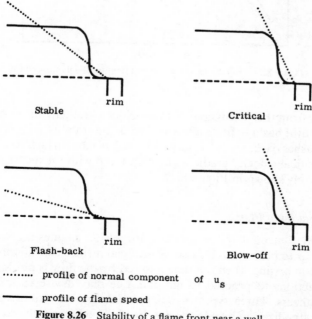

········· profile of normal component of u_s

——— profile of flame speed

Figure 8.26 Stability of a flame front near a wall

close to the wall is shown since it is here that the flame is anchored to the burner. When the flow is feeble, u_0 is greater than u_{sn} in almost the entire cross section of the barrel; therefore, the flame propagates into the barrel to bring about flash-back. On the other hand, when the flow is very strong, u_{sn} is greater than u_0 in the entire cross section to result in a blow-off. The critical criterion for blow-off arises when the u_0 and u_{sn} profiles are tangential.

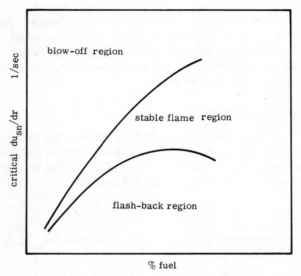

Figure 8.27 Effect of the velocity gradient on the flame stability

It is clear from this discussion that the velocity gradient (at the barrel wall near its mouth) has a definite effect on the flame stability; if it is too small, the flame flashes back; if it is too large, blow-off occurs. In fact, measurements show the critical velocity gradient correlates well with the fuel oxidant ratio as qualitatively shown in Figure 8.27.

(b) *Stabilization of Flames*

The velocities encountered in modern propulsion engines are so high that the flame has to be stabilized by some artificial means. Considering blow-off as a situation arising when the time allowed by the flow is not long enough for the reactions to proceed to ignition, one may devise various possible flame stabilizers. Three types of flame stabilizers are extensively known. These are: stabilization by pilot flames, by bluff bodies, and by recirculation.

Flames in Premixed Gases

By Pilot Flames

Suppose a pilot flame is held adjacent to the cold reactant mixture flow issuing in the form of a high velocity jet. Heat and mass are transferred across the boundary of the two streams by diffusion and mixing. The reaction rate in the cold reactant mixture is thus enhanced. Marble and Adamson dealt with this situation by solving the two-dimensional problem shown in Figure 8.28. We will not elaborate their solution here but make a rough estimation of the criterion of flame stabilization.

The mass flow rate through the jet of area A is $A\rho u_s$ gm/sec. If $\overline{W}_F'''$ is some sort of a mean reaction rate corresponding to a volume V of the jet, the rate of consumption due to reaction is $V\overline{W}_F'''$ gms/sec. Blow-off would occur if the flow rate is greater than the reaction rate. Thus there occurs a sustained flame if

$$A\rho u_s \leq V\overline{W}_F'''$$

The blow-off velocity u_{BO} is given by the limiting supply velocity when the two rates are equal. Thus,

$$u_{BO} = \frac{V}{A}\frac{\overline{W}_F'''}{\rho}$$

According to Eq. 8.15,

$$\overline{W}_F''' = \frac{\rho_s^2 u_0^2 \Lambda C^2 (T_f - T_s)}{K \Delta H}$$

Hence

$$u_{BO} = \frac{\Lambda C^2 (T_f - T_s)}{K \Delta H} \cdot \frac{V}{A} \cdot \rho_s u_0^2 \tag{8.20}$$

This equation shows that the blow-off velocity is expected to be proportional to the characteristic dimension of the system, to the pressure and the square of the flame speed. Experiments show that this is a valid prediction.

By Bluff Bodies

When a blunt body is placed in a high velocity reactant stream, flame stabilization may occur by two mechanisms. The flow is greatly slowed down at the forward stagnation point (see Figure 8.29) to give ample opportunity for reactions to proceed to ignition. Chambré analyzed this type of a problem. One major disadvantage of solid bluff bodies is the drag they exert on the flow and the resultant loss of thrust. Cambel overcomes this drawback by employing an opposing gaseous jet in the reactant stream. A stagnation point

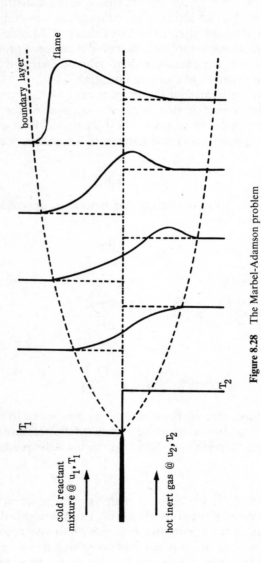

Figure 8.28 The Marbel-Adamson problem

Flames in Premixed Gases 309

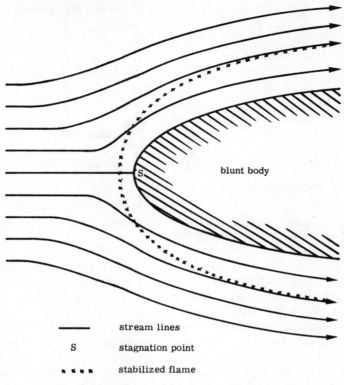

Figure 8.29 Flame stabilization by a blunt body

evolves as the opposing jet is slowed down and the flame is anchored as schematically shown in Figure 8.30. The blow-off velocity is increased by increasing the injection pressure of the opposing jet, by increasing the temperature of the opposing jet gas and by choosing a combustible mixture for the opposing jet gas.

The second mechanism by which bluff bodies stabilize the flames is described below.

By Recirculation

When the solid bluff body discussed above is of finite length in the direction of flow, the pressure distribution prevents the high velocity flow from keeping attached to the solid surface. Increasing pressure separates the boundary layer and causes eddy shedding in the "wake." Under sufficiently fast flow conditions a (symmetric) recirculation pattern of flow is established behind the blunt body as shown in Figure 8.31. The recirculation zone provides

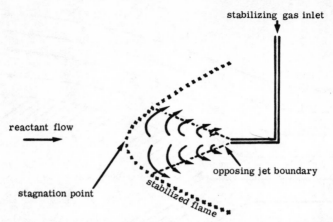

Figure 8.30 Flame stabilization by an opposing jet (adapted from: A. B. Campbell and B. H. Jennings, *Gas Dynamics*, McGraw-Hill, (1958))

a station where reactions can take place. Shown in Figure 8.32 is a hypothetical model of the longitudinal mixer and reactor associated with the recirculation regime. $\dot{m}_s$ (gms/sec) of the fresh reactants at a temperature T_s enter into the mixer to mix well (but not react) with $\dot{m}_r$ (gms/sec) of the reactor output at T_r. The resultant mixture of $\dot{m}_s + \dot{m}_r$ at T_m splits into two branches one of which is fed into the reactor. Focusing attention on the reactor, heat is lost by it to the mixture at a rate which is a linear function of the temperature difference $(T - T_s)$ where T is a characteristic mean temperature ($\approx (T_m + T_f)/2$) of the reactor contents. Figure 8.33 shows this straight line

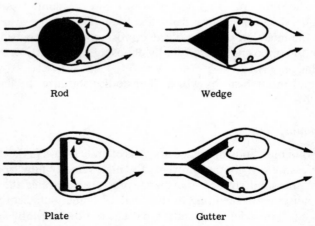

Figure 8.31 Flame stabilization by recirculation

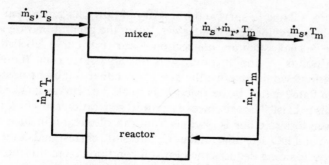

Figure 8.32 Hypothetical model for recirculation/mixing

heat loss rate relation. The slope of this line is determined by the relative magnitudes of $\dot{m}_s$ and $\dot{m}_r$, and those of the concerned specific heats. The rate of heat generation in the reactor is directly proportional to the reaction rate which is given by, say, an Arrhenius type rate equation. The curves of this rate for various kinetic constants are also drawn in Figure 8.33.

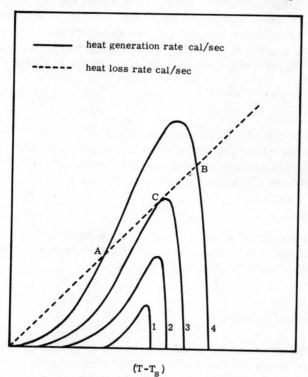

Figure 8.33 Thermal Imbalance at extinction

Now, for a flame to survive the severity of flow, the generation rate curve and the loss rate line should intersect. When the generation rate is high and loss rate is weak, such an intersection occurs (see curve labelled 4) at two nonzero points, A and B. Point A is unstable since even the slightest of perturbations would remove the system to either 0 (absent flame state) or B (stable flame state). If the reaction is weak (i.e., curves labelled 1 and 2), flame fails to exist. The existence of a critical generation rate curve (labelled 3) for a given loss rate line is obvious from this discussion in such a way that the curve and line are tangential at C. This state corresponds to the blow-off criticality. Spalding deduces the blow-off condition from this picture as

$$\frac{u_{BO}}{dPu_0^2} = \text{a constant} \tag{8.21}$$

where d is the characteristic linear dimension of the recirculation zone. Equation 8.21 is consistent with Eq. 8.20.

In the preceding rough analysis, loss of heat to the bluff body is not considered. Addition of this extra loss term makes the loss line steeper and thus blow-off, easier. Keeping the bluff body at a higher temperature would raise the blow-off velocity. Likewise, increased turbulence in the reactant mixture increases the $\dot{m}_B$ entering into our imaginary mixer and therefore results in a steeper line for loss. The effect once again is to lower the blow-off velocity.

Other Models

Equation 8.21 may also be obtained by considering the lateral mixing of the eddy-contained hot products by the model developed in Section 4.9(c).

Toong studied ignition of a cold reactant mixture flowing along a hot flat plate—a problem of relevence to the flame stabilization.

Other Methods

Since separated flows always contain recirculation to some degree, sudden increases in the cross sectional area of a duct or steps and recesses in the walls are often useful for flame stabilization.

8.9 Acoustical Instabilities

Practical flames often spontaneously become oscillatory. The flame in a pulse jet engine is inherently oscillatory. The instability of flames is desired to be minimized in view of the mechanical integrity of engines. Combustion with different degrees of instability is called by different names. If the frequency of instability is lower than about 100 cycles/sec, "chugging" or "chuffing" combustion arises. The fuel feed system may be capable of inducing the chugging type instability. If the frequency of oscillations is of the order of hundreds or thousands of cycles per second, "screaching" or "screaming" combustion arises. The interaction of the chamber boundaries with the sonic waves in the combustion gases is responsible for this mode. All too often the "screaching" is annoyingly noisy.

8.10 Concluding Remarks

Propagation of a flame in premixed gases offers a very stimulating problem for both experimental as well as analytical work. Quite extensive experimental data is available in such treatises as those of Lewis and von Elbe, Jost, Gaydon and Wolfhard, Khitrin, etc. The survey paper of Evans and the book of Williams are excellent references for the theory of propagation. The Combustion Symposia and the Journal of *Combustion and Flame* provide more recent advances in the field. The book of Fristrom and Westenberg presents the fascinating world of a skillful experimentalist. N.A.C.A. 1300 (published in 1959) gives a thorough account of the science of hydrocarbon combustion. For a comprehensive but a many-sided look at the experimental methods Princeton volume IX of High Speed Aerodynamics and Jet Propulsion Series is highly recommended. This volume also contains a concise summary of the then (1954) available flame speed data.

References

Bartholome, E., *Naturweissenschaften*, **36**, p. 171 (1949).
Becker, R., *Z. Physik*, **4**, p. 393 (1921); **8**, p. 321 (1922).
Botha, J. P. and Spalding, D. B., *Proc. Roy. Soc. A*, **225**, p. 71 (1954).
Bunsen, R., *Poggendorff's Annals*, **131**, p. 161 (1866)
Cambel, A. B. and Jennings, B. H., *Gas Dynamics*, McGraw-Hill, New York (1958).
Chambré, P. L., *J. Chem. Phys.*, **25**, p. 417 (1956).
Chapman, D., *Phil. Mag.*, **47**, p. 90 (1899).
Combustion Processes, Vol. II of the Princeton Series. Editors: Lewis, Pease, and Taylor (1956).
Courant, R. and Friedrichs, K. O., *Supersonic Flow and Shock Waves*, Interscience, New York (1948).
Damkohler, G., *Z. Elektrochem.*, **46**, p. 60 (1940), English translation, N.A.C.A. Tech. Memo. 1112 (1947).
Dugger, G. L., Heimel, S., and Weast, R. C., *I. and E.C.*, **47**, p. 114 (1955).
Egerton, A., Limits of Inflammability, (A Survey), p. 4, *Fourth Symposium* (*International*) *on Combustion* (1953).
Egerton, A. and Powling, J., *Proc. Roy. Soc. A*, **193**, p. 172 and p. 190 (1948).

Evans, M. W., Current Theoretical Concepts of Steady State Flame Propagation, *Chemical Reviews*, **51**, p. 363 (1952).
Fiock, E. F., Soap Bubble Method, *Chem. Rev.*, **21**, p. 367 (1937).
Frank-Kamenetski, D. A., *Diffusion and Heat Exchange in Chemical Kinetics*, translated by N. Thon, Princeton University Press, Princeton (1955).
Friedman, R. and Burke, E., *J. Chem. Phys.*, **21**, p. 710 (1953).
Fristrom, R. M. and Westenberg, A. A., *Flame Structure*, McGraw-Hill, New York (1964).
Gaydon, A. G., Influence of Diffusion on Flame Propagation, *Proc. Roy. Soc. A.*, **196**, p. 105 (1949).
Gaydon, A. G. and Wolfhard, H. G., *Flames*, Chapman and Hall, London (1960).
Gerstein, M., Levine, O., and Wong, E. L., N.A.C.A. R.M. E50G24 (1950), E51J05 (1951), and *J. Am. Chem. Soc.*, **73**, p. 418 (1951), and also, *I. and E.C.*, **43**, p. 2770 (1951).
Gouy, G., *Ann. Chim. et Phys.*, **5**, No. 18, p. 27 (1879).
Hartmann, E., Karlsruhe Thesis (1931).
Hilsenrath, J., *Tables of Thermal Properties of Gases*, N.B.S. Circular 564, Govt. Printg. Office, Washington, D.C. (1955).
Jahn, G., "Der Zundvorgang in Gasgemischen." Oldenberg, Berlin (1934).
Jahn, G., Karlsruhe Thesis (1934).
Jost, W., *Explosion and Combustion Processes in Gases*, McGraw-Hill, New York (1946).
Jouget, E., *Jour. of Math.*, p. 347 (1905) and p. 6 (1906).
Jouget, E., *Mechanics des Explosifs*, Doin, Paris (1917).
Khitrin, L. N., *The Physics of Combustion and Explosion*, N.S.F. Translation from Russian by the Israel Program for Scientific Translations (1962).
Lafitte, P. and Patrie, *Compt. Rend.*, **191**, p. 1335 (1930).
Lewis, B., Discussion in *Selected Combustion Problems*, AGARD, p. 177 Butterworths, London (1954).
Lewis, B. and Friauf, J., *Journal of Am. Chem. Soc.*, **52**, p. 3905 (1930).
Lewis, B. and von Elbe, G., *Combustion, Flames and Explosions of Gases*, Second Edition, Academic, New York (1951).
Linnett, J. W., Methods of Measuring Burning Velocities, p. 20, *Fourth Symposium on Combustion* (1953).
Marble, F. E. and Adamson, T. C., *Jet Propulsion*, **24**, p. 85 (1954).
Maxwell, J. B., *Data Book on Hydrocarbons*, Van Nostrand, New York (1950).
Michelson, *Wiedemanns Ann.*, **37**, p. 1 (1889).
N.A.C.A. 1300, *Basic Considerations in the Combustion of Hydrocarbon Fuels with Air* (1959).
Physical Measurements in Gas Dynamics and Combustion, Series of High Speed Aerodynamics and Jet Propulsion, Vol. IX, Ladenberg, R. W., Lewis, B., Pease, R. N., and Taylor, H. S., editors, Princeton University Press, Princeton (1954).
Powling, J., *Fuel*, **28**, p. 25 (1949).
Powling, J., The Flat Flame Burner, Section 2.2.1, p. 14 of *Experimental Methods in Combustion Research*, AGARD (1961).
Sachsse, H., *Z. Physik. Chem., A.*, **180**, p. 305 (1937).
Sachsse, H. and Bartholome, E., *Z. Electrochem.*, **53**, p. 183 (1949).
Simon, D. M., Diffusion Processes as Rate-Controlling Steps, p. 59 in *Selected Combustion Problems*, AGARD (1954).
Simon, D. M., Flame Propagation Active Particle Diffusion Theory, *Ind. and Engg. Chem.* **43**, p. 2718 (1951).
Spalding, D. B., *Combustion and Flame*, **1**, p. 287 and 296 (1957).
Spalding, D. B., *Some Fundamentals of Combustion*, Butterworths, London (1955).
Spalding, D. B., Theoretical Aspects of Flame Stabilization, *Aircraft Engg.*, **25**, p. 264 (1953).

Stevens, F. W., N.A.C.A. Report 305 (1929); 372 (1930).
Tanford, C., The Role of Free Atoms and Radicals in Burner Flames, *Third Symposium on Combustion*, p. 140 (1949).
Tanford, C. and Pease, R. N., *J. Chem. Phys.*, **15**, p. 431 (1947) and p. 861 (1947).
Tanford, C. J., *Chem. Phys.*, **15**, p. 433 (1947).
Toong, T. Y., *Sixth Symposium on Combustion*, p. 532 (1957).
Weinberg, F. J., *Optics of Flames*, Butterworths, London (1963).
Williams, F. A., *Combustion Theory*, Addison-Wesley, Reading (1964).
Zeldovich, Ia. B., *Theory of Flame Propagation*, English translation, N.A.C.A. Tech. Memo. 1282 (1948).
Zeldovich, Ia. B. and Kompaneets, A. S., *Theory of Detonation*, Academic, New York (1960).
Zukoskii, E. E. and Marble, F. E., *Comb. Research and Reviews*, p. 167 (1955).
Zukoskii, E. E. and Marble, F. E., *Proceedings of the Gas Dynamics Symposium*, p. 205, Northwestern University, Evanston (1956).

Postscript

Having thus introduced the topic of combustion to the students and researchers in engineering sciences, it is appropriate to close this book with a bibliography for further study. Listed below are some books, series, symposia and journals. Complete by no means, this list can not avoid being biased by the author's limited interests and unlimited ignorance.

I Books

Abramovich, G. N., *Theory of Turbulent Jets*, MIT Press (1963).
A.P.I., *Selected Values of Properties of Hydrocarbons and Related Compounds*, American Petroleum Institute, Washington, D.C.
Barnett, H. C. and Hibbard, R. R. (eds.), *Basic Considerations in the Combustion of Hydrocarbon Fuels with Air*, National Advisory Committee for Aeronautics, Report No. 1300 (1959).
Benson, S. W., *Thermochemical Kinetics*, John Wiley (1968).
Bird, R. B., Stewart, W. E., and Lightfoot, E. N., *Transport Phenomena*, Wiley (1962).
Bone, W. A. and Townend, D. T. A., *Flame and Combustion in Gases*, Longmans (1927).
Campbell, A. B. and Jennings, B. H., *Gas Dynamics*, McGraw-Hill (1958).
Dainton, F. S., *Chain Reactions*, Methuen (1956).
Daniels, F. and Alberty, R. A., *Physical Chemistry*, Wiley (1955).
Eckert, E. R. G. and Drake, R. M., *Heat and Mass Transfer*, McGraw-Hill (1959).
Egloff, G., *Physical Constants of Hydrocarbons*, ACS Monograph 78, The American Chemical Society (1940).
Faraday, M., *Chemical History of a Candle*, Viking Press (1960).
Fenimore, C. P., *Chemistry in Premixed Flames*, Pergamon (1964).
Frank-Kamenetski, D. A., *Diffusion and Heat Exchange in Chemical Kinetics*, Princeton University Press (1955).
Fristrom, R. M. and Westenberg, A. A., *Flame Structure*, McGraw-Hill (1965).
Gaydon, A. G., *Spectroscopy and Combustion Theory*, Chapman-Hall (1948).
Gaydon, A. G. and Wolfhard, H. G., *Flames, Their Structure, Radiation and Temperature*, Chapman-Hall (1953).
Goldsmith, S. (ed.), *Modern Developments in Fluid Dynamics*, Clarendon Press, Oxford (1938).
Hinshelwood, C. N., *Kinetics of Chemical Change*, Oxford (Clarendon) Press (1940).
Hinze, J. O., *Turbulence*, McGraw-Hill (1962).

Hirchfelder, J. O., Curtis, O. F., and Bird, R. B., *Molecular Theory of Gases and Liquids*, Wiley (1954).
Janz, G. J., *Thermodynamic Properties*, Academic (1958).
Jost, W., *Explosion and Combustion Processes in Gases*, Academic (1946).
Kays, W. M., *Convective Heat and Mass Transfer*, McGraw-Hill (1966).
Khitrin, L. N., *The Physics of Combustion and Explosion*, Israel Program for Scientific Translations, OTS-61-31205, Jerusalem (1962).
King, E. L., *How Chemical Reactions Occur*, Benjamin (1964).
Lewis, B. and von Elbe, G., *Combustion, Flames and Explosions of Gases*, Academic (1951).
Literature of the Combustion of Petroleum, A.C.S. Advances in Chemistry Series No. 20 (1958).
Maxwell, J. B., *Data Book on Hydrocarbons*, Van Nostrand (1950).
Moore, W. J., *Physical Chemistry*, 3rd Printing, Prentice-Hall (1963).
Mortimer, C. T., *Reaction Heats and Bond Strengths*, Pergamon (1962).
NAS-NRC, *Consolidated Index of Selected Property Values*, Physical Chemistry and Thermodynamics, National Academy of Sciences, National Research Council, Publication No. 976, Washington, D.C. (1962).
Palmer, K. N., *Dust Explosions and Fire*, Chapman-Hall (1973).
Penner, S. S., *Introduction to the Study of Chemical Reactions in Flow Systems*, Butterworths (1955).
Penner, S. S., *Chemical Rocket Propulsion and Combustion Research*, Gordon and Breach (1962).
Penner, S. S., *Chemistry Problems in Jet Propulsion*, Pergamon (1957).
Rodiguin, N. M. and Rodiguina, E. N., *Consecutive Chemical Reactions*, Van Nostrand (1964).
Schlichting, H., *Boundary Layer Theory*, McGraw-Hill, 6th edition (1968).
Semenov, N., *Chemical Kinetics and Chain Reactions*, Oxford (1935).
Semenov, N., *Some Problems of Chemical Kinetics and Reactivity*, Vols. 1 and 2, Pergamon (1959).
Shchelkin, K. I. and Troshin, Y. K., *Gas Dynamics of Combustion*, Mono Book Co., Baltimore (1965).
Sokolik, A. S., *Self-ignition, Flame and Detonation in Gases*, Israel Program for Scientific Translations, NASA TT F-125 OTS-63-11179, Jerusalem (1963).
Spalding, D. B., *Some Fundamentals of Combustion*, Butterworths (1955).
Steacie, E. W. R., *Atomic and Free Radical Reactions*, American Chemical Society Monograph Series Nos. 102 and 125, Reinhold (1946) and (1954).
Strehlow, R. A., *Fundamentals of Combustion*, International Textbook Co. (1968).
Stull and Sinke, *Thermodynamic Properties*, American Chemical Society, Advances in Chemistry Series, No. 18 (1956).
Thring, M. W., *Sciences of Flames and Furnaces*, Chapman-Hall (1952).
Vulis, L. A., *Thermal Regimes of Combustion*, McGraw-Hill (1961).
Weinberg, F. J., *Optics of Flames*, Butterworths (1963).
Williams, F. A., *Combustion Theory*, Addison-Wesley (1965).

II Series and Symposia

1. Princeton High Speed Aerodynamics and Jet Propulsion Series:
    Vol. I: *Thermodynamics and Physics of Matter*, Rossini, F. D. (ed.)
       II: *Combustion Processes*, Lewis, B. et al. (ed.).
       III: *Fundamentals of Gas Dynamics*, Emmons, H. W. (ed.).
       V: *Turbulent Flows and Heat Transfer*, Lin, C. C. (ed.).
       IX: *Physical Measurements in Gas Dynamics and Combustion*, Landenberg, R. W., et al. (ed.).

2. American Rocket Society, Progress in Astronautics and Rocketry Series (Academic Press):
    Vol. 1. *Solid Propellant Rocket Research*, Summerfield, M. (ed.) (1960).
        2. *Liquid Rockets and Propellants*, Bollinger, L. E., et al. (ed.) (1960).
        15. *Heterogeneous Combustion*, Wolfhard, H. G., et al. (eds.) (1964).
3. The *Combustion* Institute *Biennial* International Symposia Proceedings (since about 1947).
4. The North Atlantic Treaty Organization AGARD Publications.
    AGARDograph 4: *Spontaneous Ignition of Liquid Fuels*, Mullins, B. P., Butterworths (1955).
    AGARDograph 31: *Explosions, Detonations, Flammability and Ignition*, Mullins, B. P. and Penner, S. S., Pergamon (1959).
    AGARDograph 9: *Combustion Researches and Reviews*, Butterworths (1955).
    AGARDograph 15: *Combustion Researches and Reviews*, Butterworths (1957).
    AGARDograph 75: *Nonsteady Flame Propagation*, Markstein, G. H., (ed.), Pergamon (1964).
    Selected Combustion Problems, I: Fundamentals and Aeronautical Applications, Hawthorne, W. R., et al. (ed.), Butterworths (1954).
    Selected Combustion Problems, II: Butterworths (1956).
    Supersonic Flow, Chemical Processes and Radiative Transfer, Olfe, D. B. and Zakkay, V. (eds.), Pergamon (1964).
    Experimental Methods in Combustion Research, A Manual, Surugue, J. (ed.), Pergamon (1961).
    Combustion and Propulsion, 3rd AGARD Colloquium Proceedings, Thring, M. W., et al. (eds.), Pergamon (1958).
    Combustion and Propulsion, 5th AGARD Colloquium Proceedings, Hagerty, R. P., et al. (eds.), Pergamon (1963).
    The Chemistry of Propellants. AGARD Combustion and Propulsion Panel Meeting Proceedings, Penner, S. S. and Ducarme, J. (eds.), Pergamon (1960).
5. *Proceedings of the Heat Transfer and Fluid Mechanics Institute*, Annual meetings (since about 1947).
6. Ducarme, J., et al. (eds.), *Progress in Combustion Science and Technology, Vol. 1*, Pergamon (1960).
7. *Proceedings of the Joint Conference on Combustion*, The Institution of Mechanical Engineers, and the American Society of Mechanical Engineers (1955).
8. *Proceedings of Joint Symposia on Combustion Chemistry*, Divisions of Petroleum Chemistry, and Gas and Fuel Chemistry, American Chemical Society (1951).
9. *Annual Bulletin of Thermodynamics and Thermochemistry*, International Union of Pure and Applied Chemistry (since 1958).
10. I. & E.C., March Issues of every year, *Industrial and Engineering Chemistry*, American Chemical Society.
11. Cornelius, W. and Agnew, W. G. (eds.), *Emissions from Continuous Combustion Systems*, Plenum (1972).
12. I.C.R.P.G. *Combustion Conferences* of 1960's.
13. *Proceedings of the Gas Dynamics Symposium on Aerothermochemistry*, Northwestern University, Evanston (1956).
14. *Proceedings of the Third Midwestern Conference on Fluid Mechanics*, Minneapolis, Minnesota (March, 1953).

III Periodicals

1. *Combustion and Flame*, The Combustion Institute.
2. *Fuel*, The Institute of Fuel.

3. *Combustion Science and Technology*, Gordon and Breach Science Publishers.
4. *Fire Technology*, National Fire Protection Association.
5. *Fire and Flammability*, Technomic Publishing.
6. *Industrial and Engineering Chemistry*, American Chemical Society.
7. *Journal of Chemical Physics*.
8. *Fire Research Abstracts and Reviews*, U.S. National Academy of Sciences.
9. *A.I.A.A. Journal*, American Institute of Aeronautics and Astronautics.
10. *Journal of Chemical Society*, London.
11. *Journal of American Chemical Society*.
12. *Transactions of the Faraday Society*.
13. *ASME Journal of Heat Transfer*.

APPENDIX A

Review of Thermodynamics of Gases

Three basic laws form the core of thermodynamics. These are the gas law, first and second laws. The gas law stipulates the thermodynamic state of the system, the first law expresses conservation of energy, and the second law describes the conditions of equilibrium. A total of twelve basic properties (six fundamental and six derived) are usually encountered in the field of thermodynamics. Three of the fundamental properties—pressure, volume and temperature—arise in connection with the equation of state and three more—internal energy, heat and work—arise in connection with the first law. Of the six derived properties, enthalpy, specific heat at constant pressure and specific heat at constant volume arise from the first law whereas entropy, Helmholtz free energy and Gibbs free energy find their origin in context with the second law. Table A.1 summarizes the origin and classification of the twelve basic properties enumerated above. When mixtures of gases are concerned, an extra variable describing the composition is introduced. Any property of a thermodynamic system may be expressed as a function of any two other properties.

Consider, for example, a mixture. The laws of equilibrium stipulate the composition of an initial mixture at a given temperature and pressure. From

TABLE A.1

Twelve Basic Thermodynamic Properties

Nature of the property	Property arises in context with		
	Equation of state	First law	Second law
Fundamental	Pressure	Internal energy	—
	Volume	Heat	
	Temperature	Work	
Derived	—	Enthalpy	Entropy
		Specific heat	Helmholtz free energy
		Specific heat	Gibbs free energy

the knowledge of various thermodynamic properties of each of the components, the average properties of the initial mixture may be calculated. If the mixture now is allowed to react and the products are allowed to achieve an equilibrium state, thermodynamic laws again enable us to calculate the composition, and the various properties of the final mixture. Note must be made that whereas thermodynamics does not give any indication of the *rate* or *path* of transformation of the initial reactant mixture to the final product mixture, it does evaluate the limitations imposed on the transformation of various properties such as energy from one form to other as the state of the gases varies from the reactants to products. For this reason it seems more appropriate to call thermodynamics by the name thermostatics. If the transformation is complete and the final products attain equilibrium, the net change in properties is given by that part of thermodynamics which is called *thermochemistry*. The *rates* of transformation are themselves given by the laws of that branch of science which is known as *chemical kinetics*.

A *thermodynamic system* is a region of space or material set aside for thermodynamic investigation. A system can either be closed or open, depending on whether there is loss and gain of mass through the boundaries of the system. An isolated system is that which is isolated from the changes taking place in the surroundings. A system could also be labelled either as simple or complex, depending on whether the properties are uniform or nonuniform in its space. An equilibrium system is one whose state tends to remain fixed with respect to time.

A.1 Equation of State (Gas Law)

The thermodynamic state of a system containing a single pure gas species is fully described by the equation of state, a relation between the pressure P, volume V and temperature T. If the gas is such that the individual molecules occupy no volume and exert no mutual attractions or repulsions, it is known as an ideal gas.* Many gases may be considered as ideal at low pressures and high temperatures. For an ideal gas the pressure, volume and temperature are related through the total number of moles n by the equation of state.

$$\frac{PV}{nT} = R, \text{ a constant} \tag{A.1}$$

The constant R is known as the universal gas constant. Its value is given in various units in Table A.2. A *mole* is a convenient unit for the measure of a substance. If M is the molecular weight, a (gram-) mole of the gas weighs

---

* Only ideal gases are discussed in the present work.

Review of Thermodynamics of Gases

TABLE A.2

Universal Gas Constant R in Various Units

$R =$	1.987 BTU per lb-mole per °F
	1.987 cal per gm-mole per °K
	82.05 cm^3 atm per gm-mole per °K
	$8.314 \cdot 10^7$ gm cm^2 per sec^2 per gm-mole per °K
	$8.314 \cdot 10^3$ kgm m^2 per sec^2 per kgm-mole per °K
	$4.968 \cdot 10^4$ lb$_m$ft^2 per sec^2 per lb-mole per °K
	$1.544 \cdot 10^3$ ft lb per lb-mole per °R
	3.40709 ft lb per gm-mole per °F
	847.81 meter gm per gm-mole per °K
	8311.5 international joule per gm-mole per °K
	$8.3141 \cdot 10^{10}$ ergs per gm-mole per °K

M grams. One mole of a gas occupies 22.4 liters of volume at 0 °C temperature and 1 atm pressure. The density of a gas, therefore, at 0 °C and 1 atm is $M/22.4$ grams/liter. *Avagadro's hypothesis* states that equal volumes of all gases at the same temperature and pressure contain the same number of molecules. By careful experiments Avagadro determined that 22.4 liters of *any* gas at 0 °C and 1 atm (i.e., 1 mole of any gas) contains $6.023 \cdot 10^{23}$ molecules. This is known as Avagadro's number. When the pressure, temperature and volume of a thermodynamic system are changed from (P_1, V_1, T_1) to (P_2, V_2, T_2), the system is said to undergo a (process of)change of state. Such change of state is usually accomplished by keeping one or more of the thermodynamic properties fixed and thus *prescribing* the "path" from the initial state (P_1, V_1, T_1) to the final state (P_2, V_2, T_2). Those properties, changes in which depend upon the particular path taken by the system in its transformation from state 1 to state 2, are known as *path* properties (or path functions). Work and heat are examples of path functions. Those properties, changes in which depend only on the initial and final states but not on the particular path connecting these states, are known as *point* (or state) properties. All the properties in Table A.1, excluding work and heat, are state properties.

A.2 First Law of Thermodynamics

The first law invokes the concepts of internal energy, work and heat. Before proceeding to state and discuss the first law, let us define these three properties.

Energy is the capacity for causing a change of state in a system. Internal energy is that energy which is associated with the translational, vibrational and rotational activity of individual molecules. Temperature is, on a statistical average, a measure of the internal energy. In fact, a small change in the temperature dT may be related to the resultant small change in internal energy, du according to

$$du \equiv C_V \, dT \qquad (A.2)$$

where C_V is a (derived) property known as the *specific heat* (*at constant volume*). The units of energy are cal/gm-mole (or cal/gm) so that those of specific heat are cal/gm-mole/°K (or cal/gm/°K). C_V itself may be a function of temperature. Integrating Eq. A.2, we can obtain the change in internal energy when the temperature of the system is changed from T_1 to T_2.

$$u_2 - u_1 = \int_{T_1}^{T_2} C_V(T) \, dT \qquad (A.3)$$

If C_V is independent of temperature, Eq. A.3 simplifies to $u_2 - u_1 = C_V(T_2 - T_1)$.

Examination of Eq. A.3 indicates that if a system starting from temperature T_1 is taken through various states (i.e., various combinations of pressure and volume) to finally bring back to its original state, the change in internal energy is zero. In other words, the internal energy being a property of the state of the medium, does not depend on the particular fashion of changing the pressure and volume from one state to another. Changing the states of a medium in such a way that the final state is same as the initial is called a *cycle* of operation. The change in internal energy for a complete cycle is denoted by a circle on the integral sign as below.

$$\oint du = 0$$

If a system is isolated from all displacements (and hence from all volume changes) but is forced to experience a change of state, then *heat* is said to flow *across the boundaries* of the system. Heat is what flows from one region to another in a body by virtue of differences in temperature. A system is said to be in thermodynamic equilibrium if there are no variations in temperature (and other properties) and hence no heat flow from region to region.*

If, on the other hand, a system is isolated thermally (i.e., no heat is allowed to flow *into* or *out of* the system) but is forced to experience a change of state, then *work* is said to flow *across the boundaries* of the system. When work is done on a body it is imperative that it underwent a spatial displacement

---

* The zeroeth law of thermodynamics says that if two bodies are in their own turn in thermal equilibrium with a third body, then the two bodies themselves mutually are in thermal equilibrium.

Review of Thermodynamics of Gases

due to the action of a net force. For example, work is done when a weight is pulled up with a pulley, when water is lifted up to the overhead tank by a pump and when a balloon is blown with a gas. Work may be defined as the product of the net force resolved in the direction of the displacement and the displacement itself.

Consider now a thermodynamic system consisting of a cylinder filled with a fluid, a piston holding the fluid in the cylinder. Let P be the fluid pressure in the cylinder and A be the area of the piston. Let the piston be displaced by the fluid through a distance dx in a small time interval. Then by the definition, work done by the fluid on the piston is equal to the force acting on the piston (PA) and the displacement dx of the piston.

$$đW = PA\, dx = P\, dV \qquad (A.4)^{*\dagger}$$

where $dv = A\,dx$ is the volume of displacement. This is called P–V (or mechanical) work. Situations in various engineering problems may be present where magnetic, electrical and other kinds of forces perform work. For most part of the present engineering interest, we will ignore these nonmechanical contributions to work. As a gas is expanded, its volume increases and pressure is decreased according to the equation of state. If the expansion is under isothermal (i.e., no change in temperature) conditions, P is inversely proportional to the volume V. When such a prescribed path is incorporated into Eq. A.4, an integration yields the work performed as the volume is changed isothermally from V_1 to V_2. Thus,

$$W = \int_{V_1}^{V_2} P\, dV = \int_{V_1}^{V_2} \frac{RT\, dV}{V} = RT \ln \frac{V_2}{V_1} \qquad (A.5)\ddagger$$

Work done in a constant volume process of change of state is obviously zero. Work done in a constant pressure process, on the other hand, is simply $P(V_2 - V_1)$.

It is quite easy to see that calculations of heat and work require a statement regarding the path followed by the system from one state to the other.

---

* Whenever we talk about differential quantities of work and heat, we will use the differential sign $đ$ rather than d to distinguish inexact differentials (path functions) from exact differentials (point functions).

† If the pressure on the piston is released at an infinitesimally slow rate, the expansion process is called *reversible*. Work obtained in a reversible process is the maximum that the system can ever yield. Hence, Eq. A.4 should be written for correctness as $đW \leq P\, dV$. See section A.3 for further discussion on this topic.

‡ We will follow the customary notation of denoting the natural logarithm by the symbol "ln" and the logarithm to base ten by the symbol "log."

Therefore, both of these properties are path functions which when integrated around a cycle can be nonzero. That is,

$$\oint du = 0$$

$$\oint dw \neq 0$$

$$\oint dQ \neq 0$$

We are now equipped with the tools to state the first law of thermodynamics which, in essence, implies that *the energy of a system is conserved.**
With this implication the first law is often the starting point to derive the energy conservation equation for all engineering systems.

The internal energy of a thermodynamic system is increased by the quantity of heat dQ *absorbed* by it (i.e., entering the system) and decreased by the external work dW *performed* by it (i.e., exciting from the system). In mathematical terms the energy conservation may be expressed by the following equation.

$$du = dQ - dW \quad (A.6)$$

For an isolated system, $dQ = 0$ and $dW = 0$ so that the change in internal energy is zero. If heat is supplied to the system just to do work and not to raise the internal energy (i.e., for an isothermal expansion process),

$$dQ = dW$$

If a fraction of the heat supplied is utilized to raise the temperature (and hence u) of the system, naturally the amount of work performed is reduced by that fraction. If heat is supplied to the system with fixed impervious boundaries, Eq. A.6 gives $du = dQ$. This means, heat supplied in a constant volume process is equal to the increase in the internal energy. By Eq. A.2, then,

$$C_V = \left.\frac{dQ}{dT}\right|_V \quad (A.7)$$

---

* The following statements of first law may prove helpful in the understanding of our discussion.

Fermi: The variation of energy of a system during any transformation is equal to the amount of energy that the system receives from its environment.

Obert: Energy can neither be created nor destroyed but only converted from one form to another.

Zemansky: If a system is caused to change from an initial state to a final state by adiabatic means only, the work done is the same for all adiabatic paths connecting the two states.

Review of Thermodynamics of Gases

C_V, thus, is the heat to be added to the system at constant volume to raise its temperature by one degree.

Enthalpy is a property derived from the internal energy, pressure and volume by the following definition.

$$h \equiv u + PV \qquad (A.8)$$

Since all the three fundamental properties on the right hand side of Eq. A.8 are state functions, the derived property h is also a state function. Specification of enthalpy instead of internal energy is conventionally preferred in engineering practice since it stipulates the pressure-volume conditions as well as the temperature.

By the first law applied to a constant pressure reversible process, ($dW = P\,dV$),

$$du = dQ - P\,dV. \qquad (A.9)$$

Differentiating Eq. A.8 for a constant pressure process,

$$dh = du + P\,dV$$

Substituting Eq. A.9 to eliminate internal energy,

$$dh = dQ.$$

That is, heat supplied to a system in a constant pressure process is equal to the increase in the enthalpy. Define then the *specific heat at constant pressure* as

$$dh \equiv C_P\,dT.$$

C_P is the heat to be added to the system at constant pressure to raise its temperature by one degree.

The difference between the specific heat at constant pressure and that at constant volume is given by

$$C_P - C_V = \frac{dh}{dT} - \frac{du}{dT} = \frac{d(h-u)}{dT}$$

By Eq. A.8, $h - u = PV$ which by the equation of state is equal to RT. Hence,

$$C_P - C_V = R \qquad (A.10)$$

Table A.3 gives C_P (cal/gm-mole/°K) as a function of temperature for some gases of relevance in combustion. Knowing the universal gas constant as ≈ 2 cal/gm-mole/°K, the values of C_V may be obtained readily from this table with the help of Eq. A.10.

For a monatomic gas species kinetic theory of gases indicates that $C_V = 3R/2$ and hence $C_P = 5R/2$. Since $R \approx 2$ cal/gm-mole/°K, this C_P is ≈ 5

TABLE A.3
Molar Heat Capacity*

$$C_p = a + bT + cT^2 \text{ cal/gm-mole/°K}$$

Gas	a	b	c	a	b	c
H_2	6.1830	4.7107×10^{-3}	-10.9215×10^{-6}	4.1033	3.9818×10^{-3}	-1.4265×10^{-6}
H	4.9680	0	0	4.9680	0	0
O_2	7.3611	-5.3696×10^{-3}	20.5418×10^{-6}	8.4391	-0.3765×10^{-3}	0.6217×10^{-6}
O	5.9741	-4.2419×10^{-3}	7.9312×10^{-6}	4.7434	0.4809×10^{-3}	-0.3647×10^{-6}
N_2	7.7099	-5.5039×10^{-3}	13.1214×10^{-6}	5.6492	3.5790×10^{-3}	-1.7943×10^{-6}
N	4.9665	0.0115×10^{-3}	-0.0333×10^{-6}	4.8457	0.0808×10^{-3}	0.0879×10^{-6}
H_2O	7.9888	-1.5063×10^{-3}	6.6661×10^{-6}	3.4019	9.4330×10^{-3}	-4.0674×10^{-6}
NH_3	7.0405	1.2091×10^{-3}	18.3300×10^{-6}	20.5150	-14.0320×10^{-3}	13.2660×10^{-6}
NO	8.4623	-10.4067×10^{-3}	27.5488×10^{-6}	6.5902	2.6042×10^{-3}	-1.2912×10^{-6}
NO_2	6.6101	5.4313×10^{-3}	12.7251×10^{-6}	9.9490	4.4936×10^{-3}	-2.3473×10^{-6}
N_2O	4.8267	20.1393×10^{-3}	-22.1361×10^{-6}	9.8739	5.6103×10^{-3}	-2.9067×10^{-6}
SO	5.9203	4.6547×10^{-3}	0.0476×10^{-6}	7.8755	1.2776×10^{-3}	-0.6382×10^{-6}
SO_2	5.9262	14.4706×10^{-3}	-7.3970×10^{-6}	11.0304	3.4772×10^{-3}	-1.7638×10^{-6}
SO_3	3.9208	37.8380×10^{-3}	-41.6000×10^{-6}	15.0470	5.6292×10^{-3}	-3.0148×10^{-6}
CO	7.8122	-6.6683×10^{-3}	17.2830×10^{-6}	5.9665	3.2889×10^{-3}	-1.6605×10^{-6}
CO_2	4.3249	20.8089×10^{-3}	-22.9459×10^{-6}	8.1530	8.4114×10^{-3}	-4.7952×10^{-6}
CH_4	7.9184	-11.4172×10^{-3}	63.7345×10^{-6}	2.3689	24.0868×10^{-3}	-11.7510×10^{-6}
C_2H_2	1.6509	47.7840×10^{-3}	-81.0300×10^{-6}	8.2241	12.4790×10^{-3}	-5.6128×10^{-6}
C_2H_4	1.5087	31.9260×10^{-3}	-5.8508×10^{-6}	6.5490	26.0090×10^{-3}	-12.7360×10^{-6}
C_2H_6	1.3750	41.8520×10^{-3}	-13.8270×10^{-6}	—	—	—
C_3H_8	6.8008	28.7100×10^{-3}	62.3490×10^{-6}	2.5532	55.1590×10^{-3}	-20.8220×10^{-6}
C_4H_{10}	−39.7720	412.9000×10^{-3}	-991.7400×10^{-6}	8.5028	56.4720×10^{-3}	-23.0920×10^{-6}
C_6H_6	1.6505	68.4460×10^{-3}	-2.9566×10^{-6}	9.1653	57.2620×10^{-3}	-22.3540×10^{-6}
C_6H_{14}	7.3130	104.9060×10^{-3}	-32.3970×10^{-6}	—	—	—
Cl_2	7.5755	1.3133×10^{-3}	-0.9650×10^{-6}	—	—	—
	⟵ 300 °K < T < 2,000 °K ⟶			⟵ 2,000 °K < T < 6,000 °K ⟶		

Review of Thermodynamics of Gases

cal/gm-mole/°K. Note from Table A.3 that this argument is true for atomic H, O and N. For polyatomic gases, C_V is greater than $3R/2$ because the molecules need more energy to raise the temperature through 1°K due to the vibrational and rotational modes of energy storage.

A.3 Second Law of Thermodynamics

The second law of thermodynamics attributes, to every thermodynamic system, a characteristic state property called entropy. *Entropy* is a derived fundamental property which may be defined as the measure of availability of heat to flow per degree of temperature. Mathematically,

$$ds \equiv \frac{dQ}{T} \tag{A.11}$$

Having units same as those of specific heat, it may be calculated by imagining that the state of the system is changed from an arbitrarily chosen reference state to the actual state through a sequence of equilibrium states and by summing up the quotients of quantities of heat dQ introduced at each step and the absolute temperature T at that step. Thus the change in entropy from one state to another is given by integrating Eq. A.11.

$$s_2 - s_1 = \int_1^2 \frac{dQ}{T} \tag{A.12}$$

The second law of thermodynamics states that *for all real processes taking place in an isolated system, the entropy would only increase with time.** The concept of entropy and the second law of thermodynamics are very powerful tools in defining and describing the limitations of real working cycles.

When the entropy of the isolated system does not increase with time it remains constant (but never decreases). This temporally constant entropy state is known as equilibrium state in which all the thermodynamic properties of the contents of the system tend to be spatially uniform. In Appendix C we will show that when a constant volume system is in equilibrium, if entropy is specified to be constant, the internal energy assumes its minimum value

---

* The following statements of the second law may prove helpful in understanding our discussion.
Clausius: Heat cannot, of itself, pass from a lower to a higher temperature.
Planck: It is impossible to construct an engine that will work in a complete cycle and produce no effect except the raising of a weight and the cooling of a heat reservoir.

and if the internal energy is specified to be constant, the entropy assumes its maximum value.

If neither u nor s is fixed for the considered constant volume system at equilibrium, they both adjust together in such a way $(u - Ts)$ is a minimum. This is the origin of the derived property called *Helmholtz free energy, a*.

$$a \equiv u - Ts$$

Following the parallellism between enthalpy and internal energy, *Gibbs free energy, f*, is defined as

$$f \equiv h - Ts = a + PV$$

Because u, T, s and h are state functions, so also are the free energies. In order to consolidate the physical significance of the free energy, differentiate the definition of f to get

$$df = dh - T\,ds - s\,dT.$$

By the definition of enthalpy,

$$df = du + P\,dV + V\,dP - T\,ds - s\,dT$$

For a reversible process with only mechanical work considered, $du + P\,dV = dQ$ which by the definition of entropy is equal to $T\,ds$. Hence

$$df = V\,dP - s\,dT \qquad (A.13)$$

Equation A.13 indicates the f is a function of pressure and temperature alone for an ideal gas. Let $f = f(P, T)$. Then

$$df = \left.\frac{\partial f}{\partial P}\right|_T \cdot dP + \left.\frac{\partial f}{\partial T}\right|_P \cdot dT \qquad (A.14)$$

Comparison of Eqs. A.13 and A.14 gives

$$\left.\frac{\partial f}{\partial P}\right|_T = V; \qquad \left.\frac{\partial f}{\partial T}\right|_P = -s \qquad (A.15)$$

Table A.4 presents some further relationships between the properties P, V, T, u, Q, s, h, f, C_P and C_V.

Integration of Eq. A.14 under specific constraints results in calculation of the free energy changes. The free energy change associated with a constant temperature compression of one mole of an ideal gas ($V = RT/P$) is given by integrating $df = V\,dP = RT\,dP/P$.

$$f_2 - f_1 = RT \ln \frac{P_2}{P_1}$$

Likewise, the free energy change associated with a constant pressure process is obtained by knowing that $ds = dQ/T = dh/T = C_P(T) \, dT/T$ so that

$$s - s_0 = \int_{T_0}^{T} C_P(T) \, dT/T$$

and hence from $df|_P = -s \, dT|_P$,

$$f_2 - f_1 = - \int_{T_1}^{T_2} \left[s_0 + \int_{T_0}^{T} C_P(T) \frac{dT}{T} \right] dT$$

Similar integrations may be performed on some other relations given in Table A.4.

For extensive numerical data of enthalpy, entropy, and energy, the student is referred to the C.R.C. Handbook, JANAF and McBride Tables which are readily available in most of the engineering libraries. Dreisbach's book provides a wealth of thermodynamic data for hydrocarbon fuels.

Let us briefly pause here to explain what we mean by a reversible process. On a pressure-volume diagram, a *reversible path* is that which is obtained by connecting all the intermediate *equilibrium* states. A process that follows a reversible path is called a reversible process. In order to make gas expand reversible in a cylinder, the force on the piston has to be released so slowly that at any instant the pressure at every point in the working fluid is same (i.e., the system is in equilibrium). At any instant thus, the state of the system is represented by a unique point. Connecting all these equilibrium points, we obtain a reversible process.

Instead of infinitely slowly, suppose now that the force on the piston is released very suddenly. Next to the piston, the fluid will be at a rather low pressure whereas away from the piston, the pressure will be higher. The fluid will be in nonequilibrium due to variations in the pressure and as a result of this, the fluid will rush towards the piston to fill the newly made vacuum. In such a case, the thermodynamic state of the total system cannot be represented at any instant by a single point on the P–V diagram. The process is then called irreversible.*

In nature, reversible processes are very rare because of the slow release of the forces required to obtain a persistent equilibrium. For example, it is impossible for an engineer to conceive of an automobile engine in which the piston moves practically at zero speed. If one attempts to run the engine slower and slower, he approaches a reversible cycle but never achieves it. That is, a reversible process is a mathematical limit of the process as the speed of the piston approaches zero. In spite of the fact that it is impossible to

---

* Since our concern in this book is combustion, it is fitting to note that all real combustion processes are irreversible.

TABLE A.4
Thermodynamic Relations*

Differential energy functions:

$$du = Tds - PdV$$
$$dh = Tds + VdP$$
$$df = VdP - sdT$$

Maxwell Relations:

$$\left.\frac{\partial T}{\partial V}\right|_s = \left.\frac{\partial P}{\partial s}\right|_V \qquad \left.\frac{\partial T}{\partial P}\right|_s = \left.\frac{\partial V}{\partial s}\right|_P$$

$$\left.\frac{\partial s}{\partial V}\right|_T = \left.\frac{\partial P}{\partial T}\right|_V \qquad \left.\frac{\partial s}{\partial P}\right|_T = \left.\frac{\partial V}{\partial T}\right|_P$$

Energy Function Derivatives:

$$\left.\frac{\partial u}{\partial s}\right|_V = \left.\frac{\partial h}{\partial s}\right|_P = T \qquad \left.\frac{\partial u}{\partial V}\right|_s = -P$$

$$\left.\frac{\partial h}{\partial P}\right|_s = \left.\frac{\partial f}{\partial P}\right|_T = V \qquad \left.\frac{\partial f}{\partial T}\right|_P = -s$$

Thermal Capacity Relations:

$$C_P = \left.\frac{dQ}{\partial T}\right|_P = T\left.\frac{\partial s}{\partial T}\right|_P \qquad C_V = \left.\frac{dQ}{\partial T}\right|_V = T\left.\frac{\partial s}{\partial T}\right|_V$$

$$C_P - C_V = -T\left(\left.\frac{\partial P}{\partial T}\right|_V\right)^2 \cdot \left.\frac{\partial V}{\partial P}\right|_T = -T\left(\left.\frac{\partial V}{\partial T}\right|_P\right)^2 \cdot \left.\frac{\partial P}{\partial V}\right|_T$$

$$\frac{C_P}{C_V} = \gamma = -\left.\frac{\partial V}{\partial T}\right|_P \cdot \left.\frac{\partial T}{\partial P}\right|_V \cdot \left.\frac{\partial P}{\partial V}\right|_s$$

$$\left.\frac{\partial C_P}{\partial P}\right|_T = -T\left.\frac{\partial^2 V}{\partial T^2}\right|_P \qquad \left.\frac{\partial C_V}{\partial V}\right|_T = T\left.\frac{\partial^2 P}{\partial T^2}\right|_V$$

Dependence of u, h, and f on P, V, and T:

$$\left.\frac{\partial u}{\partial V}\right|_T = T\left.\frac{\partial P}{\partial T}\right|_V - P \qquad \left.\frac{\partial u}{\partial T}\right|_V = C_V$$

$$\left.\frac{\partial h}{\partial P}\right|_T = V - T\left.\frac{\partial V}{\partial T}\right|_P \qquad \left.\frac{\partial h}{\partial T}\right|_P = C_P$$

$$\left.\frac{\partial \Delta f/T}{\partial T}\right|_P = -\frac{\Delta h}{T^2}$$

---

* By the permission of John Wiley & Sons. From Hougen, O. A., and Watson, K. M., *Chemical Process Principles*: *Part II*: Thermodynamics. Table XXIV, pp. 472–473.

achieve a reversible process in reality, its concept makes it possible to calculate the higher limits of the thermodynamic performance of the system.

Because, by the gas law, the pressure is a function of volume alone at any given constant temperature, a system can be reversibly taken from one state (P_1, V_1) to another (P_2, V_2) only by one particular path at a given temperature. The area under such a reversible (P, V) curve gives the work done in the reversible process which is the maximum that could be extracted from that system. This is obvious since in a reversible process the expansion is carried out against the maximum possible pressure.

A.4 Thermodynamics of Nonreacting Gaseous Mixtures

When dealing with a system containing a mixture of gases, its thermodynamic state is fully described by stating the mixture composition in addition to the usual pressure, volume and temperature conditions.

(a) Gibbs-Dalton Law

In 1802, John Dalton discovered that any gas is a vacuum to any other gas mixed with it. In 1875, J. Willard Gibbs reformulated and expanded Dalton's law to read as following:

1. The pressure of a gaseous mixture is the sum of the pressures which each component would exert if it alone occupied the volume of the mixture at the temperature of the mixture. The pressure of each component is known as the *partial pressure* of the component.
2. The internal energy, enthalpy and entropy of a gaseous mixture are respectively equal to the sum of the internal energies, enthalpies and entropies which each component of the mixture would have, if each by itself occupied the volume of the mixture.

This is called the *Gibbs-Dalton law*.

Suppose a homogeneous non-reacting mixture of n_A molecules of A, n_B molecules of B, ..., n_i molecules of i (where $A, B, ..., i$ are ideal gases) is enclosed in a vessel of volume V. If equilibrium prevails in the system, the pressure and temperature would be uniform in the vessel. Let them be respectively designated by P and T. For this system, the equation of state is given by $PV = nRT$ where n is the total number of molecules in the system.

$$n = n_A + n_B + \cdots n_i.$$

Rewriting the equation of state,

$$P = nRT/V$$
$$= (n_A + n_B + \cdots + n_i)RT/V \qquad (A.16)$$
$$= n_A RT/V + n_B RT/V + \cdots + n_i RT/V$$

The pressure exerted by the jth component when occupying the whole volume V is given by,

$$P_j = n_j RT/V \qquad (A.17)$$

P_j is the partial pressure of the jth species. Hence Eq. A.16 gives the total pressure as the sum of all the partial pressures.

$$P = P_A + P_B + \cdots + P_i \qquad (A.18)$$

This is just a mathematical representation of the first part of Gibbs-Dalton law.

Note here that n/V is the number of moles in a unit volume, that is, *concentration* or *number density* C. Equation A.18 can be written in terms of concentrations as,

$$P = C_A RT + C_B RT + \cdots + C_i RT = (C_A + C_B + \cdots + C_i)RT = CRT$$

We obtain partial pressure as

$$P_j = n_j RT/V = \frac{n_j RT}{nRT/P} = \frac{n_j}{n} P \qquad (A.19)$$

The ratio (n_j/n) is called the *mole fraction* and denoted by X_j. It can also be written as the ratio of concentration of component i to the total concentration

$$X_j = \frac{C_j}{C}$$

With the definition of X_j, Eq. A.19 may be written in a simpler form as following.

$$P_j = X_j P \qquad (A.20)$$

It may quickly be seen that the sum of all the mole fractions is unity.

$$X_A + X_B + \cdots + X_i = n_A/n + n_B/n + \cdots + n_i/n = 1 \qquad (A.21)$$

It is convenient sometimes to express the composition of a mixture by *mass fractions*. The mass fraction Y_j of species j in a mixture is the ratio of mass of species j in a given volume to the total mass of the mixture in that volume.

Review of Thermodynamics of Gases

Considering a unit volume,

$$Y_j = \frac{n_j M_j}{nM} = X_j M_j / M = C_j M_j / CM = \frac{P_j M_j RT}{PMRT} \equiv \frac{P_j M_j}{PM} = \frac{\rho_j}{\rho} \quad (A.22)$$

where M_j and M are molecular weights of species j and the mixture respectively. ρ_j is called *partial density* or concentration of species j.

The definition of mass fraction given above is useful to derive a simple expression for the mean molecular weight of the mixture. Noting that the total mass of the mixture is the sum of the masses of all the components,

$$Y_A + Y_B + \cdots + Y_i = 1$$

$$\frac{n_A M_A}{nM} + \frac{n_B M_B}{nM} + \cdots + \frac{n_i M_i}{nM} = 1 \quad (A.23)^*$$

$$M = X_A M_A + X_B M_B + \cdots + X_i M_i$$

This type of a linear relationship is good for not only the mean molecular weight but also for all intrinsic properties of the thermodynamic system.

(b) *Implications of Gibbs-Dalton Law*

Implicit in the formulation of the Gibbs-Dalton law are the following inferences. If T is the temperature of the mixture,

$$T = T_A = T_B = \cdots = T_i$$

If V is volume occupied by mixture,

$$V = V_A = V_B = \cdots = V_i$$

If the density is designated by ρ, then,

$$\rho = \rho_A + \rho_B + \cdots + \rho_i \quad (A.24)$$

Let W be the total mass of mixture of volume V and $W_A, W_B, \ldots$, etc. be the mass of species $A, B, \ldots$, etc. respectively. In what follows below, the capitalized symbols refer to the total quantities in the volume V whereas script symbols refer to the quantities per unit mass or per unit gm-mole. With this notation, Gibbs-Dalton law in its second part implies the following:

(i) *Internal Energy:*

$$U = U_A + U_B + \cdots + U_i = W \cdot u = W_A u_A + W_B u_B + \cdots + W_i u_i$$

$$u = Y_A u_A + Y_B u_B + \cdots + Y_i u_i \quad (A.25)$$

---

* Mean density divided by the mean molecular weight of a mixture has units of concentration.

$$\frac{\rho}{M} [=] \text{gm-mole/cm}^3$$

(ii) *Enthalpy:*

$$H = H_A + H_B + \cdots + H_i = W \cdot h = W_A h_A + W_B h_B + \cdots + W_i h_i$$
$$h = Y_A h_A + Y_B h_B + \cdots + Y_i h_i \tag{A.26}$$

(iii) *Entropy:*

$$S = S_A + S_B + \cdots + S_i = W \cdot s = W_A s_A + W_B s_B + \cdots + W_i s_i$$
$$s = Y_A s_A + Y_B s_B + \cdots + Y_i s_i \tag{A.27}$$

(iv) *Specific Heats:*

From the definition of specific heats at constant volume and constant pressure, and from Eqs. A.25 and A.26,

$$C_V = Y_A C_{VA} + Y_B C_{VB} + \cdots + Y_i C_{Vi} \tag{A.28}$$
$$C_P = Y_A C_{PA} + Y_B C_{PB} + \cdots + Y_i C_{Pi} \tag{A.29}$$

Example 1

Gases are usually collected in the laboratory by letting them bubble through a water tower. A gas, say hydrogen, so collected over water at 25 °C becomes saturated with water vapor. The volume of the mixture of hydrogen and water vapor is 0.2 liters at a total pressure of 760 mm of mercury. Because it is a saturated mixture, the partial pressure of water vapor at 25 °C is simply read in the steam tables to be 23.8 mm of mercury. If the gas and vapor behave as ideal gases, calculate the volume occupied by dry hydrogen at a pressure of 2 atmospheres.

Solution

The partial pressure of the dry gas is equal to the difference between the total pressure and the partial pressure of water vapor.

$$P_{H_2} = P - P_{H_2O} = 760 - 23.8 = 736.2 \text{ mm}$$

By Dalton's law, this is the pressure that the dry H_2 would exert when occupying the total volume of 0.2 liters by itself. If temperature is kept constant, from the ideal gas law,

$$P_1 V_1 = nRT = P_2 V_2; \quad 736.2 \times 0.2 = 2 \times 760 \times V_2$$
$$\therefore \quad V_2 = 0.0969 \text{ liters}$$

Example 2

A and B are two gases of molecular weights, respectively 28 and 32. A mixture of A and B contains 5 gms of A and 1.569 gms of B. The total pressure of the

Review of Thermodynamics of Gases

system is 760 mm of mercury. Calculate (a) the partial pressures of A and B, and (b) the molecular weight of the mixture.

Solution

No. of moles of gas $A = 5/28 = 0.1785$
No. of moles of gas $B = 1.569/32 = 0.0491$
Total no. of moles in the mixture $= 0.1785 + 0.0491 = 0.2276$
Mole fraction of $A = 0.1785/0.2276 = 0.786$
Mole fraction of $B = 0.0491/0.2276 = 0.214$

(a) By Eq. A.20,

$$P_A = X_A P; \qquad P_B = X_B P$$
$$\therefore \quad P_A = 0.786 \times 760 = 598 \text{ mm of mercury}$$
$$P_B = 0.214 \times 760 = 162 \text{ mm of mercury}$$

(b) By Eq. A.23, the molecular weight of the mixture is,

$$M = X_A M_A + X_B M_B$$
$$= 0.786 \times 28 + 0.214 \times 32 = 28.85$$

If A and B are nitrogen and oxygen, the primary components of atmospheric air, it is useful to tabulate its composition approximately as in Table A.5 for future reference.

TABLE A.5
Composition of Air

	O_2	N_2	Argon
Composition of dry air by volume %	20.99	78.03	0.98
Composition of dry air by weight %	23.19	75.46	1.35
Mean molecular weight of air	28.97		

Example 3

The specific heats of nitrogen and oxygen are given by Table A.3. Find the specific heats C_V and C_P of the mixture of Example 2 at 298 °K. If one mole of air is heated from 298 °K to 350 °K to expand against a constant pressure standard atmosphere, find the change in enthalpy and internal energy. What is the quantity of heat added? Assume that the specific heats are independent of temperature.

Solution

From Table A.3, at 298 °K,

$$C_{P_{N_2}} = 6.94 \text{ cal/°K/mole} \quad C_{V_{N_2}} = 4.95 \text{ cal/°K/mole}$$
$$C_{P_{O_2}} = 7.05 \text{ cal/°K/mole} \quad C_{V_{O_2}} = 5.05 \text{ cal/°K/mole}$$

From example 2, the mole fractions of the mixture components are,

$$X_{N_2} = 0.786$$
$$X_{O_2} = 0.214$$

By equations A.28 and A.29,

$$C_V = 0.786 \times 4.95 + 0.214 \times 5.05 = 4.97 \text{ cal/°K/mole}$$
$$C_P = 0.786 \times 6.94 + 0.214 \times 7.05 = 6.98 \text{ cal/°K/mole}$$

Changes in enthalpy and energy are

$$dh = C_P \, dT = 6.98 \text{ (cal/°K/mole)} \times (350 - 298)°K = 364 \text{ cal/mole}$$
$$du = C_V \, dT = 4.97 \text{ (cal/°K/mole)} \times (350 - 298)°K = 258 \text{ cal/mole}$$

Because this is a constant pressure process, the heat added according to Eq. A.9 is $dQ = dh = 364$ cal/mole. The work done, on the other hand, is $P \, dV = dh - du = 106$ cal/mole.

Example 4

10 moles of a mixture at 0 °C and 760 mm contains CO_2, O_2 and N_2 by 1.31, 0.77, 7.92 moles respectively. Calculate the total volume and partial pressures.

Solution

$$X_{CO_2} = 1.31/10 = 0.131; \quad X_{O_2} = 0.077; \quad X_{N_2} = 0.792$$
$$P_{CO_2} = X_{CO_2} \cdot P = 0.131 \times 760 = 99.56 \text{ mm of mercury}$$
$$P_{O_2} = 0.077 \times 760 = 58.52 \text{ mm of mercury}$$
$$P_{N_2} = 0.792 \times 760 = 601.92 \text{ mm of mercury}$$
$$PV = nRT$$
$$V = nRT/P = 10 \times R \times 273/760 = 3.59 \text{ cubic feet}$$

A.5 Concluding Remarks

Several important thermodynamic concepts are reviewed in this appendix. Thermodynamic laws of nonreacting gases are discussed. Various methods of

Review of Thermodynamics of Gases

describing the composition of a mixture of gases are suggested. Laws to compute the average properties of a mixture of known composition are formulated on the basis of Gibbs-Dalton Law. For further study, excellent text books on thermodynamics are available in the literature. Some of these are given in the reference list of this appendix.

In Appendix B, we apply the first law of Thermodynamics to reacting mixtures in order to compute the heat liberated or absorbed due to a chemical transformation of the reactant mixture to product mixture.

References

Bosnjakovic, F. and Blackshear, P. L., *Technical Thermodynamics*, Holt, Rinehart and Winston, New York (1965).
CRC *Handbook of Physics and Chemistry*, Chemical Rubber Co., Cincinnati, Ohio.
Daniels, F. and Alberty, R. A., *Physical Chemistry*, Wiley, New York (1955).
Dreisbach, R. R., *Physical Properties of Chemical Compounds*, No. 15 of the series Advances in Chemistry, American Chemical Society (1955).
Fermi, E., *Thermodynamics*, Dover, New York (1956).
Glasstone, S., *The Elements of Physical Chemistry*, Van Nostrand, New York (1958).
Hall, N. A. and Ibele, W. E., *Engineering Thermodynamics*, Prentice-Hall, Englewood (1960).
Hougen, O. A. and Watson, K. M., *Chemical Process Principles—Part II; Thermodynamics*, Wiley, New York (1956).
Huff, V. N., et al., N.A.C.A. Report 1037 (1951).
McBride, et al., *Thermodynamic Properties to 6000 °K for 210 Substances involving the First 18 Elements*. National Aeronautics and Space Administration, Special Publication 3001 (1963).
Moor, W. J., *Physical Chemistry*, Prentice-Hall, Englewood (1956).
Obert, E. F., *Concepts of Thermodynamics*, McGraw-Hill, New York (1960).
Saad, M. A., *Thermodynamics for Engineers*, Prentice-Hall, Englewood (1966).
Spalding, D. B. and Cole, E. H., *Engineering Thermodynamics*, McGraw-Hill, New York (1959).
Tribus, M., *Thermostatics and Thermodynamics*, Van Nostrand, New York (1961).
Vay Wylen, G. J., *Thermodynamics*, Wiley, New York (1962).
Wilson, F. D. and Ries, H. C., *Principles of Chemical Engineering Thermodynamics*, McGraw-Hill, New York (1956).
Zemansky, M. W., *Heat and Thermodynamics*, McGraw-Hill, New York (1957).

APPENDIX B

Thermochemistry
(First Law of Thermodynamics Applied to Chemically Reacting Systems)

Any system possesses energy due to the virtue of its motion (kinetic energy), its position relative to the surroundings (potential energy), mutual attraction and repulsion of atoms in a molecule (chemical energy), etc. Some of the common forms of energy are: (a) kinetic (b) potential (c) electric (d) magnetic (e) thermal (f) surface (g) chemical and (h) gravitational.

Any classification such as this is quite arbitrary since energy, by first law, can be transformed from one form to another. For example, the potential energy of water at the top of the waterfall is converted into kinetic energy as it reaches the bottom of the fall. The thermal energy of steam is converted into kinetic energy by a steam engine. The chemical energy of a fuel is converted into thermal energy by combustion.

The second concept to be elucidated here is regarding the energetic nature of the chemical reactions. Let a vessel contain a mixture of a fuel and an oxidant. If reaction occurs, atoms exchange partners by breaking some old bonds (in the reactants) and making some new ones (in the products). If the new bonds contain less chemical energy than the old bonds, the balance is released (in the thermal or light form) during the reaction. Thermal energy so released will partly raise the temperature of the contents of the reaction vessel and is partly lost in the form of work and heat through the boundary of the vessel.

Suppose now that the system is kept at constant temperature while the reaction takes place. Consequently, no fraction of the thermal energy released in the reaction is used to heat up the contents. All of it is removed across the boundaries in the form of either heat or work. For a constant volume system, work term being zero, the energy released by a reaction is equal to the heat that flows out of the system. If the system, instead, is at constant pressure, work would flow out of the boundaries in addition to heat.

Thermochemistry

Thermochemistry is that branch of chemistry which with the help of the law of conservation of energy (i.e., the first law of thermodynamics) explains the changes of energy occurring from chemical to thermal form due to chemical reactions. In this appendix we will explore methods of calculation of enthalpy of formation of compounds, enthalpy of reactions and enthalpy of combustion.

When a chemical compound is formed from its elements, chemical energy is transformed to thermal energy (or vice versa). This transformed energy is called the *enthalpy of formation* of the compound (cal per mole) which characterizes the compound for thermochemical calculations. When several compounds (and elements) react with one another to form products, (say, at constant pressure) heat is either released or absorbed. The magnitude of this release or absorption is called *enthalpy of reaction* (cal). It characterizes the particular reaction from the energetic point of view. The enthalpy of reaction in which one mole of a particular reactant, which is termed "fuel," reacts with an oxidant is called *enthalpy of combustion*. Enthalpies of reactions can be computed from the enthalpies of formation of the participating chemicals as illustrated later in this appendix.

In addition to exploring the methods of computations of these enthalpy changes in chemical reactions, we also will consider in this appendix two important thermochemical laws (Lavoisier–Laplace law and Hess' law).

B.1 Enthalpy of Formation of Compounds

The standard enthalpy of formation of a compound is defined as the increase in enthalpy when one mole of the compound is formed at constant pressure from its elements, which are in their standard states at 25 °C and one atmosphere pressure. It is denoted by $\Delta h^\circ_{f_{298}}$, the subscript 298 referring to standard temperature and the superscript $^\circ$ referring to the standard pressure of 1 atm. The enthalpy of formation of a compound characterizes the compound for thermochemical calculations. It can be seen from Table B.1 that the standard enthalpy of formation of carbon dioxide is -94 kcal per mole. Consider the following examples.

$$CO(g) + \tfrac{1}{2}O_2(g) \longrightarrow CO_2(g), \quad \Delta h = -67.63 \text{ kcal/mole of } CO_2$$

-67.63 kcal/mole of CO_2 is *not* the enthalpy of formation of $CO_2(g)$ because the reactant CO is not an element.

$$C(s) + \tfrac{1}{2}O_2(g) \longrightarrow CO(g), \quad \Delta h^\circ_{f_{298}} = -26.42 \text{ kcal/mole of } CO$$

TABLE B.1

Enthalpy of Formation for Several Substances at 1 Atmosphere and Specified Temperature*

Substance	Formula	State	Enthalpy of formation kcal/gm-mole	Temperature °C
Carbon monoxide	CO	g	-26.42	25
Carbon dioxide	CO_2	g	-94.05	25
Methane	CH_4	g	-17.89	25
Acetylene	C_2H_2	g	54.23	25
Ethylene	C_2H_4	g	12.56	25
Benzene	C_6H_6	g	19.82	25
Benzene	C_6H_6	l	11.72	25
Octane	C_8H_{18}	g	-49.82	25
n-Octane	C_8H_{18}	l	-59.74	25
n-Octane	C_8H_{18}	g	-49.82	25
Calcium oxide	CaO	Crystalline	-151.80	25
Calcium carbonate	$CaCO_3$	Crystalline	-289.50	25
Oxygen	O_2	g	0	25
Nitrogen	N_2	g	0	25
Carbon (graphite)	C	Crystalline	0	25
Carbon (diamond)	C	Crystalline	0.45	25
Water	H_2O	g	-57.80	25
Water	H_2O	l	-68.32	25
Ethane	C_2H_6	g	-20.24	25
Propane	C_3H_8	g	-24.82	25
n-Butane	C_4H_{10}	g	-29.81	25
i-Butane	C_4H_{10}	g	-31.45	25
n-Pentane	C_5H_{12}	g	-35.00	25
n-Hexane	C_6H_{14}	g	-39.96	25
n-Heptane	C_7H_{16}	g	-44.89	25
Propylene	C_3H_6	g	4.88	25
Formaldehyde	CH_2O	g	-27.70	25
Acetaldehyde	C_2H_4O	g	-39.76	25
Methanol	CH_3OH	l	-57.02	25
Ethanol	C_2H_6O	l	-66.36	25
Formic acid	CH_2O_2	l	-97.80	25
Acetic acid	$C_2H_4O_2$	l	-116.40	25
Oxalic acid	$C_2H_2O_4$	s	-197.60	25
Carbon tetrachloride	CCl_4	l	-33.30	25
Glycine	$C_2H_5O_2N$	s	-126.33	25
Ammonia	NH_3	g	-11.00	18
Hydrogen bromide	HBr	g	-8.60	18
Hydrogen iodide	HI	g	6.00	18

* Abstracted from NBS Circular 500.

Thermochemistry

−26.42 kcal/mole of CO *is* the enthalpy of formation of CO(g) because both C and O_2 are elements in their natural state (i.e., carbon is in solid state and oxygen in molar gaseous state at 25 °C and 1 atm).

$$\tfrac{1}{2}H_2(g) + \tfrac{1}{2}I_2(s) \longrightarrow HI(g), \quad \Delta h°_{f_{298}} = 6.00 \text{ kcal/mole of HI}$$

Since one mole of hydrogen iodide is formed from half a mole each of $H_2(g)$ and $I_2(s)$ and since both H_2 and I_2 are elements in their standard state, the enthalpy of formation of hydrogen iodide molecule *is* 6.00 kcal/mole of HI. The standard enthalpies of formation of a number of substances are given in Table B.1.

B.2 Enthlapy of Reaction

(a) *Enthalpy—Temperature Diagram*

In combustion, we encounter generally very rapid chemical reactions taking place with liberation of heat and, sometimes, light. This evolution and transfer of heat are described to a great extent by the first law of thermodynamics. The total enthalpy contained in the products of a combustion reaction is less than that in the reactants, the balance being the heat liberated during combustion. Such reactions are called *exothermic*. Reactions that absorb energy instead, are called *endothermic*.

For an exothermic reaction the variation of enthalpy with temperature is illustrated by Figure B.1. These plots are curves instead of straight lines because the specific heats are functions of temperature. At any temperature, the difference between the enthalpy of products and reactants is the heat evolved due to the reactions and this is called *the enthalpy of reaction*. It is negative for exothermic reactions.

$$\Delta H°_{R_T} = \sum_{i=\mathscr{P}} n_i \, \Delta h°_{fTi} - \sum_{j=\mathscr{R}} n_j \, \Delta h°_{fTj} \quad (B.1)$$

The enthalpy change and mass (or mole) balances are generally represented in the form of a *stoichiometric equation*. For carbon burning with oxygen to form CO_2, for instance,

$$C(s) + O_2(g) \longrightarrow CO_2(g), \quad \Delta h°_{R_{298}} = -94 \text{ kcal}$$

This means that 12 gms of carbon react with 32 gms of oxygen to yield 44 gms of CO_2 and 94,000 calories of heat at constant pressure. The symbols s, g, and l are used to identify the natural states of the chemicals, as solid, gas, or liquid respectively. The enthalpy of a compound is its heat of formation Δh_f which for elements in their natural standard states is, as before, zero.

344   *Introduction to Combustion Phenomena*

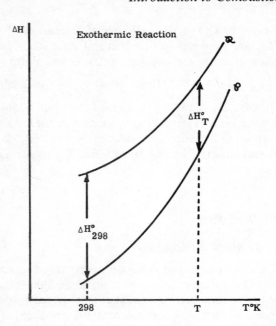

ΔH°_{298} is called standard enthalpy of reaction

Figure B.1 Definition of enthalpy of reaction

It should be understood that since thermochemical calculations are always meant to obtain differences in properties, it is quite immaterial what the arbitrarily chosen basis of reference is, as long as it is always the same in a calculation. The basis used above for the elements is simple and convenient but it by no means implies that the enthalpy of the elements at 25 °C and 1 atm indeed is zero.

Let us here review the two points we have established so far. Firstly, we stipulated a datum reference by setting the enthalpy of the elements in their standard form as zero. Secondly, the enthalpy of a reaction in which the reactants are natural elements and the product is one mole of any compound is equal to the enthalpy of formation of the compound.

Let us calculate, for illustration, enthalpy of the reaction of methane oxidation to form CO_2 and H_2O.

$$CH_4(g) + 2O_2(g) \longrightarrow CO_2(g) + 2H_2O(l)$$

Total enthalpy of the products (from Table B.1) is $1(-94.0) + 2(-68.3)$ kcal and that of the reactants is $1(-17.9) + 2(0.0)$ kcal so that the increase in

Thermochemistry

enthalpy due to the reaction, by Eq. B.1, is $-94.00 - 136.6 - (-17.9) = -212.7$ kcal.

(b) *Reactions at Constant Volume or Constant Pressure*

Reactions may take place either at constant volume (such as in a combustion bomb) or at constant pressure (such as in a Bunsen burner or a candle). While constant volume reactions involve no work flow, constant pressure reactions do. If the heat evolved by the reaction at constant pressure is greater than that evolved by the reaction at constant volume, the system would contract in volume. In such a system, work is done by the environment on the system. If the volume change between the reactants and products is zero, there is no work done on or by the system and the heat of reaction at constant pressure is same as the heat of reaction at constant volume.

Here we shall consider reactions in which there is volume change between the reactants and products (as such are most interesting reactions of combustion) and examine how the heat of reaction differs between the constant pressure case and the constant volume case. From the first law, $du = dQ - dW$ which at constant volume becomes $du = dQ$ implying that all the heat supplied by combustion in a system at constant volume is expended in raising the internal energy u of the system. This is the principle commonly used to determine the "energy" of combustion of a substance by the bomb calorimeter method. But if reactions are to take place in an open vessel, and the component species are allowed to expand or contract as they wish, the heat released is expended to raise the enthalpy h of the system, i.e., $dh = dQ$. Examination of these relations shows that heats evolved (or absorbed) in constant pressure and constant volume reactions differ merely by the work term. For an ideal gas system, $PV = nRT$ so that if the temperature is kept unchanged, any changes in volume could be related to the changes in the number of moles as the reaction proceeds from the reactants to the products at constant pressure. If (dn) is such a change, $P\,dV = (dn)RT$ so that $dh - du = (dn)RT$.

$$\Delta H_R - \Delta U_R = \Delta n_R \cdot RT. \qquad (B.2)$$

The measurements are made at constant pressure for most gaseous reactions, whereas they are made at constant volume for liquids and solids. For most combustion reactions, enthalpy and energy of reaction are of the order of tens of kilocalories whereas the change in number of moles Δn_R due to reaction is such that $\Delta n_R \cdot RT$ is only of the order of 200 calories. Often, hence, one finds in literature that the difference between "enthalpy of reaction" and "energy of reaction" is neglected. In this sense, use of the name "heat of reaction" does not introduce any significant misunderstandings.

B.3 Enthalpy of Combustion

The heat release accompanying the complete combustion of *one* mole of a compound is known as its *heat of combustion*, or more rigorously the "enthalpy" of combustion if combustion took place at constant pressure. (Note, the compound of our interest here is a reactant whereas in the calculation of enthalpy of formation the interest is in the product.) We have already calculated the enthalpy of combustion of methane as -212.7

TABLE B.2

Heats of Combustion at 25 °C*
(Products: N_2, $H_2O(l)$ and CO_2)

Fuel	State	Formula	ΔH_c kcal/mole
Carbon (graphite)	s	C	-93.9
Hydrogen	g	H_2	-68.3
Carbon monoxide	g	CO	-67.6
Methane	g	CH_4	-210.8
Ethane	g	C_2H_6	-368.4
Propane	g	C_3H_8	-526.3
Butane	l	C_4H_{10}	-686.1
Pentane	l	C_5H_{12}	-833.4
Heptane	l	C_7H_{16}	$-1,149.9$
Octane	l	C_8H_{18}	$-1,302.7$
Dodecane	l	$C_{12}H_{26}$	$-1,943.7$
Hexadecane	s	$C_{16}H_{34}$	$-2,559.1$
Ethylene	g	C_2H_4	-337.3
Ethyl alcohol	l	C_2H_5OH	-327.6
Methyl alcohol	l	CH_3OH	-170.9
Benzene	l	C_6H_6	-782.3
Cycloheptane	l	C_7H_{14}	$-1,087.3$
Cyclopentane	l	C_5H_{10}	-783.6
Acetic acid	l	$C_2H_4O_2$	-209.4
Benzoic acid	s	$C_7H_6O_2$	-771.2
Ethyl acetate	l	$C_4H_8O_2$	-536.9
Naphthalene	s	$C_{10}H_8$	$-1,232.5$
Sucrose	s	$C_{12}H_{22}O_{11}$	$-1,349.6$
Camphor	s	$C_{10}H_{16}O$	$-1,411.0$
Styrene	l	C_8H_8	$-1,047.1$
Toluene	l	C_7H_8	-934.2
Xylene	l	C_8H_9	$-1,091.7$
Urethane	s	$C_3H_7NO_2$	-397.2

* Abstracted from NBS Circular 500.

kcal/mole of CH_4. The values for many other fuels could either be calculated from the known enthalpies of reactions and stoichiometry or be found in literature. Some are listed in Table B.2.

B.4 Calculation of Entlaphy of Reaction from Bond Energies

It is found that the energy needed to break a particular type of bond between two atoms is approximately the same regardless of the molecule in which the bond occurs. In other words, different pairs of atoms have different affinities between one another and different amounts of energy have to be applied to break them apart. It is called bond energy and denoted by ε_{i-j}. Table B.3 is a compact list of such bond energies for bonds between various atom combinations.

Suppose one desires to find ΔH_R of a reaction involving a compound whose enthalpy of formation is not known. Perhaps the compound has never been synthesized or is not suitable for burning in a calorimeter. The thermodynamic methods described so far are useless to predict the enthalpy of reaction. At this point, one turns to the bond structure of the molecules

TABLE B.3

Mean Bond Energies*
(kilocalories/bond)

Bond	Energy	Bond	Energy
C—C	85	O—H	109
C=C	143	O—N	150
C≡C	198	N—H	88
C—H	98	P—P	48
C—O	86	S—S	50
C=O	173	Cl—Cl	57
C—N	81	Br—Br	46
C≡N	210	I—I	36
C—Cl	78	F—F	36
C—Br	67	H—Cl	103
C—I	64	H—Br	88
C—F	102	H—I	72
C—S	64	H—F	135
O—O	33	H—P	76
O=O	117	H—S	81
N—N	60	P—Cl	78
N≡N	225	P—Br	64
H—H	103	S—Cl	60

* Abstracted from CRC Handbook.

to see if the bond energies can be of any help to determine the enthalpy of formation or reaction.

Consider the reaction $C_2H_4 + H_2 \rightarrow C_2H_6$. In molecular synbols

$$\begin{array}{c} H \ \ H \\ |\ \ \ | \\ C=C \\ |\ \ \ | \\ H\ \ H \end{array} + H-H \longrightarrow \begin{array}{c} H\ \ H \\ |\ \ \ | \\ H-C-C-H \\ |\ \ \ | \\ H\ \ H \end{array}$$

In this reaction, one C=C bond and one H—H bond are broken, whereas one C—C bond and two C—H bonds are made. The bond energy of the bonds "broken" minus the bond energy of bonds "made" is equal to Δu_R and to a good approximation to ΔH_R. Thus from Table B.3, bond energy "in" = $\varepsilon_{C=C} + \varepsilon_{H-H}$ = 143 + 103 = 246 kcal and bond energy "out" = $\varepsilon_{C-C} + 2\varepsilon_{C-H}$ = 85 + 196 = 281 kcal so that the enthalpy of the above reaction is given by ΔH_R = bond energy "in" − bond energy "out" = 246 − 281 = −35 kcal i.e., 35 kcal of heat is evolved (the minus sign shows that the "new" bonds contain less energy than the "old" ones) when 1 mole of ethylene is hydrogenated.

From Table B.1, the enthalpies of formation of ethylene and ethane are respectively 12.56 kcal and −20.24 kcal. Hence for the reaction under scrutiny, by Eq. B.1, $\Delta H_R = 1(-20.24) - (1 \cdot 12.56 + 1 \cdot 0) = -32.8$ kcal. This is in rough agreement with the value calculated from bond energy considerations.

For a further example, consider simple straightforward oxidation of methane to form carbon dioxide and water vapor. The stoichiometric equation in skeleton form will be

$$\begin{array}{c} H \\ | \\ H-C-H \\ | \\ H \end{array} + O-O + O-O$$

$$\longrightarrow O-C-O + H-O-H + H-O-H$$

showing that, in the process, four C—H bonds and two O—O bonds are broken whereas two C—O bonds and four O—H bonds are made. To break four C—H bonds and two O—O bonds, we have to put "into" our system $(4 \cdot 98 + 2 \cdot 33) = 458$ kcal of energy. However, when two C—O bonds and four O—H bonds are made in the products, energy is released by the amount of $(2 \cdot 86 + 4 \cdot 109) = 608$ kcal. That is, atoms produced by dissociation cost us 458 kcal of energy. These atoms when recombined to form the final products yield 608 kcal. Thus the excess energy $(608 − 458) = 150$ kcal is released in the reaction thermally and/or radiationally. Hence for the

Thermochemistry

oxidation of methane, the heat of reaction is approximately 150 kcal. Obviously, even if the bond energy calculations are rather crude and approximate, they are of great use when the enthalpies of formation of compounds are not available.

B.5 Thermochemical Laws

In the calculation of the thermochemical properties described above, it is implicitly assumed that energy is conserved by way of the first law. On the same basic foundation of conservation of energy, two important thermochemical laws evolve.

(a) Lavoisier–Laplace Law

The thermal energy which must be supplied to decompose a compound into its elements is equal to the thermal energy which is evolved when that compound is formed from its elements.

In other words, the enthalpy of decomposition of a compound is equal and opposite in sign to its enthalpy of formation. This is in agreement with the first law, for if not, one would be able to create energy by synthesizing a compound from its elements and then decomposing it, or vice versa—this would be the ultimate in the energy production history. Lavoisier–Laplace Law enables us to write thermochemical equations in reverse order. Consider the familiar reaction of carbon and oxygen to give carbon monoxide and 26.42 kcal of heat at constant pressure.

$$C(s) + \tfrac{1}{2}O_2(g) \longrightarrow CO(g), \qquad \Delta h_f = -26.42 \text{ kcal}$$

Lavoisier–Laplace Law indicates that if by some means carbon monoxide can be broken at constant pressure into carbon solid and oxygen gas, the heat needed to accomplish this is 26.42 kcal

$$CO(g) \longrightarrow C(s) + \tfrac{1}{2}O_2(g), \qquad \Delta h_d = 26.42 \text{ kcal}$$

Consider the second example—that of hydrogen iodide formation

$$\tfrac{1}{2}H_2(g) + \tfrac{1}{2}I_2(s) \longrightarrow HI(g), \qquad \Delta h_f = 6 \text{ kcal}$$

The enthalpy of decomposition (which is an exothermic process) of HI(g) is -6.00 kcal.

$$HI(g) \longrightarrow \tfrac{1}{2}H_2(g) + \tfrac{1}{2}I_2(s), \qquad \Delta h_d = -6 \text{ kcal}$$

Similarly, the following two sets of equations further illustrate the Lavoisier–Laplace postulate.

$$S(s) + O_2(g) \longrightarrow SO_2(g), \quad \Delta h_f = -70.9 \text{ kcal}$$
$$SO_2(g) \longrightarrow S(s) + O_2(g), \quad \Delta h_d = 70.9 \text{ kcal}$$
$$CO_2(g) + H_2(g) \longrightarrow CO(g) + H_2O(g), \quad \Delta H_R = -11.2 \text{ kcal}$$
$$CO(g) + H_2O(g) \longrightarrow CO_2(g) + H_2(g), \quad \Delta H_R = 11.2 \text{ kcal}$$

(b) *Hess' Law of Summation*

The net heat evolved or absorbed in a chemical reaction is the same whether it takes in one or many steps.

The statement simply means that in considerations of energy conversion, it is the initial and final states of the system that are important and that the overall process is not affected by the intermediate states that may occur between the reactants and products. Hess' Law is of extreme practical use because implicit in it is that thermochemical equations can be added or subtracted algebraically. Thus, the enthalpies of reactions that are difficult to measure directly can be calculated from other thermochemical data obtained from the literature or relatively easier experiments. Consider the following examples:

Carbon monoxide is formed from carbon and oxygen.

$$C(s) + \tfrac{1}{2}O_2(g) \longrightarrow CO(g), \quad \Delta h_f = ?$$

The heat of formation of CO is to be calculated by using Hess' Law. Suppose we can conduct an experiment or find the data in literature for the formation of carbon dioxide as described by the following two equations:

$$C(s) + O_2(g) \longrightarrow CO_2(g), \quad \Delta h_f = -93.91 \text{ kcal}$$
$$CO(g) + \tfrac{1}{2}O_2(g) \longrightarrow CO_2(g), \quad \Delta H_R = -67.62 \text{ kcal}$$

Subtracting one equation from the other,

$$C(s) + O_2(g) - CO(g) - \tfrac{1}{2}O_2(g)$$
$$\longrightarrow CO_2(g) - 93.91 \text{ kcal} - CO_2(g) + 67.62 \text{ kcal}$$

results in

$$C(s) + \tfrac{1}{2}O_2(g) \longrightarrow CO(g), \quad \Delta h_f = -26.29 \text{ kcal}$$

Let us illustrate the principle to calculate the enthalpy of formation of ethanol from the following information.

$$C_2H_5OH(l) + 3O_2(g) \longrightarrow 2CO_2(g) + 3H_2O(l), \quad \Delta H_R = -327 \text{ kcal}$$
$$C(s) + O_2(g) \longrightarrow CO_2(g), \quad \Delta h_f = -93.91 \text{ kcal}$$
$$H_2(g) + \tfrac{1}{2}O_2(g) \longrightarrow H_2O(l), \quad \Delta h_f = -68.31 \text{ kcal}$$

Thermochemistry

Multiply the second equation by 2, the third by 3 and subtract the first equation from their sum to get

$$2C(s) + 3H_2(g) + \tfrac{1}{2}O_2(g) \longrightarrow C_2H_5OH(l), \qquad \Delta h_f = -65.15 \text{ kcal}$$

Another application of Hess' Law is to calculate enthalpy of a reaction from known enthalpies of combustion of all the reactants and products. Suppose we need to find the enthalpy of reaction of ethylene and hydrogen.

$$C_2H_4(g) + H_2(g) \longrightarrow C_2H_6(g), \qquad \Delta H_R = ?$$

The heats of combustion of ethylene, hydrogen, and ethane from Table B.2 are respectively -337.3, -68.3, and -368.4 kcal at standard temperature. The following three combustion equations are written.

$$C_2H_4(g) + 3O_2(g) \longrightarrow 2CO_2(g) + 2H_2O(l), \qquad \Delta h_C = -337.3 \text{ kcal}$$
$$H_2(g) + \tfrac{1}{2}O_2(g) \longrightarrow H_2O(l), \qquad \Delta h_f = -68.3 \text{ kcal}$$
$$C_2H_6(g) + \tfrac{7}{2}O_2(g) \longrightarrow 2CO_2(g) + 3H_2O(l), \qquad \Delta h_C = -368.4 \text{ kcal}$$

If the third equation is subtracted from the sum of the first two,

$$C_2H_4(g) + H_2(g) \longrightarrow C_2H_6(g), \qquad \Delta H_R = -37.2 \text{ kcal}$$

The student at this point can make up his own examples for the use of Hess' Law.

B.6 The Effect of Physical State on Enthalpy of Reaction

When oxygen and hydrogen gases react, water is formed. Depending on the temperature, the product water may be either in gaseous form or in liquid form. Table B.1 shows two values for the enthalpy of formation of water for its liquid state and gaseous state. Similarly, some other items in the Table B.1 show the dependency of Δh_f on the state of the product. Also Table B.2 indicates that Δh_C of a fuel depends on its state.

Consider

$$H_2(g) + \tfrac{1}{2}O_2(g) \longrightarrow H_2O(l), \qquad \Delta h_f = -68.31 \text{ kcal}$$

If instead, water in the product is in vapor state, we have to add the molar enthalpy of evaporation to the enthalpy of formation of $H_2O(l)$

$$H_2O(l) \longrightarrow H_2O(g), \qquad \Delta h_{vap} = 10.52 \text{ kcal}$$

Hence, by Hess' Law,

$$H_2(g) + \tfrac{1}{2}O_2(g) \longrightarrow H_2O(g), \qquad \Delta h_f = -57.79 \text{ kcal}$$

Different forms of crystalline solid compounds involve different energies of crystalline transformations. For example, the two allotropic states of carbon burning with oxygen lead to

$$C(\text{Diamond}) + O_2(g) \longrightarrow CO_2(g), \quad \Delta h_f = -94.5 \text{ kcal}$$
$$C(\text{Graphite}) + O_2(g) \longrightarrow CO_2(g), \quad \Delta h_f = -93.91 \text{ kcal}$$

Therefore,

$$C(\text{Diamond}) \longrightarrow C(\text{Graphite}), \quad \Delta h = -0.59 \text{ kcal}$$

Thus 0.59 kcal of heat is evolved when 12 gms of diamond is converted into graphite at constant pressure.

B.7 Temperature Dependency of Heat of Reaction

Kirchoff's Law

The numerical values of enthalpies of reactions that are used thus far are qualified as "standard" (i.e., they are determined and tabulated at a pressure of 1 atmosphere and 25 °C). They are denoted by a subscript 298 for temperature and a superscript ° for pressure, as, for example $\Delta h^{\circ}_{f_{298}}$.

Enthalpy being independent of pressure for an ideal gas, enthalpy of reaction also would behave independent of pressure. On the other hand, enthalpy change associated with any constant temperature process usually varies with the level of temperature in a manner typical of Figure B.1.

Conservation of energy (first law) correlates this variation with the other macroscopic properties of the system.

Consider a simple case where r moles of a reactant $\mathcal{R}$ transforms by physical or chemical process into p moles of a product $\mathcal{P}$.

$$r\mathcal{R} \longrightarrow p\mathcal{P}$$

The enthalpy of reaction ΔH_R is equal to the decrease in the enthalpy of the system as it goes from $\mathcal{R}$ to $\mathcal{P}$;

$$\Delta H_R = \Delta H_{\mathcal{P}} - \Delta H_{\mathcal{R}} = p\,\Delta h_{\mathcal{P}} - r\,\Delta h_{\mathcal{R}} \tag{B.3}$$

The variation of ΔH_R with temperature is given by

$$\left.\frac{d\,\Delta H_R}{dT}\right|_P = p\left.\frac{d\,\Delta h_{\mathcal{P}}}{dT}\right|_P - r\left.\frac{d\,\Delta h_{\mathcal{R}}}{dT}\right|_P \tag{B.4}$$

By the definition of specific heat at constant pressure,

$$\left.\frac{d\,\Delta H_R}{dT}\right|_P = pC_{P\mathcal{P}} - rC_{P\mathcal{R}} \tag{B.5}$$

Thermochemistry

This result implies that the rate of change of enthalpy of a reaction with temperature is equal to the difference between the specific heats of the reactants and products at constant pressure. This is called Kirchoff's Law of variation of enthalpy of reaction with temperature. If one has to find the change in enthalpy of reaction between two temperatures, one can integrate the above equation.

$$\Delta H_{R2} - \Delta H_{R1} = \int_{T_1}^{T_2} (pC_{P\mathscr{P}} - rC_{P\mathscr{R}})\, dT \tag{B.6}$$

Here ΔH_{R2} and ΔH_{R1} are enthalpies of reaction at temperatures T_2 and T_1 respectively. Specific heats of $\mathscr{R}$ and $\mathscr{P}$ will vary with temperature as discussed in Appendix A. If $C_{P\mathscr{R}}$ and $C_{P\mathscr{P}}$ are considered independent of temperature,

$$\Delta H_{R2} - \Delta H_{R1} = (pC_{P\mathscr{P}} - rC_{P\mathscr{R}})(T_2 - T_1) \tag{B.7}$$

Furthermore, Kirchoff's Law may be applied to systems where more than one reactant and one product are present; i.e.,

$$A + B + C + \cdots \longrightarrow M + N + O + \cdots$$

In this case, the specific heats will be the average values for products and reactants as calculated by Eq. A.29.

$$C_{P\mathscr{R}} = X_A C_{PA} + X_B C_{PB} + \cdots$$
$$C_{P\mathscr{P}} = X_M C_{PM} + X_N C_{PN} + \cdots$$

Hence, if the enthalpy of reaction at a temperature is known, (as given in Tables B.1 and B.2), it may be computed at any other temperature by Eq. B.7.

Example 1

Heat of formation of water at 25 °C is given in Table B.1 as -68.32 kcal. Calculate it at 90 °C. Assume that the mixture is stoichiometric initially and that $C = a + bT$, a and b being given in Table A.3.

Solution

Initial mixture is $H_2(g) + \frac{1}{2}O_2(g)$; $r = \frac{3}{2}$
Final mixture is $H_2O(l)$; $p = 1$

$$X_{H_2} = \frac{1}{3/2} = \frac{2}{3}, \qquad X_{O_2} = \frac{1/2}{3/2} = \frac{1}{3}$$

$$C_{P\mathscr{R}} = \tfrac{2}{3}(6.94 - 0.2 \cdot 10^{-3}T) + \tfrac{1}{3}(6.09 + 3.25 \cdot 10^{-3}T)\,\text{cal/mole/°K}$$
$$= 6.64 - 0.00024T \,\text{cal/mole/°K}$$

$$C_{P\mathscr{P}} = C_{P_{H_2O(l)}} = 18 \,\text{cal/mole/°K}$$

Note that the temperature here must be in °K.

$$pC_{P\mathscr{P}} - rC_{P\mathscr{R}} = 1 \cdot 18 - \tfrac{3}{2}(6.64 - 0.00024T) = 8.04 + 0.00036T$$

Substituting now the known quantities in Eq. B.6

$$\Delta H_{R2} - (-68.32) = \int_{298}^{363} 10^{-3} \cdot (8.04 + 0.00036T)\, dT \text{ kcal}$$

$$\Delta H_{R2} = -67.782 \text{ kcal}$$

In a generalized form Eq. B.6 can be rewritten as

$$\Delta H_{RT} = \Delta H_{RT_1} + \int_{T_1}^{T} (pC_{P\mathscr{P}} - rC_{P\mathscr{R}})\, dT \tag{B.6}$$

$(pC_{P\mathscr{P}} - rC_{P\mathscr{R}})$ can be written as a polynomial of temperature as shown in the example above.

$$pC_{P\mathscr{P}} - rC_{P\mathscr{R}} = a' + b'T + C'T^2 + \cdots$$

so that

$$\Delta H_{RT} = \Delta H_{RT_1} + [a'T + b'T^2/2 + C'T^3/3 + \cdots]_{T_1}^{T}$$

If the enthalpy of reaction is measured at one particular temperature, T_1, and if specific heat capacities of all components are known, ΔH_{RT} can thus be computed at any temperature.

Example 2

Compute the enthalpy of the reaction for

$$C(s) + H_2O(g) \longrightarrow H_2(g) + CO(g), \quad \Delta H_R = ?$$

from the following information

$$C(s) + \tfrac{1}{2}O_2(g) \longrightarrow CO(g), \quad \Delta h_f = -26.42 \text{ kcal}$$
$$H_2(g) + \tfrac{1}{2}O_2(g) \longrightarrow H_2O(g), \quad \Delta h_f = -57.80 \text{ kcal}$$

Solution

By Hess' Law, subtracting the second from the first equation,

$$C(s) + \tfrac{1}{2}O_2(g) - H_2(g) - \tfrac{1}{2}O_2(g)$$
$$\longrightarrow CO(g) - 26.42 \text{ kcal} - H_2O(g) + 57.80 \text{ kcal}$$

Therefore,

$$C(s) + H_2O(g) \longrightarrow H_2(g) + CO(g) \quad \Delta H_R = 31.38 \text{ kcal}$$

Thermochemistry

Example 3

What is the enthalpy of reaction of

$$CO(g) + H_2(g) \longrightarrow C(s) + H_2O(g) \qquad \Delta H_R = ?$$

By Lavoisier–Laplace Law, the result of Example 2 gives ΔH_R for this reaction as -31.38 kcal.

Example 4

Consider the burning of methane in air. Compute the enthalpy of combustion assuming that all the available methane is consumed.

Solution

First we have to write the stoichiometric equation.

$$aCH_4 + bnO_2 + 2.73bnN_2 \longrightarrow cCO_2 + dH_2O + 2.73bnN_2$$

For a balance, the atoms of carbon, oxygen, nitrogen and hydrogen should be conserved in the process of transforming the reactant mixture to the product mixture.

For carbon, $a = c$. For hydrogen, $4a = 2d$. For oxygen, $2bn = 2c + d$. For nitrogen, $2 \cdot 2.73bn = 2 \cdot 2.73bn$. Solve the first 3 equations for c, d, and b by assuming an arbitrary value of unity to a.

$$a = c = 1 \qquad d = 2 \qquad bn = 2$$

so that

$$CH_4 + 2O_2 + 5.46N_2 \longrightarrow CO_2 + 2H_2O + 5.46N_2$$

Enthalpy of combustion

$$\Delta h_C = \sum_{i=\mathscr{P}} n_i \Delta h_{fi} - \sum_{j=\mathscr{R}} n_j \Delta h_{fj}$$

From Table B.1

$$\Delta h_{f\,CO_2(g)} = -94.05 \text{ kcal/mole}$$

$$\Delta h_{f\,H_2O(g)} = -57.80 \text{ kcal/mole}$$

$$\Delta h_{f\,CH_4(g)} = -17.89 \text{ kcal/mole}$$

What are the enthalpies of formation of O_2 and N_2? By our definition, the enthalpy of formation of all elements in their natural form at 25 °C and 1 atm is zero. Hence

$$\Delta h_C = [1 \cdot (-94.05) + 2 \cdot (-57.80) + 5.46 \cdot 0.0] - [1 \cdot (-17.89) + 2.0 \cdot 0.0 + 5.46 \cdot 0.0]$$
$$= -191.75 \text{ kcal/mole of } CH_4$$

What is the difference between this answer and that calculated in the example of oxidation of methane in Section B.4? Why does it occur?

B.8 Concluding Remarks

In this appendix, we have elucidated how chemical energy is stored in the bonds between various atoms. From this concept, the enthalpy of formation of a compound is defined as the change in enthalpy when the compound is formed from its elements in their standard state at 25 °C and 1 atm. Arbitrary datum has been chosen in such a way that all elements in their standard state at 25 °C and 1 atm have an enthalpy of formation equal to zero. Concepts then are derived for the enthalpy of reaction and of combustion. Emphasis might here be put again that the units of enthalpy of formation are kcal per mole of the compound. Units for enthalpy of reaction are simply kcal, whereas enthalpy of combustion is in kcal per mole of the fuel. Two important thermochemical laws are demonstrated to be useful in calculations. Kirchoff's Law is then obtained to calculate $\Delta H°_{RT}$ at any temperature from a knowledge of its value at any other temperature. JANAF, CRC, McBride, Huff and Dreisbach tables given in Appendix A provide comprehensive data on the heats of formation and combustion.

References

Daniels, F. and Alberty, R. A., *Physical Chemistry*, Wiley, New York (1955).
Glasstone, S., *Elements of Physical Chemistry*, Van Nostrand, New York (1958).
Moore, W. J., *Physical Chemistry*, Prentice-Hall, Englewood (1956).
Tribus, M., *Thermostatics and Thermodynamics*, Van Nostrand, New York (1961).
Rossini, F. D., et al. (eds.), *Selected Values of Chemical Thermodynamic Properties*, National Bureau of Standards Circular No. 500 (1952).
Rossini, F. D., et al., *Tables of Selected Values of Physical and Thermodynamic Properties of Hydrocarbons and Related Compounds*, Carnigie Press (1953).

APPENDIX C

Equilibrium
(Application of the Second Law of Thermodynamics to Chemically Reacting Systems)

Equilibrium, according to Webster's Dictionary, is a state of balance, or even adjustment, between opposing influences, interests, etc. It is also a state of balance between forces or actions, either "static" as in the case of a body acted on by forces whose resultant is zero or "dynamic" as in a reversible chemical reaction when the velocities in both directions are equal.

The simple experience of a marble in a bowl is a familiar example to illustrate the principle of mechanical equilibrium. Suppose a marble is placed in a bowl as shown in Figure C.1a. As the marble settles down at the bottom of the bowl "O," and if by chance it is slightly displaced to either "A" or "B," it will seek its stable position of "O" again. "O" is thus a stable equilibrium state of the marble-bowl system. Now consider Figure C.1b in which a marble is shown sitting on the apex of the inverted bowl. The point "O" here is an unstable equilibrium state because any small perturbations in the position of

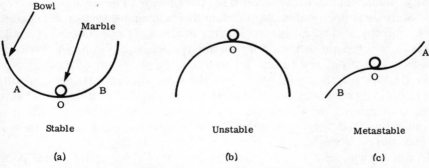

Figure C.1   Three different equilibrium (mechanical) states

the marble would displace it away from this point. Consider now a third situation. Imagine a worn-out staircase in which the landing between two consecutive steps is simply a point of zero in slope. Such a point in mathematical terms is called the *inflexion point* or saddle point. If the marble sitting at the saddle point is slightly displaced upwards, say to "A," it returns to "O" but if it crosses (due to inertia) the point of inflexion by any chance, it displaces itself farther and farther away from "O." O is then known as a metastable equilibrium state.

C.1 Concept of Minimum Energy for Equilibrium

In the preceding simple example the following three equilibrium characteristics of a mechanical system are clear. (1) The marble at stable equilibrium is at its minimum potential energy state. (2) Any process by which a change is attempted in the state of the marble at stable equilibrium is a reversible process. (3) The marble at stable equilibrium would refuse to change its state.

Consider now a thermodynamic system consisting of a vessel of volume V isolated from the surroundings and containing a gas at pressure P and temperature T. Ths system is said to be in thermodynamic equilibrium if the thermodynamic properties of the system refuse to change. This situation is quite analogous to the marble-in-the-bowl experiment.

The concept of entropy introduced in Appendix A becomes helpful to describe the thermodynamic equilibrium. The second law states that any spontaneous change in an isolated system would only result in an increase of entropy. By the first law, the energy in an isolated system could be neither destroyed nor created. Furthermore, because the volume of the system is kept constant, no work is transferred between the system and the surroundings. Thus, for an isolated system, the energy is constant. The second law may, hence, be rephrased as: in a system of constant energy and volume, any spontaneous changes would cause the entropy to increase.

Since no property of the equilibrium thermodynamic system will change, the entropy is at its maximum. Further analysis of these statements would result in a general definition for thermodynamic equilibrium. A system in which the entropy and volume are constant, the energy must be a minimum at thermodynamic equilibrium—just the same concept attained with the marble in its minimum potential energy for mechanical equilibrium. That is, at equilibrium, for $V = $ constant with a given $s = $ constant, then u is minimum; and for $V = $ constant with a given $u = $ constant, then s is maximum.

But what if neither u nor s remain constant? It is exactly at this point the free energy comes to help us. In Appendix A the free energy of a constant

Equilibrium

pressure system is defined as $f = h - Ts$. The work function, "a," (Helmholtz free energy) is defined on a parallel basis as $a = u - Ts$. "a" and "f" are both state functions because the properties composing them are all state functions. If a system is made to change its state in a constant temperature process, as in a bomb calorimeter, the changes in the free energy and the work function could be written as

$$df = dh - T\,ds, \quad P = \text{constant} \quad (C.1)$$

$$da = du - T\,ds, \quad V = \text{constant} \quad (C.2)$$

Equations C.1 and C.2 indicate that the change in free energy between the two states is equal to the change in enthalpy minus a quantity ($T\,ds$) which we will call the *unavailable energy*; and that the change in the work function is equal to the change in internal energy minus the unavailable energy.

In the vicinity of the equilibrium state, the system could be perturbed in either direction with equal case, i.e., near equilibrium any process the system could be subjected to, is a reversible process. For a reversible process, the net work done by a constant-pressure-constant-temperature system is equal to the decrease in free energy. Hence at equilibrium,

$$df = -dW_{net} \quad (C.3)$$

In common practice, dW_{net} in many processes is negligible so that the necessary conditions for thermodynamic equilibrium are

$$df = 0 \text{ at equilibrium}, P = \text{constant}, T = \text{constant} \quad (C.4)$$

Whereas, Eq. C.4 describes the condition for thermodynamic equilibrium, we show in section C.5 that the free energy of reaction ΔF_R (to be defined in a manner similar to the enthalpy of reaction) has to the zero to realize chemical equilibrium.

Equation C.4 implies that when a system is in thermodynamic equilibrium, at constant pressure and temperature, any changes in the system properties are such that the free energy would remain constant and minimum. It can readily be seen from Eq. C.1 that either a decrease in enthalpy or an increase in entropy would reduce the free energy. Thus, for a system tending towards equilibrium, the free energy continuously decreases till it assumes a constant value, complying with both the maximum entropy and minimum energy stipulations. By a similar argument, the work function is minimum at constant volume and temperature, such as in a bomb, at equilibrium.

C.2 Free Energy

(a) *Variation of Free Energy with Pressure*

For an isothermal process, the first part of Eq. A.15 gives

$$df = V\, dP, \qquad T = \text{constant}.$$

In Appendix A, we have shown that, at any temperature T, the free energy varies with pressure according to

$$\Delta f_T^P - \Delta f_T^\circ = RT \int_{P_0}^{P} \frac{dP}{P} = RT \ln \frac{P}{P_0} \tag{C.5}$$

(b) *Variation of Free Energy with Temperature*

For an isobaric process, the second part of Eq. A.15 gives

$$df = -s\, dT, \qquad P = \text{constant}.$$

To integrate this equation, two avenues are available—one in which s is eliminated via the relation $-s = (f - h)/T$ and the other in which $s(T)$ is derived for the process as in Appendix A.

Suppose now that the free energy change is measured in isothermal processes by experiments conducted at a series of temperatures all at a certain specified constant pressure. The dependency of free energy change on the level of temperature is of interest to us. By Eq. A.15 and the definition of f,

$$\left.\frac{\partial \Delta f}{\partial T}\right|_P = -\Delta s = \frac{\Delta f - \Delta h}{T} \tag{C.6}$$

Equation C.6 is well known in thermodynamics as the *Gibbs-Helmholtz equation*. Multiplying both sides of this equation by $1/T$ and transposing the terms,

$$\frac{1}{T}\left.\frac{\partial \Delta f}{\partial T}\right|_P - \frac{\Delta f}{T^2} = -\frac{\Delta h}{T^2}$$

The left hand side of this equation is simply $\partial(\Delta f/T)/\partial T$ so that Gibbs-Helmholtz equation could be written in other forms as either

$$\left.\frac{\partial}{\partial T}\left(\frac{\Delta f}{T}\right)\right|_P = -\frac{\Delta h}{T^2}$$

or

$$\frac{\partial(\Delta f/T)}{\partial(1/T)} = -\Delta h.$$

Equilibrium

The last equation means that if we plot $\Delta f/T$ against $1/T$ for the series of experiments, the graph would be a straight line with a slope of $-\Delta h$ as shown in Figure C.2. Conversely, if $\Delta h(T)$ is a known function, $\Delta f(T)$ is calculable.

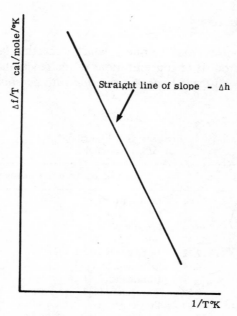

Figure C.2 Variation of free energy with temperature

C.3 Chemical Equilibrium

In Sections C.1 and C.2 we have considered mechanical and thermodynamic equilibria. In many a combustion problem, an engineer is interested in reacting mixtures of gases. When a mixture of reactants is put in a vessel and allowed to react, surprisingly enough not *all* the reactants get consumed to yield products. Consequently, analysis of the contents of the vessel after reactions cease to proceed further shows that certain fractions of the initial reactants are still present in the mixture. This state in which reactions refrain from proceeding further is called *chemical equilibrium*. The composition of the equilibrium mixture depends upon temperature and pressure.

The chemical equilibrium state is "dynamic" in its nature rather than "static." That is, the reactions really never cease to occur even at equilibrium.

While the reactants interact to form the products, simultaneously, a reverse process in which the products interact to form the original reactants would also take place. At equilibrium, the rates of the forward and backward reactions are equal.

C.4 Law of Mass Action

Law of mass action says that the rate at which a reaction takes place is proportional to the product of concentrations of the reactants, the constant of proportionality being called *specific reaction rate constant*. Denoting the

TABLE C.1

Free Energy of Formation*
Δf°_{f298} kcal/mole

Gases		Gaseous organic compounds	
H_2O	-54.64	Methane (CH_4)	-12.14
		Ethane (C_2H_6)	-7.86
O_3	39.06	Propane (C_3H_8)	-5.61
HCl	-22.77	n-Butane (C_4H_{10})	-3.75
HBr	-12.72	Isobutane (C_4H_{10})	-4.30
HI	0.31	n-Pentane (C_5H_{12})	-1.96
SO_2	-71.79	Isopentane (C_5H_{12})	-3.50
SO_3	-88.52	Neopentane (C_5H_{12})	-3.60
H_2S	-7.89	Ethylene (C_2H_4)	16.28
N_2O	24.90	Acetylene (C_2H_2)	50.00
NO	20.72	1-Butene (C_4H_8)	17.22
NO_2	12.39	Cis-2-Butane (C_4H_8)	15.74
NH_3	-3.97	Trans-2-Butene (C_4H_8)	15.05
CO	-32.81	Isobutene (C_4H_8)	13.88
CO_2	-94.26	1, 3-Butadiene (C_4H_6)	36.01
		Methyl Chloride (CH_3Cl)	-14.00
Gaseous atoms		Liquid organic compounds	
H	48.58	Methanol (CH_3OH)	-39.73
F	14.20	Ethanol (C_2H_5OH)	-41.77
Cl	25.19	Acetic acid ($C_2H_4O_2$)	-93.80
Br	19.69	Benzene (C_6H_6)	31.00
I	16.77	Chloroform ($CHCl_3$)	-17.10
C	160.85	Carbon tetra chloride (CCl_4)	-16.40
N	81.47		
O	54.99		

* Abstracted from NBS Circular 500.

Equilibrium

concentration of any component j by C_j, if k_f is the forward reaction rate constant, and k_b is the backward reaction rate constant, at equilibrium, for the reaction

$$aA + bB \underset{k_b}{\overset{k_f}{\rightleftarrows}} cC + dD \tag{C.7}$$

the law of mass action gives

$$k_f C_A^a C_B^b = k_b C_C^c C_D^d.$$

By rearrangement,

$$\frac{k_f}{k_b} = \frac{C_C^c C_D^d}{C_A^a C_B^b}$$

The quotient k_f/k_b denoted by K, is called the *equilibrium constant*. A subscript C is also given to K to indicate that its definition is based upon concentrations.

$$K_C \equiv \frac{k_f}{k_b} = \frac{C_C^c C_D^d}{C_A^a C_B^b} \tag{C.8}$$

Equation C.8 is sometimes known as *Guldberg and Waage's law* of chemical equilibrium.

C.5 Free Energy and Chemical Equilibrium

Analogous to our definition of the standard enthalpy reaction for a chemical reaction a standard *free energy of reaction* may also be defined as

$$\Delta F_{R298}^\circ = \sum_{i=\mathscr{P}} n_i \Delta f_{f298i}^\circ - \sum_{j=\mathscr{R}} n_j \Delta f_{f298j}^\circ \tag{C.9}$$

where Δf_{f298k}° is standard free energy of formation or species k some values of which are given in Table C.1.

A positive ΔF_R would mean that work has to be put into the system to keep the reaction going. If ΔF_R is negative, the reaction proceeds spontaneously by itself and in this process releases net work to the surroundings. The free energy of reaction thus acts as a driving force for the reaction. With detailed arguments it can be shown that when the free energy of reaction is zero, then the reaction is in chemical equilibrium.

C.6 Equilibrium Constant and Standard Free Energy of Reaction

In all the different kinds of equilibria considered so far, we found that the system tends to take the lowest potential energy state. It is the free energy of reaction that enables us to find whether or not equilibrium would exist in chemical reactions. In this section we will establish a quantitative relationship between the equilibrium constant K and the free energy of a reaction.

Consider once again the chemical reaction represented by Eq. C.7. If the reactants in their standard state, assumed ideal, are converted by the reaction to the products in standard state, the standard free energy of reaction by Eq. C.9 is

$$\Delta F_R^\circ = c\Delta f_{f_C}^\circ + d\Delta f_{f_D}^\circ - a\Delta f_{f_A}^\circ - b\Delta f_{f_B}^\circ$$

At any arbitrary total pressure of P,

$$\Delta F_R^P = c\Delta f_{f_C}^P + d\Delta f_{f_D}^P - a\Delta f_{f_A}^P - b\Delta f_{f_B}^P.$$

The change in the free energy of reaction due to the pressure change is

$$\Delta F_R^P - \Delta F_R^\circ = c(\Delta f_{f_C}^P - \Delta f_{f_C}^\circ) + d(\Delta f_{f_D}^P - \Delta f_{f_D}^\circ) - a(\Delta f_{f_A}^P - \Delta f_{f_A}^\circ) - b(\Delta f_{f_B}^P - \Delta f_{f_B}^\circ)$$

Substituting expressions for $\Delta f_{f_i}^\circ$ from Eq. C.5,

$$\Delta F_R^P - \Delta F_R^\circ = RT(c \ln P_C + d \ln P_D - a \ln P_A - b \ln P_B) = RT \ln \frac{P_C^c P_D^d}{P_A^a P_B^b}$$

P_C, P_D, P_A, and P_B are any specified pressures of C, D, A, and B. At the standard state, trivially $\Delta F_R^P - \Delta F_R^\circ = 0$, so that

$$P_C^c P_D^d = P_A^a P_B^b$$

This is true as all the pressures individually at their standard state are equal to 1 atm.

Now at equilibrium, according to the extension of Eq. C.4, $\Delta F_R^P = 0$ so that

$$\ln\left(\frac{P_C^c P_D^d}{P_A^a P_B^b}\right) = -\frac{\Delta F_R^\circ}{RT}$$

Let us now define a new equilibrium constant K_P based on partial pressures

$$K_P \equiv \frac{P_C^c P_D^d}{P_A^a P_B^b}$$

Equilibrium

so that

$$\ln K_P = -\frac{\Delta F_R^\circ}{RT} \quad \text{or} \quad K_P = \exp(-\Delta F_R^\circ/RT) \quad (C.10)$$

Equation C.10 relates the equilibrium constant to the standard free energy of the reaction through the temperature. Since ΔF_R° is a constant, K_P is a constant at a given temperature.

Equation C.10 is sometimes derived by the following slightly different method. Consider the box shown in Figure C.3. Into this box "a" moles of

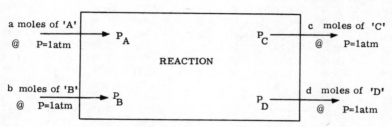

Figure C.3 Equilibrium box method

A at 1 atmosphere pressure and 298 °K and "b" moles of B at 1 atmosphere pressure and 298 °K are pumped in. Their pressures then are reduced to the equilibrium partial pressures P_A and P_B keeping the temperature constant. The change in free energy due to this pressure reduction is given by

$$aRT \ln \frac{P_A}{1} + bRT \ln \frac{P_B}{1}$$

A and B which are at P_A and P_B respectively are mixed and made to react at constant temperature of 298 °K, to yield c moles of C and d moles of D at their respective equilibrium partial pressures P_C and P_D. The free energy change involved with this equilibrium conversion is

$$\Delta F|_{\text{at equilibrium}} = 0$$

---

* At constant temperature, it is easy to show that the standard free energy, enthalpy and entropy of a reaction are related according to

$$\Delta F_R^\circ = \Delta H_R^\circ - T \Delta S_R^\circ.$$

Upon calculation of the free energy of reaction either by this equation from known (tabulated) values of ΔH_R° and ΔS_R°, or by Eq. C.9 from known free energies of formation, one can use Eq. C.10 to obtain the equilibrium constant. Refer to Tables (JANAF, McBride, Huff, Dreisbach, etc.) to familiarize yourself with the important thermochemical properties of various substances and their reactions.

TABLE C.2

$T\,°K$	\multicolumn{8}{c}{$\log_{10} K_P$ for reactions}
	1	2	3	4	5	6	7	8
298.2	11.91	−15.04			3.70			
400	7.68	−11.07			1.07			
500	5.21	−8.74			−0.45			
600	3.57	−7.20			−1.41			
700	2.37	−6.07			−2.11			−15.75
800	1.47	−5.11			−2.63	−20.40		−13.26
900	0.78	−4.58	−11.06	−9.95	−3.05	−17.70		−11.45
1000	0.22	−4.06	−9.67	−8.65	−3.39	−15.59	−21.15	−10.01
1100	−0.23	−3.62	−8.45	−7.55	−3.64	−13.80	−18.60	−8.82
1200	−0.59	−3.29	−7.46	−6.66	−3.86	−12.49	−16.52	−7.85
1300	−0.92	−2.99	−6.60	−5.90	−4.05	−11.10	−14.75	−6.98
1400	−1.19	−2.71	−5.91	−5.25	−4.21	−10.06	−13.29	−6.27
1500	−1.42	−2.47	−5.29	−4.69	−4.35	−9.18	−11.98	−5.68
1600	−1.61	−2.27	−4.75	−4.19	−4.47	−8.37	−10.81	−5.14
1700	−1.81	−2.09	−4.25	−3.75	−4.59	−7.67	−9.79	−4.67
1800	−1.98	−1.94	−3.83	−3.37	−4.68	−7.06	−8.93	−4.25
1900	−2.11	−1.82	−3.44	−3.02	−4.76	−6.49	−8.11	−3.87
2000	−2.25	−1.70	−3.10	−2.74	−4.83	−5.98	−7.40	−3.52
2100	−2.37	−1.58	−2.78	−2.44	−4.89	−5.52	−6.73	−3.20
2200	−2.48	−1.47	−2.53	−2.20	−4.95	−5.10	−6.12	−2.92
2300	−2.57	−1.38	−2.29	−1.97	−5.01	−4.72	−5.57	−2.67
2400	−2.66	−1.29	−2.06	−1.75	−5.07	−4.38	−5.07	−2.45
2500	−2.75	−1.21	−1.84	−1.55	−5.12	−4.06	−4.62	−2.25
2600	−2.83	−1.14	−1.63	−1.36	−5.17	−3.77	−4.19	−2.06
2700	−2.90	−1.07	−1.44	−1.19	−5.21	−3.49	−3.79	−1.87
2800	−2.97	−1.01	−1.26	−1.03	−5.25	−3.23	−3.42	−1.70
2900	−3.03	−0.95	−1.09	−0.89	−5.29	−3.00	−3.08	−1.54
3000	−3.09	−0.90	−0.93	−0.76	−5.32	−2.79	−2.77	−1.39
3100	−3.14	−0.85	−0.78	−0.63	−5.36	−2.60	−2.47	−1.25
3200	−3.19	−0.80	−0.63	−0.51	−5.39	−2.41	−2.19	−1.12
3300	−3.24	−0.76	−0.50	−0.40	−5.42	−2.22	−1.93	−1.00
3400	−3.28	−0.71	−0.38	−0.30	−5.45	−2.04	−1.69	−0.89
3500	−3.32	−0.67	−0.26	−0.21	−5.47	−1.86	−1.46	−0.78
3600	−3.36	−0.63	−0.15	−0.11	−5.49	−1.69	−1.24	−0.68
3700	−3.40	−0.59	−0.05	−0.02	−5.51	−1.53	−1.03	−0.58
3800	−3.44	−0.56	0.05	0.07	−5.53	−1.39	−0.83	−0.49
3900	−3.47	−0.53	0.14	0.15	−5.55	−1.26	−0.63	−0.41
4000	−3.50	−0.50	0.23	0.23	−5.57	−1.14	−0.44	−0.33
4100	−3.53	−0.47	0.32	0.31	−5.59	−1.03	−0.26	−0.25
4200	−3.56	−0.44	0.40	0.38	−5.61	−0.92	−0.09	−0.17
4300	−3.59	−0.41	0.47	0.44	−5.62	−0.82	0.07	−0.10
4400	−3.62	−0.39	0.54	0.50	−5.63	−0.72	0.22	−0.03
4500	−3.65	−0.37	0.61	0.56	−5.64	−0.63	0.36	0.03
4600	−3.67	−0.35	0.67	0.62	−5.66	−0.54	0.49	0.09
4700	−3.69	−0.33	0.72	0.67	−5.68	−0.46	0.61	0.15
4800	−3.71	−0.31	0.77	0.72	−5.69	−0.38	0.72	0.21
4900	−3.73	−0.29	0.82	0.77	−5.70	−0.30	0.82	0.27
5000	−3.75	−0.28	0.86	0.81	−5.71	−0.22	0.91	0.33

Equilibrium

Equilibrium Constants for 17 Gas Reactions

	\multicolumn{9}{c}{$\log_{10} K_p$ for reactions}								
$T\ °K$	9	10	11	12	13	14	15	16	17
298.2		−4.50	−20.52			16.02	11.00	4.62	27.02
400		−2.90	−13.02			10.12	6.65	0.35	16.77
500		−2.02	−8.64			6.62	4.08	−2.15	10.70
600		−1.43	−5.69	−20.11		4.26	2.36	−3.81	6.60
700	−16.60	−1.00	−3.59	−16.59	−13.73	2.59	1.12	−5.02	3.71
800	−14.06	−0.67	−1.98	−13.93	−11.63	1.31	0.20	−5.92	1.51
900	−12.07	−0.41	−0.74	−11.86	−10.02	0.33	−0.53	−6.63	−0.20
1000	−10.50	−0.22	0.26	−10.23	−8.72	−0.48	−1.05	−7.20	−1.53
1100	−9.22	−0.07	1.08	−8.89	−7.67	−1.15	−1.49	−7.63	−2.64
1200	−8.14	0.06	1.74	−7.79	−6.78	−1.68	−1.91	−8.02	−3.59
1300	−7.22	0.17	2.30	−6.81	−6.02	−2.13	−2.24	−8.37	−4.37
1400	−6.45	0.27	2.77	−6.01	−5.40	−2.50	−2.54	−8.64	−5.06
1500	−5.78	0.35	3.18	−5.33	−4.84	−2.83	−2.79	−8.87	−5.64
1600	−5.20	0.42	3.56	−4.73	−4.35	−3.14	−3.01	−9.08	−6.15
1700	−4.66	0.48	3.89	−4.19	−3.94	−3.41	−3.20	−9.27	−6.61
1800	−4.21	0.54	4.18	−3.71	−3.56	−3.64	−3.36	−9.44	−7.00
1900	−3.79	0.59	4.45	−3.27	−3.20	−3.86	−3.51	−9.59	−7.37
2000	−3.49	0.64	4.69	−2.88	−2.88	−4.05	−3.64	−9.72	−7.69
2100	−3.07	0.69	4.91	−2.54	−2.61	−4.22	−3.75	−9.84	−7.97
2200	−2.79	0.73	5.10	−2.24	−2.37	−4.37	−3.86	−9.95	−8.23
2300	−2.52	0.76	5.27	−1.96	−2.14	−4.51	−3.96	−10.05	−8.47
2400	−2.27	0.79	5.43	−1.69	−1.92	−4.64	−4.06	−10.14	−8.70
2500	−2.03	0.82	5.58	−1.43	−1.72	−4.76	−4.15	−10.22	−8.91
2600	−1.81	0.85	5.72	−1.21	−1.53	−4.87	−4.23	−10.30	−9.10
2700	−1.60	0.87	5.84	−1.00	−1.35	−4.97	−4.30	−10.47	−9.27
2800	−1.41	0.89	5.95	−0.81	−1.18	−5.06	−4.37	−10.44	−9.43
2900	−1.24	0.91	6.05	−0.63	−1.04	−5.14	−4.43	−10.50	−9.57
3000	−1.07	0.93	6.16	−0.46	−0.91	−5.23	−4.49	−10.56	−9.72
3100	−0.92	0.95	6.25	−0.30	−0.79	−5.30	−4.55	−10.61	−9.85
3200	−0.78	0.97	6.33	−0.15	−0.68	−5.37	−4.61	−10.66	−9.98
3300	−0.64	0.99	6.41	−0.01	−0.57	−5.44	−4.66	−10.71	−10.10
3400	−0.51	1.01	6.49	0.12	−0.47	−5.50	−4.71	−10.76	−10.21
3500	−0.28	1.02	6.56	0.24	−0.37	−5.56	−4.75	−10.81	−10.31
3600	−0.26	1.03	6.63	0.35	−0.28	−5.62	−4.78	−10.85	−10.40
3700	−0.13	1.04	6.71	0.36	−0.19	−5.67	−4.81	−10.89	−10.48
3800	−0.04	1.05	6.78	0.56	−0.11	−5.72	−4.84	−10.93	−10.56
3900	0.05	1.06	6.85	0.65	−0.03	−5.78	−4.87	−10.96	−10.65
4000	0.13	1.07	6.91	0.74	0.05	−5.83	−4.90	−10.99	−10.73
4100	0.21	1.08	6.97	0.83	0.12	−5.88	−4.93	−11.02	−10.81
4200	0.29	1.09	7.03	0.92	0.19	−5.93	−4.96	−11.05	−10.89
4300	0.37	1.10	7.08	1.00	0.25	−5.97	−4.99	−11.08	−10.96
4400	0.44	1.11	7.13	1.08	0.31	−6.01	−5.02	−11.11	−11.03
4500	0.51	1.12	7.17	1.15	0.37	−6.05	−5.05	−11.14	−11.10
4600	0.58	1.13	7.21	1.22	0.43	−6.08	−5.08	−11.16	−11.16
4700	0.64	1.14	7.25	1.29	0.48	−6.11	−5.11	−11.18	−11.22
4800	0.70	1.15	7.28	1.36	0.53	−6.13	−5.14	−11.20	−11.27
4900	0.76	1.16	7.31	1.43	0.58	−6.15	−5.17	−11.22	−11.32
5000	0.82	1.17	7.34	1.50	0.63	−6.17	−5.19	−11.24	−11.36

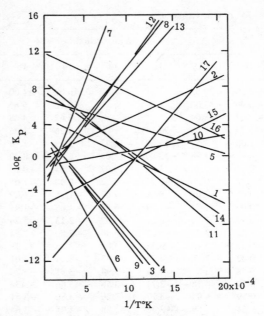

Figure C.4 Equilibrium constants for 17 gas reactions (from Hawkins and Smith, Purdue University Engineering Bulletin, Research Series 108, Volume 33, No. 3, 1949)

Following are the reactions and the corresponding numbers, they are shown by, in Table C.2 and Figure C.4.

1. $SO_2 + \tfrac{1}{2}O_2 \rightleftharpoons SO_3$
2. $\tfrac{1}{2}O_2 + \tfrac{1}{2}N_2 \rightleftharpoons NO$
3. $\tfrac{1}{2}O_2 \rightleftharpoons O$
4. $\tfrac{1}{2}H_2 \rightleftharpoons H$
5. $\tfrac{1}{2}N_2 + \tfrac{3}{2}H_2 \rightleftharpoons NH_3$
6. $\tfrac{1}{2}N_2 \rightleftharpoons N$
7. $NO \rightleftharpoons N + O$
8. $H_2O \rightleftharpoons H_2 + \tfrac{1}{2}O_2$
9. $H_2O \rightleftharpoons \tfrac{1}{2}H_2 + OH$
10. $CO_2 + H_2 \rightleftharpoons CO + H_2O$
11. $CO_2 + C \rightleftharpoons 2CO$
12. $CO_2 \rightleftharpoons CO + \tfrac{1}{2}O_2$
13. $2C + H_2 \rightleftharpoons C_2H_2$
14. $H_2 + CO \rightleftharpoons C + H_2O$
15. $C + 2H_2 \rightleftharpoons CH_4$
16. $CO + 2H_2 \rightleftharpoons CH_3OH$
17. $CO + 3H_2 \rightleftharpoons CH_4 + H_2O$

Now the products C and D at P_C and P_D and 298 °K are isothermally expanded to the pressures of 1 atmosphere. The free energy change associated with this isothermal expansion is

$$cRT \ln \frac{1}{P_C} + dRT \ln \frac{1}{P_D}$$

Hence the net free energy change caused by converting the reactants at 1 atm and 298 °K to products at 1 atm and 298 °K is the sum of the three

Equilibrium

changes listed above.

$$\Delta F°_{R298} = aRT \ln P_A + bRT \ln P_B$$
$$+ \Delta F|_{\text{at equilibrium}}$$
$$+ cRT \ln \frac{1}{P_C} + dRT \ln \frac{1}{P_D}$$

Rewriting,

$$-\frac{\Delta F°_{R298}}{RT} = \ln \left[\frac{P_C^c P_D^d}{P_A^a P_B^b} \right]$$

which is the same as Eq. C.10.

The equilibrium constants as functions of temperature are tabulated along with thermochemical data for various combustion reactions in the literature. Figure C.4 and Table C.2 illustrate some of these data.

C.7 Relation between the Equilibrium Constants Defined on the Basis of Concentrations and Partial Pressures

We ask ourselves the question, how are K_P and K_C related? By ideal gas law,

$$C_A = \frac{n_A}{V} = \frac{P_A}{RT}.$$

Therefore,

$$K_C = \left(\frac{P_C}{RT}\right)^c \left(\frac{P_D}{RT}\right)^d \bigg/ \left(\frac{P_A}{RT}\right)^a \left(\frac{P_B}{RT}\right)^b = \left(\frac{1}{RT}\right)^{\Delta n_R} \cdot K_P \quad \text{(C.11)}$$

where R is in liter-atmospheres per °K per mole and Δn_R is the change in the number of moles of gas due to the reaction.

$$\Delta n_R = \sum_{i=\mathscr{P}} n_i - \sum_{j=\mathscr{R}} n_j = (c+d) - (a+b).$$

If the reaction is such that there is no change in the number of moles due to reaction, then, $K_P = K_C$. The reaction $\frac{1}{2}H_2 + \frac{1}{2}I_2 \rightleftharpoons HI$ is an example of such a case.

Example 1

Calculate the equilibrium composition at 1,300 °K and a total pressure of 1 atmosphere, for the reaction

$$\tfrac{1}{2}Cl_2(g) + \tfrac{3}{2}F_2(g) \rightleftharpoons ClF_3(g)$$

Solution

Neglecting dissociation, there will be 3 components in the equilibrium mixture, namely Cl_2, F_2 and ClF_3. We need three equations to solve for these three unknowns. These are equilibrium constant relation, mass conservation of atoms of various kinds, and Dalton's law of partial pressures. The first equation can be obtained as

$$K_P(T) = \frac{P_{ClF_3}}{P_{Cl_2}^{1/2} P_{F_2}^{3/2}}.$$

This could either be referred to in the tables cited above corresponding to $T = 1,300\,°K$ or could be computed from

$$K_P = \exp(-\Delta F_R^\circ/RT)$$

provided the standard free energy change due to the reaction can be computed from the thermodynamic tables that give the free energy of the components. From the tables, at $1,300\,°K$,

$$\log_{10} K_P = -0.3307 \quad \text{i.e.,} \quad K_P = 0.467 \tag{a}$$

The conservation of atoms is given as below for the atoms of Cl and F. Total number of chlorine atoms is

$$n_{tCl} = 1 = 2n_{Cl_2} + n_{ClF_3}. \tag{b}$$

Total number of fluorine atoms is

$$n_{tF} = 3 = 3n_{ClF_3} + 2n_{F_2}. \tag{c}$$

Equations (b) and (c), by ideal gas law, become

$$\frac{n_{tCl}}{n} = \frac{1}{n} = 2X_{Cl_2} + X_{ClF_3} \tag{d}$$

and

$$\frac{n_{tF}}{n} = \frac{3}{n} = 3X_{ClF_3} + 2X_{F_2} \tag{e}$$

where n is the total number of moles in the system and X_i is mole fraction of i. From (d) and (e),

$$RT/PV = 2X_{Cl_2} + X_{ClF_3}$$
$$3RT/PV = 3X_{ClF_3} + 2X_{F_2}.$$

Combining these two expressions to eliminate X_{ClF_3} and RT/PV, $3X_{Cl_2} = X_{F_2}$. Noting $X_i = P_i/P$ and that the total pressure $P = 1$ atm.,

$$3P_{Cl_2} = P_{F_2}. \tag{f}$$

Equilibrium

The third equation is Dalton's law.

$$P_{Cl_2} + P_{F_2} + P_{ClF_3} = P = 1 \text{ atm.} \tag{g}$$

Equations (a), (f), and (g) are solved for the three unknown partial pressures to obtain the equilibrium composition.

An alternate method convenient for simple reactions is called the "extent of reaction method." At any stage of completion λ, the stoichiometric equation can be written as

$$\tfrac{1}{2}Cl_2(g) + \tfrac{3}{2}F_2(g) \longrightarrow (1-\lambda)[\tfrac{1}{2}Cl_2(g) + \tfrac{3}{2}F_2(g)] + \lambda[ClF_3(g)]$$

thus satisfying mass conservation implicitly. Note that when $\lambda = 0$, reaction is yet to start, whereas at $\lambda = 1$, reacton is complete. We can tabulate the information as below:

i	n_i	$X_i = n_i/\sum n_i$	$P_i = X_i P$
Cl_2	$(1-\lambda)/2$	$(1-\lambda)/2(2-\lambda)$	$(1-\lambda)P/2(2-\lambda)$
F_2	$3(1-\lambda)/2$	$3(1-\lambda)/2(2-\lambda)$	$3(1-\lambda)P/2(2-\lambda)$
ClF_3	λ	$\lambda/(2-\lambda)$	$\lambda P/(2-\lambda)$

Hence

$$K_P = \frac{[\lambda P/(2-\lambda)]}{[(1-\lambda)P/2(2-\lambda)]^{1/2}[3(1-\lambda)P/2(2-\lambda)]^{3/2}} = \frac{4}{3^{3/2}} \frac{\lambda(2-\lambda)}{P(1-\lambda)^2}$$

Knowing K_P at the desired temperature from tables, for any given total pressure P, this equation can be solved for λ which gives the equilibrium composition. The student must be able to discover where in this method we have incorporated Dalton's Law.

C.8 Influence of Temperature and Pressure on the Equilibrium Constant

(a) *Temperature Dependence*

The Gibb's–Helmholtz equation at standard pressure is

$$\frac{\Delta F° - \Delta H°}{T} = \frac{d\Delta F°}{dT} \tag{C.12}$$

The equilibrium constant is related to $\Delta F°$ by Eq. C.10 which when differentiated with respect to temperature yields

$$-\frac{d\Delta F°}{dT} = R \ln K_P + RT \frac{d \ln K_P}{dT}. \tag{C.13}$$

Combining Eqs. C.12 and C.13,

$$\Delta H° - \Delta F° = RT \ln K_P + RT^2 \frac{d \ln K_P}{dT}.$$

By substituting Eq. C.10 in this expression,

$$\Delta H° = RT^2 \frac{d \ln K_P}{dT}. \qquad (C.14)$$

Equation C.14 is very useful since it enables calculation of the standard heat of reaction from the known equilibrium constant at various temperatures.

Separating variables and integrating Eq. C.14 once,

$$d \ln K_P = (\Delta H°/RT^2) \, dT$$
$$\ln K_P = -\frac{\Delta H°}{RT} + \text{constant} \qquad (C.15)$$

This equation, named after *Van't Hoff*, implies that if the equilibrium constants determined by experiments at constant temperatures are plotted as in Figure C.5, the graph would be a straight line with a slope of $\Delta H°/R$.

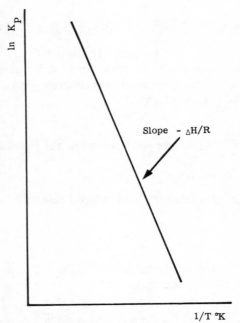

Figure C.5 The slope of $\ln K_p$ versus $1/T$ gives the heat of reaction

Equilibrium

(b) Pressure Dependence

From Eq. C.10, at constant temperature,

$$\frac{\partial \ln K_P}{\partial P} = -\frac{1}{RT}\frac{\partial \Delta F^\circ}{\partial P}.$$

The right hand side of this equation is zero because free energy ΔF° is evaluated at $P = 1$ atmosphere. Hence the equilibrium constant does not depend on pressure. Care should be taken, however, to note that this statement does not mean that the equilibrium composition is independent of total pressure. As shown in example 1, the equilibrium composition *does* depend on total pressure of the system and this dependency is implicitly invoked by Dalton's Law.

C.9 Adiabatic Flame Temperature

(a) *Method of Calculation*

Consider an exothermically reacting mixture in an isolated system. If the mixture (from a specified initial pressure and temperature) is allowed to approach chemical equilibrium by a constant pressure process adiabatically, the final temperature attained by the system is called the *adiabatic flame temperature* T_f. It depends on the pressure, the initial temperature and composition of the reactants.

Since the system is adiabatic, all the heat released in the chemical transformation of the reactants to form the equilibrium products is utilized to raise the temperature of the system. If $\Delta H_\mathcal{R}$ is the total enthalpy (includes the potential chemical energy) in the reactants and $\Delta H_\mathcal{P}$ is the total enthalpy of the equilibrium products, the adiabaticity implies that

$$\Delta H_\mathcal{R} = \Delta H_\mathcal{P} \tag{C.16}$$

This equation is illustrated in Figure C.6 by the horizontal line labelled "1" on the enthalpy-temperature diagram.

The enthalpy of reactants would consist of the summation of the heat of formation for all reactants.

$$\Delta H_\mathcal{R} = \sum_{j=\mathcal{R}} n_j \Delta h_{fj}.$$

The enthalpy of the products in their final state is the sum of the enthalpies of formation of all the components plus the sensible enthalpy rise which

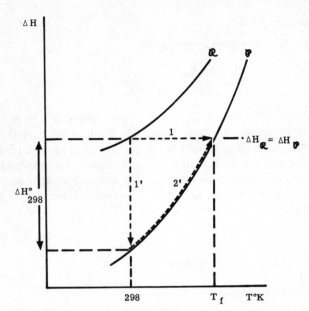

Figure C.6 Adiabaticity of an exothermic reaction

results from the heating up of the products from the standard state to their final state.

$$\Delta H_{\mathscr{P}} = \sum_{i=\mathscr{P}} n_i \, \Delta h_{fi} + \sum_{i=\mathscr{P}} n_i C_{Pi} \, dT.$$

Hence, by introducing these quantities into Eq. C.16,

$$\sum_{j=\mathscr{R}} n_j \, \Delta h_{fj} = \sum_{i=\mathscr{P}} n_i \, \Delta h_{fi} + \sum_{i=\mathscr{P}} \int_{298}^{T_f} n_i C_{Pi} \, dT \qquad (C.17)$$

Rearranging,

$$\sum_{i=\mathscr{P}} \int_{298}^{T_f} n_i C_{Pi} \, dT = \sum_{j=\mathscr{R}} n_j \, \Delta h_{fj} - \sum_{i=\mathscr{P}} n_i \, \Delta h_{fi}$$

The right hand side of this equation is already well known to us as the heat of the reaction with the sign reversed. Hence,

$$\sum_{i=\mathscr{P}} \int_{298}^{T_f} n_i C_{Pi} \, dT = - \Delta H^{\circ}_{R298} \qquad (C.18)$$

The right hand side of this equation in magnitude is simply indicated in Figure C.6 by the line marked 1', the enthalpy change at standard temperature

Equilibrium

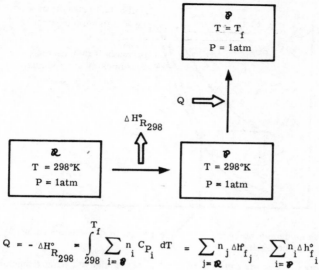

Figure C.7 Hess' law interpretation of T_f

$$Q = -\Delta H^{\circ}_{R_{298}} = \int_{298}^{T_f} \sum_{i=\mathcal{P}} n_i C_{P_i} \, dT = \sum_{j=\mathcal{R}} n_j \Delta h^{\circ}_{f_j} - \sum_{i=\mathcal{P}} n_i \Delta h^{\circ}_{f_i}$$

as the reactants transform into products. Equation C.18 hence shows that the heat thus liberated is utilized to raise the temperature of the products as indicated by the step 2' in Figure C.6. This principle is quite consistent with Hess' law and is graphically illustrated in Figure C.7.

In Eq. C.18, if we know the final products, the only unknown is the final temperature. However, at this point the big question arises, what *are* the final products? The composition of the final products, as we have seen earlier in this appendix, depends strongly on the final temperature due to equilibrium principles. We thus have two interdependent unknowns in the system—the equilibrium composition and the final temperature. We utilize an iteration technique involving the following three steps to solve Eq. C.18 under the constraint of chemical equilibrium. Figure C.8 illustrates this calculation process.

Step 1: Assume a value of T_f and determine, by the methods described earlier in this appendix, the equilibrium composition of the reacting mixture.

Step 2: For the process of transformation of the reactants to equilibrium products at the assumed T_f, from a knowledge of the heats of formation of all the components involved, calculate the heat liberated in the reaction at standard temperature and given pressure.

Step 3: All the heat so liberated by the reaction is then used to increase the sensible (i.e., $C_P \, dT$) heat of the equilibrium mixture from the standard

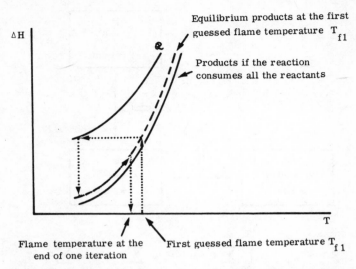

Figure C.8 Iterative procedure to calculate the adiabatic flame temperature

temperature to a new flame temperature. If this new flame temperature is not equal to the initially guessed flame temperature, make a refined guess and repeat the procedure of calculation.

(b) *Comments on Adiabatic Flame Temperature*

The calculation of the adiabatic flame temperature described above simply involves simultaneously applying the constraints of the energy conservation and equilibrium to the system. The factors which influence the final temperature of a reacting system include:

  (i) the amount of heat liberated by complete combustion of the fuel-oxidant system—i.e., the heat of combustion;
 (ii) the failure of a part of this heat of combustion to be released due to the fact that a fraction of the initial reactants are unconsumed due to equilibrium considerations;
(iii) the departure of real life systems from adiabaticity; and lastly,
(iv) diluents lower the final temperature.

Nitrogen in air is an example for such diluents. Whereas the diluents do not contribute any of their share in liberating heat, they merely take up heat in the process of heating up from room temperature to the flame temperature. As a matter of fact, looking at the situation in this viewpoint, the affect of diluents may be considered as an apparent nonadiabaticity in which a part of the heat liberated in the reaction is lost to heat the diluents.

Equilibrium

We stress the fact that the processes in most engineering situations are non-adiabatic, for to realize and simulate a perfectly adiabatic situation is impossible in real systems. A fraction of the heat is always lost to the environment and some work is done on the environment no matter how carefully the experiment is controlled. Adiabatic conditions can at most be *approached* by making the combustion process very rapid. This point makes the adiabatic flame temperature calculated by the method described above only a hypothetical upper limit of the temperature which the system can achieve under ideal conditions. In actual combustion situations, the flame temperatures will be much lower than the adiabatic flame temperature.

In the following example, we will illustrate a slightly different method of calculation of adiabatic flame temperature. Note that even though the mechanics of calculation differ, the principles involved remain unchanged.

Example 2

Assuming that the temperatures involved are low enough, and pressure high enough to preclude atomic and free radical species (H, O and OH), calculate the adiabatic flame temperature at $P = 1$ atm for the reaction

$$H_2(g) + \tfrac{1}{2}O_2(g) \rightleftharpoons H_2O(g).$$

Solution

By the method of "extent of reaction,"

$$H_2(g) + \tfrac{1}{2}O_2(g) \longrightarrow (1 - \lambda)[H_2(g) + \tfrac{1}{2}O_2(g)] + \lambda[H_2O(g)].$$

The partial pressures are given by

$$P_{H_2} = X_{H_2}P = 2(1 - \lambda)P/(3 - \lambda)$$
$$P_{O_2} = X_{O_2}P = (1 - \lambda)P/(3 - \lambda)$$
$$P_{H_2O} = X_{H_2O}P = 2\lambda P/(3 - \lambda)$$

Equilibrium constant is, therefore,

$$K_P = \frac{P_{H_2O}}{P_{H_2} P_{O_2}^{1/2}} = \frac{\lambda(3 - \lambda)^{1/2}}{P^{1/2}(1 - \lambda)^{3/2}} \tag{a}$$

P is given as 1 atm. K_P is known as a function of T. Hence Eq. (a) gives λ as a function of T.

The energy released for any value of λ is given by

$$(1 - \lambda)\Delta h^\circ_{fH_2 298} + \frac{(1 - \lambda)}{2}\Delta h^\circ_{fO_2 298} + \lambda \Delta h^\circ_{fH_2O 298} - \Delta h^\circ_{fH_2 298} - \tfrac{1}{2}\Delta h^\circ_{fO_2 298}$$

This heat is used to heat up the products from 298 °K to T, so that

$$\int_{298}^{T_f} \left[(1-\lambda)C_{PH_2} + \frac{(1-\lambda)}{2} C_{PO_2} + \lambda C_{PH_2O} \right] dT = [\Delta h^\circ_{fH_2 298} + \tfrac{1}{2}\Delta h^\circ_{fO_2 298}]$$

$$- \left[(1-\lambda)\Delta h^\circ_{fH_2 298} + \frac{(1-\lambda)}{2} \Delta h^\circ_{fO_2 298} + \lambda \Delta h^\circ_{fH_2O 298} \right]$$

But Δh°_{f298} for all elements in their natural state at 298 °K and 1 atm is zero by our datum convention. Hence

$$\int_{298}^{T_f} \left[(1-\lambda)C_{PH_2} + \frac{(1-\lambda)}{2} C_{PO_2} + \lambda C_{PH_2O} \right] dT = -\lambda \, \Delta h^\circ_{fH_2O 298} \quad \text{(b)}$$

The energy conservation thus implicitly gives the second relation between the extent of reaction λ and temperature T. We can solve Eqs. (a) and (b) for the two unknowns λ and T. It is best done either graphically or by a computer. Figure C.9 shows the two $\lambda - T$ relationships qualitatively. The intersection point of the two curves gives the equilibrium conditions.

In this problem, we assumed that the initial mixture is in stoichiometric proportions. If one of the species is in excess, the excess quantity can be treated as a third and different species and calculations performed by the usual method. The student can immediately see that the adiabatic flame temperature will be maximum when the initial mixture is stoichiometric. On the rich and lean sides, T_f will be lower due to dilution effect as indicated qualitatively in Figure C.10

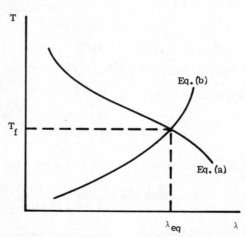

Figure C.9 Figure for example 2 solution by the method of extent of reaction

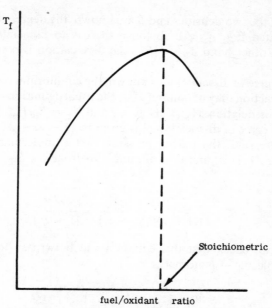

Figure C.10 Dependence of Adiabatic flame temperature on the fuel/oxidant ratio

C.10 Dissociation of Gases

When gases are intensely heated, they dissociate into atomic species. The extent to which dissociation occurs is often of interest in many combustion problems. Measurements of the density of the dissociated gas gives the degree of dissociation because due to dissociation, the number of moles present in the volume increases, thus changing the density.

Let β be the number of atoms produced when one mole of the molecular gas under consideration is dissociated. Then the total number of moles in the dissociated mixture is $(1 - \lambda) + \beta\lambda$, the extent of dissociation being denoted, as usual, by λ.

At constant pressure, the density of a gas is inversely proportional to the number of moles present. The ratio of ρ_1 (the density of the undissociated gas) to ρ_2 (that of partially dissociated gas) is, hence,

$$\frac{\rho_1}{\rho_2} = 1 + (\beta - 1)\lambda$$

Solving for the degree λ of dissociation,

$$\lambda = \frac{(\rho_1/\rho_2) - 1}{\beta - 1} \tag{C.19}$$

Therefore, if the two densities and β are known, the degree of dissociation may be calculated. If $\rho_1/\rho_2 = 1$, obviously there is no dissociation and hence $\lambda = 0$. On the other hand if $\rho_1 = \beta\rho_2$ the dissociation is complete and so $\lambda = 1$.

Once the degree of dissociation is known, the equilibrium constant for the dissociation reaction may be computed by the usual definitions. For example, for hydrogen dissociation, $H_2 \rightleftharpoons 2H$, $\beta = 2$ and $K_p = P_H^2/P_{H_2}$. Total number of moles at a degree of dissociation λ is given by $(1 - \lambda) + 2\lambda = 1 + \lambda$. If P is the total pressure, the partial pressure of the molecular hydrogen is $P_{H_2} = (1 - \lambda)P/(1 + \lambda)$; and that of atomic hydrogen is $P_H = 2\lambda P/(1 + \lambda)$. Therefore,

$$K_p = \frac{[2\lambda P/(1 + \lambda)]^2}{[(1 - \lambda)P/(1 + \lambda)]} = \frac{4\lambda^2 P}{(1 - \lambda^2)}$$

The student can check that this equation can be written for the general reaction (1 molecule $\to$ β atoms) as

$$K_p = \frac{(\beta\lambda)^\beta}{(1 - \lambda)} \cdot \left(\frac{1}{1 + (\beta - 1)\lambda}\right)^{(\beta - 1)} \cdot P^{(\beta - 1)} \qquad (C.20)$$

which is useful to compute the degree of dissociation at any given temperature if the equilibrium constant is known.

C.11 Concluding Remarks

In view of the minimum energy concept for mechanical equilibrium, criteria of maximum entropy and minimum energy are derived for thermodynamic equilibrium. The role of free energy in chemical equilibrium is then discussed. Equilibrium constant is introduced and related to the standard free energy of reaction. The temperature dependency of the equilibrium constant is derived. Methods of calculation of adiabatic flame temperature and degree of dissociation are introduced.

References

CRC, *Handbook of Physics and Chemistry*, Chemical Rubber Co., Cincinnati, Ohio.
Daniels, F. and Alberty, R. A., *Physical Chemistry*, Wiley, New York (1955).
Hall, N. A. and Ibele, W. E., *Engineering Thermodynamics*, Prentice-Hall, Englewood (1960).

Huff Tables, N.A.C.A. TR 1037.
JANAF Thermochemical Tables: Dow Chemical Company, Midland, Mich.
Moore, W. K., *Physical Chemistry*, Prentice-Hall, Englewood (1956).
Thermodynamic Tables to 6000 °K for 210 Substances including the First 18 Elements. N.A.S.A. SP-3001.
Tribus, M., *Thermostatics and Thermodynamics*, Van Nostrand, New York (1961).
Williams, F. A., *Combustion Theory*, Addison-Wesley, Reading (1965).

APPENDIX D

Transport Property Tables

In this appendix we present viscosity, thermal conductivity, mass diffusion coefficients for various gases, (vapors) and liquids of interest in combustion studies. These tables are compiled from a variety of sources available in the literature with care taken to keep the units uniform and consistent. One who wishes to obtain more information is referred to the following sources:

1. *Handbook of Physics and Chemistry*, Chemical Rubber Co., Cincinnati, Ohio. (Periodically updated and most widely available.)
2. *International Critical Tables*, McGraw-Hill Book Co., New York, N.Y. (1926–1930).
3. *Smithsonian Physical Tables*, The Smithsonian Institute, Washington, D.C.
4. *Mechanical Engineers' Handbook*, Marks, L. S., McGraw-Hill Book Co., New York, N.Y., 5th edition (1951).
5. *Tables of Thermal Properties of Gases*, Hilsenrath, J., et al., National Bureau of Standards, Circular 564 (Nov. 1955).
6. *Thermal Conductivity of Gases and Liquids*, Tsederberg, V., The M.I.T. Press, Cambridge, Mass. (1965).
7. *The Properties of Gases and Liquids*, Reid, R. C. and Sherwood, T. K., McGraw-Hill Book Co., New York, N.Y. (1958).
8. Westerberg, A. A., "Present Status of Information on Transport Properties of Applicable to Combustion Research," *Combustion and Flame*, **1**, p. 346 (1957).
9. Fristrom, R. M. and Westenberg, A. A., "Molecular Transport Properties for Flame Studies," *Fire Research Abstracts and Reviews*, **8**, p. 155 (1966).

Transport Property Tables

TABLE D.1

Viscosity ($\times 10^{-6}$) of Gases and Vapors
μ (in gms/cm/sec)

T °C	1	2	3 Air	4	5	6	7	8	9	10	11	12	13	14	15	16	17	18
0			171						83	85	91	102						
10			180		72	139	166		84	87	92	103		170				
20			183	94	76	140	170		87	91	101	109		178				
30			186		79	148	173		90	94	104	113		184				
40			190		82	153	175		92	98	108	117		188				
50		68	194		85	157	177		94	101	110	120		192				
75		82	210		91	168	180	98	100	108	120	129		203				
100	105	93	218		97	180	188	108	105	114	126	133	127	211	81		76	
150		110	242		108	190	196	124	113	127	140	148	125	229	84	63	80	
200		125	255		120	211	218	138	122	141	154	160	136	245	88	68	84	126
250		136	275		132	232	240	153	130	155	167	170	152	253	90	77	87	144
300		147	295		145	250	266	166	139	171	180	181	170	280	94	85	90	164
350		158	313		158	268	294	180	148			193	188	295	100	92	96	183
400		166	332		174	286		194	156			205	211	311	108	100	102	202
500			360			304			171			226		340	120		114	221
600			390			333			185			256		369	134		125	241
700			420			361			198			285		381	151		137	
800			442			389			211					419	170			
900			470			415			224					440				
1,000			500			440			235					460				

Legend: 1. Acetic Acid Vapor, 2. Acetone Vapor, 3. Air, 4. Acetylene, 5. Benzene Vapor, 6. Carbon Dioxide, 7. Carbon Monoxide, 8. Ethyl Alcohol Vapor, 9. Hydrogen, 10. Ethane, 11. Ethylene, 12. Methane, 13. Methyl Alcohol Vapor, 14. Nitrogen, 15. Propylene, 16. n-Octane Vapor, 17. Propane, 18. Water Vapor.

Units: 1 gm/cm/sec = 1 poise = 1 dyne sec/cm^2 = 10^2 centipoise = 10^6 micropoise.

TABLE D.2

Thermal Conductivity of Gases and Vapors

Gas	$K_0(T_0 = 273\ °K)$ cal/cm/sec/°K	Gas	$K_0(T_0 = 273\ °K)$ cal/cm/sec/°K
*Air	5.83×10^{-5}	*Helium	32.40×10^{-5}
*Argon	3.90×10^{-5}	Heptane	2.56×10^{-5}
Acetone	2.32×10^{-5}	Hexane	2.68×10^{-5}
Ammonia	5.02×10^{-5}	*Hydrogen	40.40×10^{-5}
Amyl alcohol	2.58×10^{-5}	Methane	7.33×10^{-5}
Benzene	2.20×10^{-5}	Methanol	3.05×10^{-5}
Butane	3.17×10^{-5}	Methyl bromide	1.50×10^{-5}
Butyl alcohol	2.64×10^{-5}	Methyl chloride	2.20×10^{-5}
*Carbon dioxide	3.60×10^{-5}	*Nitrogen	5.50×10^{-5}
*Carbon monoxide	5.30×10^{-5}	Octane	2.33×10^{-5}
Carbon tet	1.43×10^{-5}	*Oxygen	6.25×10^{-5}
Chloroform	1.52×10^{-5}	Pentane	2.94×10^{-5}
Cyclohexane	2.32×10^{-5}	Propane	3.64×10^{-5}
Ether	3.11×10^{-5}	Steam	3.61×10^{-5}
Ethane	4.52×10^{-5}	*Sulfur dioxide	2.00×10^{-5}
Ethanol	3.08×10^{-5}	Toluene	3.08×10^{-5}

$K = K_0(T/T_0)^n$
$n \approx 0.94$ for permanent gases (asterisked), $T_{max} < 1{,}000\ °C$
$n \approx 1.83$ for condensible gases, $T_{max} < 600\ °C$
1 cal/cm/sec/°K = 134.5 BTU/ft/hr/°F
 241.9

TABLE D.3
Viscosity and Conductivity of Mixtures

X_1	μ	K	X_1	μ	K
	$H_2 + CO_2$			$H_2 + O_2$	
0.000	13.69×10^{-5}	3.60×10^{-5}	0.000	20.45×10^{-5}	6.25×10^{-5}
0.100	13.86×10^{-5}	5.10×10^{-5}	0.034	20.44×10^{-5}	6.51×10^{-5}
0.142	13.92×10^{-5}	5.70×10^{-5}	0.250	19.94×10^{-5}	11.12×10^{-5}
0.250	14.06×10^{-5}	7.70×10^{-5}	0.500	18.55×10^{-5}	18.27×10^{-5}
0.355	14.15×10^{-5}	10.00×10^{-5}	0.750	15.88×10^{-5}	27.49×10^{-5}
0.500	14.17×10^{-5}	13.50×10^{-5}	0.947	10.88×10^{-5}	37.44×10^{-5}
0.750	13.41×10^{-5}	22.70×10^{-5}	1.000	8.87×10^{-5}	41.80×10^{-5}
0.901	11.63×10^{-5}	31.50×10^{-5}			
1.000	8.54×10^{-5}	40.40×10^{-5}			
	$H_2 + CO$			$H_2 + Ar$	
0.000	17.06×10^{-5}	5.30×10^{-5}	0.000	21.35×10^{-5}	3.90×10^{-5}
0.163	17.15×10^{-5}	8.00×10^{-5}	0.090	21.26×10^{-5}	5.50×10^{-5}
0.272	16.49×10^{-5}	10.30×10^{-5}	0.180	21.11×10^{-5}	7.30×10^{-5}
0.566	15.38×10^{-5}	18.00×10^{-5}	0.400	20.20×10^{-5}	12.60×10^{-5}
0.634	14.88×10^{-5}	20.90×10^{-5}	0.600	18.63×10^{-5}	18.70×10^{-5}
0.794	13.03×10^{-5}	27.00×10^{-5}	0.802	14.80×10^{-5}	27.00×10^{-5}
1.000	8.53×10^{-5}	40.40×10^{-5}	1.000	8.54×10^{-5}	40.40×10^{-5}
	$H_2 + N_2$			$N_2 + Ar$	
0.000	16.88×10^{-5}	5.50×10^{-5}	0.000	20.89×10^{-5}	3.85×10^{-5}
0.159	16.70×10^{-5}	8.00×10^{-5}	0.204	29.37×10^{-5}	4.17×10^{-5}
0.390	16.00×10^{-5}	12.70×10^{-5}	0.359	30.64×10^{-5}	4.44×10^{-5}
0.652	14.49×10^{-5}	19.40×10^{-5}	0.611	30.43×10^{-5}	4.90×10^{-5}
0.795	12.85×10^{-5}	25.20×10^{-5}	0.780	28.63×10^{-5}	5.24×10^{-5}
0.803	12.74×10^{-5}	25.70×10^{-5}	1.000	16.59×10^{-5}	5.66×10^{-5}
1.000	8.53×30^{-5}	40.40×10^{-5}			

X_1 is the mole fraction of the ligher component.
μ is in gm/cm/sec.
K is in cal/cm/sec/°C.
(All mixtures at $T = 273$ °K except $H_2 + O_2$ which is at 295 °K.)

TABLE D.4

Conductivity of Mixtures

X_1	K	X_1	K	X_1	K
He + Ar (273 °K)		H_2 + CO_2 (273 °K)		H_2O + Air (353 °K)	
0.000	3.89×10^{-5}	0.000	3.39×10^{-5}	0.000	6.86×10^{-5}
0.270	7.41×10^{-5}	0.170	6.08×10^{-5}	0.197	7.15×10^{-5}
0.454	10.76×10^{-5}	0.370	10.35×10^{-5}	0.306	7.08×10^{-5}
0.847	23.18×10^{-5}	0.607	17.24×10^{-5}	0.444	6.89×10^{-5}
0.946	29.40×10^{-5}	0.834	27.98×10^{-5}	0.519	6.73×10^{-5}
1.000	33.89×10^{-5}	1.000	41.66×10^{-5}	1.000	5.23×10^{-5}
NH_3 + Air (293 °K)		H_2 + CO (273 °K)		C_2H_2 + Air (293 °K)	
0.000	6.00×10^{-5}	0.000	5.30×10^{-5}	0.000	6.00×10^{-5}
0.246	6.31×10^{-5}	0.163	8.00×10^{-5}	0.141	5.96×10^{-5}
0.366	6.28×10^{-5}	0.272	10.30×10^{-5}	0.320	5.84×10^{-5}
0.608	6.09×10^{-5}	0.566	18.02×10^{-5}	0.536	5.66×10^{-5}
0.805	5.75×10^{-5}	0.634	20.91×10^{-5}	0.630	5.55×10^{-5}
1.000	5.49×10^{-5}	0.794	26.99×10^{-5}	0.900	5.30×10^{-5}
		1.000	40.43×10^{-5}	1.000	5.22×10^{-5}
NH_3 + CO (295 °K)		H_2 + C_2H_4 (278 °K)		CO + Air (291 °K)	
0.000	5.74×10^{-5}	0.000	5.28×10^{-5}	0.000	5.97×10^{-5}
0.220	5.97×10^{-5}	0.170	8.61×10^{-5}	0.108	5.95×10^{-5}
0.338	6.03×10^{-5}	0.314	11.48×10^{-5}	0.321	5.90×10^{-5}
0.620	5.94×10^{-5}	0.514	16.91×10^{-5}	0.562	5.83×10^{-5}
0.790	5.82×10^{-5}	0.611	20.61×10^{-5}	0.978	5.69×10^{-5}
1.000	5.55×10^{-5}	0.865	32.90×10^{-5}	1.000	5.68×10^{-5}
		1.000	43.73×10^{-5}		
CH_4 + Air (295 °K)					
0.000	6.04×10^{-5}				
0.076	6.13×10^{-5}				
0.390	6.49×10^{-5}				
0.700	6.87×10^{-5}				
0.880	7.08×10^{-5}				
1.000	5.98×10^{-5}				

X_1 is mole fraction of the first gas.
K is in cal/cm/sec/°C.

TABLE D.5

Binary Diffusion Coefficients*
(at $T_0 = 298\ °K$ and $P_0 = 1$ atm)

Gas pair	D_0 cm²/sec	Gas pair	D_0 cm²/sec
N_2–He	0.71	H_2–nC_4H_{10}	0.38
N_2–Ar	0.20	H_2–O_2	0.81
N_2–H_2	0.78	H_2–CO	0.75
N_2–O_2	0.22	H_2–CO_2	0.65
N_2–CO	0.22	H_2–CH_4	0.73
N_2–CO_2	0.16	H_2–C_2H_4	0.60
N_2–H_2O	0.24	H_2–C_2H_6	0.54
N_2–C_2H_4	0.16	H_2–H_2O	0.99
N_2–C_2H_6	0.15	H_2–Br_2	0.58
N_2–nC_4H_{10}	0.10	H_2–C_6H_6	0.34
CO_2–O_2	0.18	Air–H_2	0.63
CO_2–CO	0.14	Air–O_2	0.18
CO_2–C_2H_4	0.15	Air–CO_2	0.14
CO_2–CH_4	0.15	Air–H_2O	0.23
CO_2–H_2O	0.19	Air–CS_2	0.10
CO_2–C_3H_8	0.09	Air–Ether	0.08
CO_2–CH_3OH	0.09	Air–CH_3OH	0.11
CO_2–C_6H_6	0.06	Air–C_6H_6	0.08
O_2–C_6H_6	0.07	H_2O–CH_4	0.28
O_2–CO	0.21	H_2O–C_2H_4	0.20
CO–C_2H_4	0.13	H_2O–O_2	0.27

$$D = D_0(T/T_0)^m(P_0/P)$$

$m \approx 1.75$ for permanent gases
$m \approx 2.00$ for condensible gases

1 cm²/sec = 3.875 ft²/hr

Compare values of D with $\alpha = 0.187$ cm²/sec and $\nu = 0.133$ cm²/sec for air at 298 °K and 1 atm.

*Compiled partly from A. A. Westenberg, *Combustion and Flame*, **1**, p. 346 (1957), with permission of the Combustion Institute, and partly from various data sources.

TABLE D.6

Viscosity of Liquids

Liquid	μ gm/cm/sec	at T °C
Acetaldehyde	2.20×10^{-5}	20
Acetic acid	11.55×10^{-5}	25
Acetone	3.16×10^{-5}	25
Aniline	37.10×10^{-5}	25
Benzene	6.52×10^{-5}	20
Butanol (n)	29.48×10^{-5}	20
Carbon tetrachloride	9.69×10^{-5}	20
Caster oil	$9,860.00 \times 10^{-5}$	20
Chloroform	5.42×10^{-5}	25
Cyclohexane	10.20×10^{-5}	17
Dodecane	13.50×10^{-5}	25
Ethyl acetate	4.41×10^{-5}	25
Ethanol	12.00×10^{-5}	20
Ethylene glycol	199.00×10^{-5}	20
Formic acid	18.04×10^{-5}	20
Heptane	3.86×10^{-5}	25
Hexane	2.94×10^{-5}	25
Methanol	5.47×10^{-5}	25
Octane (iso)	5.40×10^{-5}	20
Octane (n)	5.42×10^{-5}	20
Phenol	127.00×10^{-5}	18
Propanol	22.56×10^{-5}	20
Propyl acetate	5.90×10^{-5}	20
Toluene	5.90×10^{-5}	20
Water	10.02×10^{-5}	20
Xylene (o)	8.10×10^{-5}	20

TABLE D.7
Conductivity of Liquids

Liquid	K_{30} cal/cm/sec/°C	ω °C^{-1}	Range °C
Acetic acid	40.56×10^{-5}	1.20×10^{-3}	15–90
Acetaldehyde	44.30×10^{-5}	2.60×10^{-3}	12–31
Acetone	37.50×10^{-5}	2.20×10^{-3}	15–50
Aniline	41.25×10^{-5}	0.45×10^{-3}	15–90
Benzene	34.58×10^{-5}	1.80×10^{-3}	15–71
Butanol (n)	36.11×10^{-5}	1.40×10^{-3}	15–90
Carbon tetrachloride	24.31×10^{-5}	1.65×10^{-3}	15–90
Caster oil	42.78×10^{-5}	0.45×10^{-3}	15–90
Chloroform	27.36×10^{-5}	1.80×10^{-3}	15–70
Cyclohexane	29.31×10^{-5}	1.80×10^{-3}	12–38
Dodecane	31.25×10^{-5}	—	—
Ethyl acetate	34.30×10^{-5}	2.10×10^{-3}	16–50
Ethanol	39.72×10^{-5}	1.40×10^{-3}	16–91
Ethylene glycol	61.11×10^{-5}	-0.75×10^{-3}	15–90
Formic acid	63.89×10^{-5}	0.30×10^{-3}	15–90
Glycerol	68.05×10^{-5}	-1.20×10^{-3}	15–90
Heptane (n)	30.00×10^{-5}	1.80×10^{-3}	13–90
Hexane (n)	29.44×10^{-5}	2.00×10^{-3}	16–60
Kerosene	35.72×10^{-5}	—	—
Lube oil	32.50×10^{-5}	0.45×10^{-3}	15–90
Methanol	47.78×10^{-5}	1.20×10^{-3}	13–51
Octane (iso)	23.61×10^{-5}	1.80×10^{-3}	16–90
Octane (n)	30.56×10^{-5}	—	—
Pentanol	32.64×10^{-5}	0.90×10^{-3}	15–90
Phenol	38.33×10^{-5}	—	—
Propanol (n)	37.50×10^{-5}	1.40×10^{-3}	15–70
Propyl acetate	33.05×10^{-5}	—	—
Toluene	31.72×10^{-5}	—	—
Water	146.50×10^{-5}	2.48×10^{-3}	0–60
Xylene (o)	31.25×10^{-5}	—	—

$K = K_{30}[1 - \omega(T - 30)]$, T in °C

APPENDIX E

Some Problems for the Student

Thermodynamics

1. Calculate the work performed by a body expanding from an initial volume of 3.1 liters to a final volume of 3.5 liters at a constant pressure of 29 atm.
2. Calculate the work performed by 10 gms of O_2 expanding isothermally at 20 °C from 1 to 0.3 atm. of pressure.
3. Calculate the pressure of 28 gms of hydrogen contained in 1 liter vessel at 25 °C.
4. Calculate the change in internal energy of a system which performs $3.2 \cdot 10^8$ ergs of work and takes in 30 cal of heat. (Note, 1 cal = $4.185 \cdot 10^7$ ergs.)
5. An ideal gas expands isothermally from $P_1 = 5$ atm to $P_2 = 3$ atm at a temperature of 25 °C. If there are 4 moles of the gas in the system, how many calories of heat are absorbed in the process?
6. A mixture of gases at 25 °C and 1 atm pressure consists of O_2, H_2, N_2, and He. If Y_{O_2}, Y_{H_2}, and Y_{N_2} are respectively 0.15, 0.18 and 0.32, calculate: (a) mole fractions of all the components, (b) specific heat of the mixture, (c) partial pressure of each component, and (d) change in enthalpy, if the mixture is heated from 50 °K to 250 °K.
7. Calculate the volume of CO_2 at 70 °F and 1 atm. when 10 lb of dry ice is sublimed.
8. A body of mass 12 lb is lifted 10 feet above an arbitrary reference plane. If the local gravity is such that $g = 10$ ft/sec², calculate the work done in lifting.
9. Steam at 200 psia when introduced into a cylinder confined by a piston of cross-sectional area of 15 square inches displaces it through 3 inches. Assuming a constant pressure process, calculate the work done on the piston by the steam.

Some Problems for the Student

10. Consider a system containing 5 lb of air. If 1,500 BTU of heat is added to this system at constant volume, calculate (a) the change in internal energy of the system, and (b) work done by the gas.
11. 7 lbs of water at 98 °C is mixed thoroughly with 2 lbs of water at 15 °C. calculate the mixture temperature.
12. The change in entropy for a constant volume process is 0.10 cal/gm/°K. If the specific heat (at constant volume) of the fluid is 0.25 cal/gm/°K, and the initial temperature is 130 °C find the final temperature of the system.
13. Calculate the entropy increase when 50 grams of ice at -10 °C at 1 atm pressure is heated to steam at 100 °C at 1 atm pressure.
14. A cylinder the open end of which is confined with a movable piston initially contains 1,500 cm^3 of a certain ideal gas at 10 atm. pressure. If the gas is allowed to expand isothermally to 1 atm. calculate the work done by the gas on the piston.

Stoichiometry and Thermochemistry

15. Balance the following chemical equations by either mass conservation or charge conservation method:

$$C_6H_{15}N + HNO_3 \longrightarrow CO_2 + H_2O + N_2$$
$$C_2F_6 + ClF_3 \longrightarrow CF_4 + Cl_2$$
$$C_3H_9N + HClO_4 \longrightarrow Cl_2 + CO_2 + H_2O + N_2$$
$$C_2H_8N_2 + HN_2O_5 \longrightarrow CO_2 + H_2O + N_2$$
$$C_2H_5OH + H_2O_2 \longrightarrow CO_2 + H_2O$$
$$CH_4 + O_2 \longrightarrow CO_2 + H_2O$$

16. Calculate the stoichiometric fuel-oxidant ratio and enthalpies of reaction for

   (a) $CO(g) + H_2O(g) \longrightarrow CO_2(g) + H_2(g)$
   (b) $C_2H_4(g) + O_2(g) \longrightarrow CO_2(g) + CO(g) + H_2O(g)$
   (c) $CH_4(g) + O_2(g) \longrightarrow CO_2(g) + H_2O(g)$
   (d) $CO(g) + O_2(g) \longrightarrow CO_2(g) + O_2(g)$
   (e) $NO(g) + O_2(g) \longrightarrow NO_2(g)$

17. A hydrocarbon fuel is burnt with air. An Orsat analysis on the products yields the following composition by volume:

CO_2	10.5%
O_2	5.3%
N_2	84.2%

   Calculate the composition of the reactant mixture on mass basis.

18. Hydrogen peroxide is used sometimes as an oxidizer in special power plants such as torpedoes and rockets. Determine the standard enthalpy of combustion per gram mass of the fuel for the following combustion reaction:

$$H_2O_2(l) + \tfrac{1}{4}CH_4(g) \longrightarrow \tfrac{1}{2}H_2O(g) + \tfrac{1}{4}CO_2(g)$$

Also compare this enthalpy of combustion with that calculated for methane burning in pure oxygen. Assume $\Delta h^\circ_{f_{298}} = -145.3$ kcal/gm of H_2O_2.

19. From the table of enthalpies of formation, calculate the standard enthalpy of combustion of (a) C_2H_6, and (b) C_2H_5OH when they burn in oxygen. From the values of enthalpies of combustion so calculated, compute the "energy" of combustion for the two cases.

20. Calculate the standard enthalpy of reaction for the following cracking reactions:

$$\begin{array}{rcl} C_2H_6 + H_2 & \longleftarrow & 2\,CH_4 \\ n\text{-}C_4H_{10} + H_2 & \longleftarrow & 4\,CH_4 \\ \text{iso-}C_4H_{10} + 3H_2 & \longleftarrow & 4\,CH_4 \end{array}$$

Equilibria

21. Obtain the equilibrium constant $K_P = P_M^m P_N^n / P_A^a P_B^b \cdots$ from the law of mass action for the reaction

$$aA + bB + \cdots \rightleftharpoons mM + nN + \cdots.$$

22. If N_2O_4 dissociates into NO_2, derive an expression for K_P in terms of degree of dissociation λ and total pressure P.

23. Repeat Problem 22 for the following reactions:

$$\begin{array}{rcl} HCl & \rightleftharpoons & \tfrac{1}{2}H_2 + \tfrac{1}{2}Cl_2 \\ H_2S & \rightleftharpoons & H_2 + \tfrac{1}{2}S_2 \\ CO_2 & \rightleftharpoons & CO + \tfrac{1}{2}O_2 \\ NH_3 & \rightleftharpoons & \tfrac{1}{2}N_2 + \tfrac{3}{2}H_2 \end{array}$$

24. A mixture at 298 °K and 1 atm. pressure consists of 1 mole of H_2 and $\tfrac{1}{2}$ mole of O_2. They are slowly heated to 2,500 °K keeping pressure constant. What is the final equilibrium composition?

25. If the initial mixture of Problem 24 consists of 1 mole each of H_2 and O_2, what would be the equilibrium composition at 2,500 °K?

Some Problems for the Student 393

26. The operating temperatures in many combustion engines will not be too much higher than 1,500 °K, a temperature at which dissociation is assumed negligible. Evaluate the validity of this assumption if H_2, O_2, H_2O, CO, CO_2, are some of the species which occur in our combustion chamber.

27. A heated tube reactor is operated at 2,500 °K. The flowing mixture initially contains 2 moles of H_2O and 1 mole each of O_2 and N_2. If the total pressure is approximately 2 atm. and the outlet equilibrium mixture contains only H_2O, O_2, H_2, N_2 and OH, calculate the outlet composition.

28. Table C.2 in our text gives K_P as a function of temperature for several reactions of importance in combustion. Choose any three of these reactions and calculate the enthalpies of reaction for these three cases from the $K_P(T)$ data. Assume no data whatsoever is available regarding enthalpies of formation.

29. PCl_5 decomposes into PCl_3 and Cl_2 at elevated temperatures. If for $PCl_5 \rightleftarrows PCl_3 + Cl_2$ at 250 °C, K_P is given as 1.78, at what pressure should the system be operated in order to obtain a 50% decomposition at 250 °C?

30. Let C dissociate into A and B according to $C \rightleftarrows A + B$. If λ is degree of decomposition and P is total pressure, K_P is a function of λ and P. In other words, while K_P is a function of temperature T at which the system is operated, so is also P for obtaining any desired conversion. Eliminating T from the two relations, for any desired degree of dissociation, K_P can be related to P. If the desired λ is 0.25, find this relation for the above reaction.

31. Repeat Problem 30 for the following reactions if λ desired is 0.50:

  (a) $2C \rightleftarrows A + B$
  (b) $C \rightleftarrows 2A + B$
  (c) $3C \rightleftarrows 3A + B$
  (d) $3C \rightleftarrows A + B$
  (e) $C \rightleftarrows 2A + 3B$

32. In a vessel containing a mixture of H_2, F_2 and Cl_2, fluorine and chlorine compete with each other for hydrogen to form HF and HCl by the following reactions:

$$\tfrac{1}{2}H_2 + \tfrac{1}{2}F_2 \longrightarrow HF \quad K_P = 15,800$$
$$\tfrac{1}{2}H_2 + \tfrac{1}{2}Cl_2 \longrightarrow HCl \quad K_P = 51.2$$

at 3,500 °K

Are we more likely to observe more HF than HCl or vice versa if the total pressure is 1 atm.?

33. Write the balanced stoichiometric equations for the following fuels and oxidizers:

Fuel	Oxidizer	Products
(a) C_2H_2	O_2	CO_2, H_2O
(b) Al	O_2	Al_2O_3
(c) C	O_2	CO_2
(d) SiH_4	O_2	SiO_2, H_2O
(e) CH_3OH	O_2	CO_2, H_2O

34. Calculate the mole and mass fractions and mean molecular weights for the stoichiometric proportions of reactants and of the products in 33(e).

35. What sign would you expect the free energy changes associated with the reaction 33(a–e) to have? Why?

36. Tell in which direction you think changes in temperature and pressure and the addition of inert diluents will shift the extent of reactions in reactions 33(a–e)?

37. CO_2 is heated to 3,000 °K. CO_2 decomposes into CO and O_2 by

$$CO_2 \rightleftharpoons CO + \tfrac{1}{2}O_2$$

If $K_P = 2.93$ for the reaction $CO + \tfrac{1}{2}O_2 \to CO_2$, and total pressure is 4 atm., assuming only CO, O_2 and CO_2 are present in the equilibrium mixture, calculate the equilibrium composition.

38. Liquid C_8H_{18} burns in an oxidant composed of O_2 and N_2 in the ratio 1 mole O_2 for every 9 moles of N_2. If the initial temperature is 25 °C and dissociation is negligible, given $C_{P\,Products} = 6.4 + 1.10^{-2}T$ (cal/mole °K), and equilibrium $\lambda = 1$, calculate the adiabatic flame temperature. Assume the following enthalpies of formation:

Species	C_8H_{18}	CO_2	$H_2O(g)$	$H_2O(l)$
$\Delta h^\circ_{f\,298}$ (kcal/mole)	−49.82	−94.05	−57.80	−66.32

39. Consider

$$A_2 + B_2 \longrightarrow 2AB$$

Let molecular weights of A_2 and B_2 be equal to 28 and $C_{PA_2} \approx C_{PB_2} \approx C_{PAB} \approx 8$ cal/mole °K, a constant. If $\varepsilon_{A-A} = 100$ kcal, $\varepsilon_{B-B} = 41.8$ kcal, and $\varepsilon_{A-B} = 100$ kcal, calculate the theoretical flame temperature. Assume that K_P for the given reaction varies with temperature roughly as $K_P = 30,000/T$.

Some Problems for the Student

40. For $2B \rightarrow A$, at $25\,°C$, $K_P = 20$. If 1 mole of A were placed in a container and kept at room temperature, find the equilibrium composition at $P = 1\,atm$.

41. The equilibrium constants for the reaction of formation are

Compound	$\log_{10} K_P$
$C_2H_5OH(g)$	29.536
$CO_2(g)$	69.000
$H_2O(g)$	41.096

    What is the K_P for the reaction

    $$C_2H_5OH(g) + 3O_2(g) \longrightarrow 2CO_2(g) + 3H_2O(g)$$

42. $H_2(g)$ and $O_2(g)$ react to form water. Calculate the adiabatic flame temperature (neglect dissociation).

43. Ethylene burns in air. Calculate T_f if the products contain only $CO_2(g)$, $H_2O(g)$ and $N_2(g)$. Would dissociation lower, keep the same, or raise T_f? Why? Make any necessary assumptions.

44. (a) The molar heat capacity of H_2, Cl_2 and HCl are given by (cal/mole/°K)

    $$C_{P_{H_2}} = 6.947 - 0.2 \times 10^{-3}T + 0.4808 \times 10^{-6}T^2$$

    $$C_{P_{Cl_2}} = 7.576 + 2.424 \times 10^{-3}T - 0.965 \times 10^{-6}T^2$$

    $$C_{P_{HCl}} = 6.732 + 0.4325 \times 10^{-3}T + 0.3697 \times 10^{-6}T^2$$

    The standard molar heat of formation of HCl at $300\,°K$ is 22,060 calories. Construct an enthalpy composition diagram for HCl. Make a plot of flame temperature vs. composition for all mixtures of $Cl_2 + H_2$ from 0 to 100% H_2 starting at $300\,°K$.

    (b) Are the flame temperatures you have obtained in this way going to be too high or too low? Why?

    (c) Hydrogen and chlorine may be burnt in a constant volume bomb or a Bunsen burner? Would you expect higher or lower flame temperature in the bomb? Why? If you cannot answer this question easily, construct whatever mass fraction diagram that would be useful to you in determining explosion temperatures.

45. A mixture containing one mole of hydrogen, one mole of chlorine and two moles of helium undergoes an adiabatic explosion in a bomb. The reaction is

    $$H_2 + Cl_2 \longrightarrow 2HCl$$

The molar heat capacity (C_P) in calories per degree Kelvin per mole of hydrogen chloride is given by

$$C_P = 6{,}732 + 0.4325 \times 10^{-3}T + 0.3697 \times 10^{-6}T^2$$

The standard heat of formation of hydrogen chloride is -22.06 kcal/gm mole at 300 °K.

(a) If the initial temperature of the mixture in the constant volume bomb is 300 °K, what will be the final temperature?

(b) If this mixture is permitted to burn in a Bunsen burner in the laboratory, what will be the flame temperature?

46. What is the adiabatic flame temperature of a methane-air mixture that contains twice as much oxygen as is necessary to burn the methane completely to carbon dioxide and water? Find the necessary thermodynamic data in the *Handbook of Chemistry and Physics*. Assume CO_2 and H_2O are the only combustion products.

47. A vessel initially contains pure molecular hydrogen. For $\frac{1}{2}H_2 \rightleftarrows H$, the $\log_{10}$ of equilibrium constant at $T = 1{,}000$ °K is given to be -8.65. If the total pressure is kept at 1 atm. calculate the equilibrium composition.

48. Define equilibrium constant in terms of concentrations, mole fractions and mass fractions and relate each of these to K_P.

49. (a) At 35 °C and 1 atm, $N_2O_4(g)$ is 27% dissociated into $NO_2(g)$. Calculate K_P.

(b) At 250 °C and 1 atm, K_P for decomposition of phosphorous pentachloride gas is 1.78. Calculate the degree of decomposition.

$$PCl_5 \rightleftarrows PCl_3 + Cl_2 \quad \text{(all are gases.)}$$

(c) Assuming all gases behave ideally, state the effect of pressure on degree of decomposition for the two examples of 49a and 49b.

50. Recalling that $K_P = e^{-\Delta F^\circ / RT}$ and that

$$\Delta F_T^\circ = \sum_{\mathscr{P}} n_i \Delta f^\circ_{f_{iT}} - \sum_{\mathscr{R}} n_j \Delta f^\circ_{f_{jT}}$$

calculate the equilibrium constant at 25 °C and 1 atm. for

$$Cl_2(g) + 2HI(g) \rightleftarrows 2HCl(g) + I_2(s)$$

$\Delta f^\circ_{fCl_2 298} = \Delta f^\circ_{fI_2 298} = 0$ by datum choice. $\Delta f^\circ_{fHCl 298} = -22.69$ kcal/mole. $\Delta f^\circ_{fHI 298} = +315$ cal/mole.

51. A mixture of 1 mole of CO_2 and 2 moles of O_2 at 25 °C and 1 atm. is slowly heated 3,000 °K in a closed tank. Determine the equilibrium composition at the final state assuming that only CO_2, CO and O_2 are present.

52. What percentage of OH is formed when H_2O is heated to 5,000 °K at 1 atm. pressure?

$$H_2O \longrightarrow \tfrac{1}{2}H_2 + OH.$$

Chemistry of Combustion

53. Justify the statement "A given exothermic chemical reaction is almost always faster than its reverse reaction" by showing why this should be so. You may like to use a diagram to make your point clear.
54. What is meant by half-life period of a reaction? How does it depend on the initial concentration of the reactants for a first order reaction?
55. In the study of a reaction at a fixed temperature, the concentration of one of the reactant species is noted as a function of time. From this data, how do you determine the order of the reaction and the specific reaction rate constant?
56. The specific reaction rate constant k is determined as a function of temperature T by conducting the experiment suggested above at several constant temperatures. From this data (i.e., $k = k(T)$, how do you determine the activation energy E and the pre-exponential frequency factor $\mathscr{A}$?
57. Adiabatic reactions often start out with an explosive self-accelerating behavior, but soon transform into non-explosive. How do you explain this?
58. Qualitatively, sketch a rate $-dC_A/dt$ vs. time t plot for a hypothetical exothermic reaction. Show on this sketch the times when the ignition temperature and flame temperature occur. On the same time scale, show the rate of change of the reaction rate $d(-dC_A/dt)/dt$ with time.
59. Explain in your own words, as you would to a layman, what you learned about chemical kenetics by studying Chapter 2. Use brief illustrations if you please but refrain from extensive use of equations. Limit your work to 400 words.
60. The activation energy E for dissociation of hydrogen iodide to H_2 and I_2 is 44.3 kcal/mole. Knowing that the enthalpy of formation of HI as -1.35 kcal/mole, determine the activation energy for the reaction $H_2 + I_2 \rightarrow 2HI$.
    (Hint: Recall the extent of reaction vs. energy diagram we talked about in the text and note the physical meaning of E and ΔH_R)
61. The decomposition of a compound A is studied in a constant temperature reaction at $T = 318$ °K. Recorded is the following information.

t (min)	0	10	20	30	40	50	60	70	80
P_A(mm)	348.4	247	185	140	105	78	58	44	33
t		90	100	120	140	160	200		
P_A		13	18	10	4	3	0		

Find the order of the reaction and the specific reaction rate constant.

62. For the reaction $H_2 + I_2 \rightarrow 2HI$, specific reaction rate constants at different temperatures are determined by a method similar to that described in Problem 61.

$T\,°K$	556	575	629
k_n	1.19×10^{-4}	3.53×10^{-4}	6.76×10^{-3}

$T\,°K$	666	700	781
k_n	3.79×10^{-2}	1.72×10^{-1}	3.58

Determine energy of activation and pre-exponential frequency factor.

63. For an over-all reaction $A + B \rightarrow C + E + F$ a mechanism as below is proposed.

$$A + B \underset{k_2}{\overset{k_1}{\rightleftharpoons}} C + D^*$$

$$D^* \xrightarrow{k_3} E + F$$

Assuming (a) that the order of any of the three reactions with respect to any of the reactions is unity and (b) that there exists a pseudo-steady state condition for the concentration of the species D^*, derive an expression for dC_A/dt in terms of C_A, C_B, C_C and k_1, k_2 and k_3.

Physics of Combustion

64. Calculate the steady state conductive heat flux across a stagnant layer of air confined between two metal walls 0.5 cm apart and maintained at 100 °C and 115 °C. Assume that the fluid is stagnant.
65. If the air of Problem 64 is now set in motion at a velocity of 100 cm/sec, calculate the heat flux.
66. Calculate the mean free path length in normal air at room temperature and pressure.
67. Calculate the mean free path length in a nitrogen cylinder at room temperature and 2,000 psig.
68. Using Eqs. 3.15–3.17 calculate the viscosity, conductivity and diffusion coefficient for carbon dioxide at atmospheric pressure, as functions of temperature (25 °C < T < 200 °C). Prepare plots for future use.
69. Compare on the same plots the same properties calculated from Eqs. 3.18–3.20.
70. Using Eq. 3.24, calculate the viscosity and conductivity of a mixture of 21 % oxygen and 79 % nitrogen at 300 °K.

71. As water at 25 °C flows over a flat plate 4 inch long at a free stream velocity of 4 ft/sec, calculate the average boundary layer thickness and friction coefficient. What is the total shear force exerted if the plate width is 4 ft?

72. A hot (1,000 °C) flat plate 1 inch wide, 1/16 inch thick and 1 inch high is suspended vertically in room air. The lower edge is smoothed out sharply. Calculate and plot the boundary layer as a function of the plate height. Calculate the heat transfer rate as a function of height. Deduce the total heat loss rate.

73. Steady state heating occurs as hot air at 100 °C is blown across a 1 cm diameter copper pipe carrying water at a steady rate of 1 cc/sec. If the pipe length is 50 cm and water inlet temperature is 25 °C, determine its outlet temperature.

74. Determine the Grashof number for the flow induced around a hot, horizontal, infinitely long cylinder. The ambient fluid is air at 20 °C. The cylinder diameter is 2 cm and its temperature is 450 °C.

75. Calculate the steady heat transfer rate in the situation described in Problem 74.

76. If the cylinder described above is porous and bears methanol, calculate the rate of methanol mass transfer to the ambience. Make any approximation deemed necessary by so stating explicitly.

Kinetically Controlled Phenomena

77. A simple thermal reaction (whose specific reaction rate is $3.53 \cdot 10^{-4}$ sec^{-1}) occurs in a 25 fps flow in a 5 ft long combustor. Assuming that the reaction is of order 1, find whether the system is kinetically controlled or diffusionally controlled.

78. Describe the Semenov theory of self ignition in combustible gases. For one particular reacting mixture (i.e., $\dot{q}_g$) how does the heat loss affect the ignition characteristics? (Use a diagram to illustrate this point. Describe all three types of situations on the diagram.) How would ignition delays be observed with this scheme?

79. Discuss the similarities and differences between thermal explosions and branched chain explosions.

80. Semenov's theory of self ignition relates the critical conditions by

$$\ln\left(\frac{P_c}{T_c^{(n+2)/n}}\right) = \ln\left(\frac{hSR^{(n+1)}}{\Delta H V k_n X_{Ac}^n E}\right)^{1/n} + \frac{E}{nRT_c}$$

For a particular reaction, $n = 3$, $\Delta H = -50$ kcal/mole, $E = 40$ kcal/mole, $k_n = 10^{10}$(cm^3/mole)^2/sec. $V/S = 2$ inches, $h = 10^{-4}$ cal/cm^2/sec/°C, and X_A is kept at 0.5. Obtain a P–T graph delineating the zones of "ignition" and "no ignition."

81. What is the basic principle underlying the self-ignition theory of Semenov? You may need to write the necessary expressions to explain the quantities under consideration.
82. Semenov's theory predicts the critical $P-T$ curve which separates the regimes of ignition and no-ignition. Show, qualitatively, for a second order reaction how the ignition limit curve is altered if:
    (i) volume/surface ratio of the reaction vessel is increased.
    (ii) activation energy is decreased by treating the reactants with additives.
    (iii) heat transfer coefficient is increased by providing efficient cooling.
    (iv) ΔH is increased by changing the reactants.
83. The following graphs show knock effects in engines (pre-mixed charge). Explain in view of Semenov's theory the reasons for the exhibited behavior. Limit your answer to one or two sentences for each case.

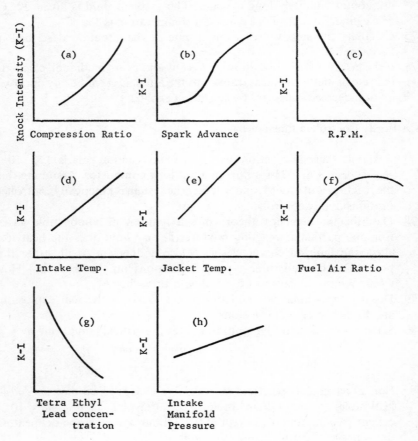

84. The specific heat of CO_2 and argon respectively are 0.27 and 0.127 cal/gm/°C. The respective thermal conductivities are $3.3 \cdot 10^{-5}$ and $3.9 \cdot 10^{-5}$ cal/cm/°C/sec. Which gas do you expect to be a better fire extinguishment agent? Why?

Diffusive Combustion of Liquid Fuels

85. Calculate "B" for the following conditions by utilizing approximate methods.
    (a) $T_\infty = 1{,}500\,°K$ $T_R = 300\,°K$ $Y_{C_8H_{18}\infty} = 0$ Use $T_\infty \gg T_{B.P.}$
    (b) $T_\infty = 300\,°K$ $T_R = 300\,°K$ $Y_{C_8H_{18}\infty} = 0$ Use $T_\infty \approx T_{B.P.}$

86. (a) Calculate the error committed in "B" because of the use of approximate methods in Problem 85 instead of the exact method.
    (b) Calculate the error in mass transfer rate.

87. Determine the wet bulb temperature when a droplet of C_8H_{18} is evaporating in air at
    (a) $1{,}500\,°K$
    (b) $1{,}000\,°K$
    (c) $500\,°K$

88. Determine the value of B in each case in 87.

89. A water film protects a porous circular cylinder in a furnace. The value of

$$B = \frac{Y_{AW} - Y_{A\infty}}{1 - Y_{AW}} = 0.1$$

is determined for the conditions in the furnace. The $Sc = v/D = 1.10$, $D_A = 2 \times 10^{-4}$ gm/sec cm, diameter $= 1$ cm, $Re = 100$.
    (a) Calculate the vaporization rate per unit area of surface.
    (b) Calculate the mass flux rate from the surface by diffusion only if $Y_{A\infty} = 0$.
    (c) What will be the effect on vaporization rate if, all other things being equal,
        (i) B is increased 4 fold
        (ii) ρD is increased 4 fold.

90. At one atmosphere total pressure the vapor pressure of water in a water-vapor-air mixture at equilibrium with liquid water can be read out from C.R.C. Tables. Estimate B if
    (i) $T = 40\,°F$, $Y_{A\infty} = 0$
    (ii) $T = 1{,}000\,°F$, $Y_{A\infty} = 0$

91. Water sits in an insulated glass in a room at temperature T. The level of the water is a distance l cm from the mouth of the glass. A gentle flow of dry room air assures that the mass fraction of H_2O at the mouth of the glass is zero. Show how you would calculate the vaporization rate of water from the glass.

92. (a) Using the curve of B vs. T_W shown below calculate the vaporization time for a droplet 1 mm in diameter.

$$L = 100 \,\frac{cal}{gram} \qquad T_\infty = 400\,°C \qquad C_g = 0.25 \,\frac{cal}{gram\,°K}$$

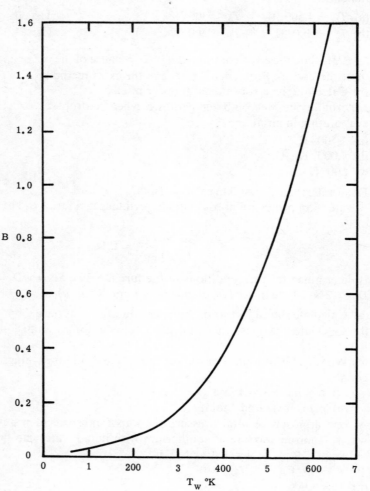

Some Problems for the Student

$$\rho_l = 1 \frac{\text{gram}}{\text{cm}^3} \qquad T_{\text{drop}} = T_W \qquad C_l = 1 \frac{\text{cal}}{\text{gram }°K}$$

$$D = \alpha_g = 0.5 \frac{\text{cm}^2}{\text{sec}} \qquad N_u = 2 \qquad \rho_g = 10^{-3} \frac{\text{gram}}{\text{cm}^3}$$

(b) at $t = 0$ what is the mass transfer rate from the surface?
(c) what is T_W?

93. What do we mean physically when we say the Lewis Number is one?
94. Let us assume we are attempting to cool the nose cone of a missile re-entering the atmosphere. We will use hydrogen as the cooling gas. The mass flow rate $\dot{W}_W''$ and mass transfer coefficient $\bar{h}°/C_g$ will be constant. $Le = 1$ and the stream (stagnation temp) $= T_\infty$. Consider two cases (i) the hydrogen does not burn, (ii) the hydrogen is allowed to burn with oxygen in the air.
Which wall temperature will be lower, T_{Wi} or T_{Wii}? Why?
95. Referring to the following figure, derive the steady state differential equation describing the rate at which a liquid A evaporates in a glass of diameter d in a room of briskly circulating gas B, and solve it with appropriate boundary conditions.

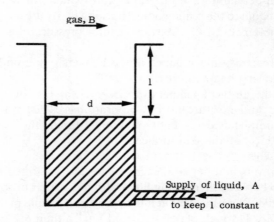

96. The hole in a wall shown below has a cross-sectional area of 1 cm². Calculate, by assuming $\rho_g D \approx 2 \cdot 10^{-4}$ gm/cm/sec,
(a) mass flux of O_2,
(b) mass flux of CH_3OH,
(c) mass fraction distributions Y_{O_2} and Y_{CH_3OH}.

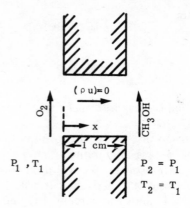

97. The front hood of a Volkswagen loses heat at a rate of 0.0003 cal/cm²/sec when traveling at 60 mph with a temperature difference of 2 °C between the hood and air. At normal temperature the B for car wax is 0.04. What would you estimate the wax loss rate of a Volkswagen hood to be at 60 mph?

98. Calculate the driving force, B, for droplets of (a) ethane (b) butane and (c) n-octane at wet bulb temperature (which with burning is approximately equal to the boiling point) burning in (i) pure oxygen and (ii) pure air to water and CO_2. Assume a total pressure of 1 atm. and $T_\infty = 300$ °K.

99. Measurements (not obtained at the University) on nude humans yield the following temperature data.
When the ambient temperature is 23° heat lost by convection is 20 cal/hour and moisture lost by evaporation is 0.020 gm/hr. If the ambient air is completely dry, what would you estimate the average mass fraction of H_2O on the skin surface to be?
(Note, when $Y_{AW} \ll 1$, $B \approx (Y_{AW} - Y_{A\infty})$)

100. Two porous infinite parallel flat plates are separated by a distance δ, the gap between them being filled with a constant property incompressible fluid. Plate 1 is kept at a temperature T_1 while plate 2 is kept at T_2 ($T_2 > T_1$). Through plate 1 is forced "in," a fluid of same physical and chemical properties as the fluid in the interspace, at a mass velocity of $\dot{W}''$ gms/cm²/sec while the same amount is sucked "out" through plate 2—thus introducing convection that opposes conduction. The energy equation and its boundary conditions are

$$K\frac{d^2T}{dx^2} - \dot{W}''C\frac{dT}{dx} = 0; \quad x = 0, T = T_1; \quad x = \delta, T = T_2$$

Introducing normalized variables

$$\theta \equiv \frac{T - T_1}{T_2 - T_1} \quad \text{and} \quad \xi \equiv x/\delta$$

simplifies the problem statement to

$$\frac{d^2\theta}{d\xi^2} - Pe\frac{d\theta}{d\xi} = 0; \quad \xi = 0, \theta = 0; \quad \xi = 1, \theta = 1$$

where Pe is a nondimensional number $\equiv \dot{W}''C\delta/K$ and is called Peclet Number. (a) solve the conservation equation to obtain $\theta = \theta(\xi, Pe)$ (b) explain physically what Pe means, (c) draw a rough sketch of $\theta(\xi)$ for several values of the parameter Pe, (d) from the solution of part (a) find an expression for the heat flux at plate 1, (e) from part (d) find the limiting heat flux as Pe tends to zero, and (f) what is the physical significance of the situation when Pe approaches zero?

101. A pan filled to brim with kerosene is burning in a hot air stream in a wind tunnel. How do you go about to determine the burning rate? (Recall that you need to know two parameters, $\bar{h}°/C_g$ and B.)
102. How would the life-time of a vaporizing droplet vary if one of the properties is changed as below while keeping all others fixed?
(i) ρ_l is doubled (ii) d_0 is halved (iii) $(T_\infty - T_W)$ is doubled.

Solid Fuels

103. Consider a porous ceramic sphere of diameter 6 cm which is soaked to saturation with water. Let the initial temperature of this sphere be 20 °C. Let the surface of the sphere be raised to 250 °C for times greater than or equal to zero. Assuming that water evaporates abruptly at $T = 100$ °C and that the diffusion coefficient of steam through the dry porous material is infinity, either Figure 6.2 or Eq. 6.3 may be used to locate the water front in the drying sphere. Calculate the mass-loss rate (gm/sec) of this sphere with time and prepare a plot. Assume that the thermal diffusivity of dry porous ceramic filled with steam is 0.01 cm²/sec. Neglect convective effects.
104. Using Eq. 6.12(a), calculate and graphically illustrate the oxygen mass fraction as a function of radial distance from a particle (of radius 1 mm.) of a material which undergoes with oxygen a surface reaction. The fuel/oxygen ratio is 0.20. The free stream oxygen mass fraction is 0.232.
105. Assuming a density of 0.001 gm/cm³ for the gas phase and 0.9 gm/cm³ for the solid phase and a diffusion coefficient of 0.2 cm²/sec, calculate the life-time of the particle of Problem 104.

106. Consider a boron particle of initial diameter 3 microns burning in air. Calculate the temperature profile in the vicinity of this particle. Assume $T_\infty = 20\,°C$ and $C_g \approx 0.23$ cal/gm/°C. Refer to the CRC Handbook of Chemistry and Physics to obtain the heat of combustion of boron.

Gas Jet Flames

107. Assuming that the potential core length of a plane jet is four times the slit width, draw the 10, 8, 6, 4, 2, and 0 cm/sec isovelocity contours for a laminar jet of air issuing at a velocity of 10 cm/sec from a slit of 1 cm width into a quiescent room. Use Eq. 7.23. See Figure 7.6 for a qualitative plot.
108. If the jet of Problem 107 were of pure propane gas the isovelocity profiles will also be isocomposition profiles. Knowing that the fuel mass fraction/oxygen mass fraction $= f$ at the flame front, locate the flame contour in the figure of Problem 107. (Refer to Eq. 7.28).
109. Equation 7.57 gives the turbulent flame hieght as a function of the port diameter, fuel/oxygen ratio and free stream oxygen concentration. Assuming the mixing constant C' to be 0.02, plot the flame height as a function of oxygen mass fraction ($0.15 \leq Y_{O\infty} \leq 1$) for two fuels—propane and methane—issuing out of a port of diameter 1 cm.

Premixed Flames

110. Consider a person walking at a speed of 3.9 mph on an excalator which itself is moving in the same direction (as the person) at a speed of 6.2 mph. What is the speed of the person relative to a fixed position? (Refer to Eq. 8.7.)
111. In the preceding example if the escalator speed is 4 mph and if the speed of the person relative to a fixed point is 1.7 mph in a direction opposing the escalator motion, what is the speed of the person relative to the escalator?
112. In a horizontal tube containing a combustible mixture, ignition is accomplished at one end by an electrical spark. The flame so initiated propagates along the tube. In one particular experiment this speed, measured by photography, is 143 cm/sec. The other end of the tube has a soap bubble provision to measure the gas escape rate. If the gas escape rate is 90 cm/sec, what is the fundamental flame speed?
113. Consider a propane-air mixture whose mean reaction rate is 1.10 moles/cm³/sec. If the flame and supply temperatures are respectively 2,300 and 400 °K, assuming the flame speed eigenvalue to be 1/2, calculate from Eq. 8.15 the fundamental flame speed.

Subject Index

Activated complex 37
Activation energy 27, 38, 84, 94, 101, 109, 138
Adiabatic flame temperature 131, 175, 211, 247, 279, 291, 297, 373, 376
Air, composition of 337
Arrhenius law 28, 90, 93, 104, 138, 200, 289
Atom conservation 355, 370
Atomization 146

Bimolecular reactions 10
Blow-off velocity 279, 282, 283, 305, 307, 312
Bluff body stabilization 307, 309, 310
Body forces 59, 237
Boiling point 130, 144, 161, 163, 172, 196
Boltzmann constant 36
Bond energy 10, 347
Boundary layer 55, 57, 59, 69, 157, 188, 266
Branched chain reactions 21, 91, 132
Bunsen burner 81, 280, 283, 305
Buoyancy 59, 247, 278
Burke–Schumann problem 260
Burning rate 145, 171, 203, 206

Capacitance spark 121
Carbon particle combustion 204, 205, 209, 211
Chain
  carriers 17, 19, 22, 280, 293, 301
  reactions 16, 21, 292
Chapman–Jouguet point(s) 275
Characteristic time 82, 83, 213
Chemical equilibrium 361, 363, 366
Chemical kinetics 8, 81, 125, 131, 135, 138, 322
Collision
  diameter 33

  integral 52
  factor 30, 33
  frequency 30
Combustion
  chamber 2, 7, 42, 137, 165, 187, 256, 313
  chemistry 8, 192
  detonative 272, 274
  deflagrative 272, 274, 276
  of gases 8, 115, 135, 217, 270
  of liquids 142
  phenomena 1, 81
  physics 43
  processes 2, 4, 81, 85, 91, 142, 195, 217, 270
  rate 84, 174
  of solids 195, 213
  of sprays 179, 213, 265
  stability 279, 305, 312
  temperature 84, 175, 211, 247, 279, 291, 297, 373, 376
Completeness of combustion 5, 361, 371, 378
Complex reactions 12, 15, 16, 21, 91, 132
Concentration 9, 334, 363
Conservation equations 73, 92, 152, 167, 205, 272, 297
Conserved property 153
Continuity equation 73, 152, 205, 221, 272
Control
  diffusional 82, 84, 142, 195, 202, 217, 270
  kinetic 81, 82, 84, 202, 270
Convective
  heat transfer 45, 57, 62, 70, 77
  mass transfer 67, 72, 78
Critical
  concentration 85, 100
  temperature 85, 100

Critical (*continued*)
  thickness 121
  velocity gradient 306

Dalton's law 333, 370, 371
Decomposition 18, 19, 196
Deflagration 272, 274, 276
Degree of Completeness 5, 361, 371, 378
Delay time, ignition 92, 107
Density 30, 323
Detonation in gases 272, 274
Diameter, molecular 33, 53
Diffusion coefficient 46, 50, 387
Diffusion flames 81, 142, 195, 217
  of gaseous jets 217
  of liquid fuels 142
  of solid fuels 195
Dilute gases 48
Displacement 279, 324
Dissociation equilibrium 379
Distance, quenching 122, 125, 131, 281
Distribution
  cumulative 147
  differential 148
  of heat 43, 77, 83, 175, 247
  of species 78, 83
Driving force, mass transfer 155, 162, 171, 173, 206, 208
Droplet
  burning 168, 178, 265
  vaporization 152, 157, 164
Dynamic viscosity 44, 50, 383, 385, 388

Eddy
  diffusivity 70, 221, 243, 303
  mixing 69, 135, 221, 230, 304
Eigenvalue, flame speed 300
Endothermic reactions 40, 343
Energy equation 77, 92, 104, 119, 221, 272, 297
Energy
  internal 323
  bond 347
  of combustion 346
  of reaction 343, 345
Enthalpy 327
  of combustion 345, 346
  of formation 341, 342
  of reaction 343
  of vaporization 150

Entropy 329
Equation of state 59, 322
Equilibrium 357
  composition 377, 379
  constant 363, 364, 368
  flame temperature 131, 138, 176, 211, 247, 289, 373, 379
Evaporation constant 166, 192, 179, 265
Exothermic reactions 38, 374, 343
Explosion 3, 21, 81, 132
  chain 21, 26, 132
  criteria 92, 95
  limits 99, 102, 126, 131
  theory 95, 104
  thermal 95, 104
Exponential, Arrhenius 26, 29, 41, 84, 93, 138, 291, 297
Extinction 85, 100, 102, 120, 122, 135, 187, 202, 279
Fick's law of diffusion 45, 72
Fire point 143, 144
First law of thermodynamics 321, 323, 340
First order reactions 10, 14
Flame
  diffusion 82, 84, 142, 195, 217, 270
  height 241, 243, 247, 262, 265
  holder 305
  kinetically controlled 81, 84
  overventilated 262
  propagation 115, 122, 131, 272, 283, 285, 296, 301
  quenching 122, 125, 131
  shape 172, 240, 247, 251, 255, 262, 278, 281, 283, 304
  speed 115, 131, 279, 282, 285, 301
  stabilization 305
  structure 115, 277
  temperature 131, 138, 175, 211, 247, 290, 297, 373, 376
  thickness 116, 124, 254, 278
  turbulent 135, 217, 229, 234, 243, 248, 301
  underventilated 262
Flammability limits 99, 103, 126, 131, 305
Flash-back 279, 305
Flash point 143, 144
Flat plate 55, 57, 59, 62, 188, 310
Flow 70, 81, 135, 217, 221, 228, 249, 257, 273, 279, 301, 305
Fourier's law of conduction 44
Free jet 218, 229, 231, 234, 236

Subject Index

Fundamental flame speed 115, 131, 279, 282, 285, 301
Furnace 1, 42, 135, 138

Gain, heat 77, 95, 113, 119, 278, 311, 326, 327
Gas
  velocity 217, 241, 272, 279, 283
  unburnt 115, 117, 279, 307
Gas constant, universal 323
Gas phase time 84, 242, 278, 305
Gaseous fuel jets 217
Gasification 146, 196, 266
Gibbs-Dalton law 333, 370, 371
Grashof number 61, 62, 157
Gravity 59, 237, 247, 278
Gutters 310

Heat
  balance 57, 77, 93, 95, 104, 136, 150
  capacity 324, 327, 328
  loss 95, 136
  of combustion 130, 144, 346
  of formation 341
  of reaction 343
  release 42, 138
  transfer 43, 95, 142, 195, 217, 270
Heating, space rate 42, 138
Hess' law 350, 375
Heterogeneous combustion 113, 142, 195
Higher order reactions 10, 14
Holding, flame 305
Homogeneous reactions 8, 81, 135, 217, 270
Hugoniot curve 273
Hydrocarbon combustion 17, 109, 125, 130, 138, 142, 144, 173, 195, 217, 270
Ideal gas law 59, 322
Ignition 81
  chain 132
  by flames 118
  by hot surface 117
  by sparks 121
  delay 92, 107
  energy 122, 125, 131
  forced 92, 112
  induction time 92, 107
  spontaneous 91, 92, 104, 110, 131
  temperature 101, 131, 298
  thermal 93, 95, 102, 104, 112

Inert additives 128, 192, 293, 376
Inhibitors 128, 294
Instability 305, 312
Integral, collision 52
Intensity 69, 135
Intermediate
  reactions 17, 19, 23, 132
  species 17, 18, 19
Internal energy 324, 340
Irreversible process 331
Isothermal processes 12, 88, 228, 250

Jet
  cylindrical 231
  enclosed 256
  entrainment in 219, 221, 230
  flames 217, 236
  flow in 217
  free 217, 218, 225, 229
  gas, combustion of 217
  laminar 217, 225
  plane 218, 225, 229
  turbulent 217, 229, 248
Jouguet–Chapman point 275

Kinematic viscosity 44
Kinetic energy 31, 340
Kinetic theory
  of gases 48
  of reaction rate 30
Kinetically controlled phenomena 81
Kinetics, chemical 8, 81
Knudsen number 49

Lag 92, 107
Laminar
  boundary layer 55, 57, 59, 61
  diffusion flames 156, 195, 217
  flame propagation 115, 279, 297
  jet flames 225, 241, 248
Law of mass action 362
  Gibbs–Dalton 333, 370 371
  Hess' 350, 375
  ideal gas 59, 322
  Lavoisier–Laplace 374
Lewis number 55, 253
Life-time 13, 14, 164, 178, 207, 208, 265
Lifted flames 248, 279
Limits, flammability 99, 103, 126, 131, 305
Liquid fuel combustion 142

Marble–Adamson problem 308
Mass fraction 334
Mass transfer 46, 67, 72, 78
   boundary condition 151
   coefficient 67, 156
   driving force 155, 162, 171, 173, 206, 208
Mean free path 37, 48
Mechanical equilibrium 357
Metal combustion 196, 200, 208
Metastable state 358
Minimum ignition energy 122, 125, 131
Mists 2, 146, 179, 213
Mixing
   combustion 82
   intensity 69
   length 70
   scale 69
   time 83
Mole fraction 334
Molecular
   collisions 33, 48, 49, 52
   diameter 33, 53
   parameters 53
   velocity 30, 34, 276
   weight 322
Molecularity of a reaction 10
Momentum equation 55, 60, 75, 219, 272

Natural convection 59, 64, 190, 278
Newton's law of viscosity 44
Number
   Grashof 62
   Knudsen 49
   Lewis 55
   Mass transfer 155, 162, 171, 173, 206, 208
   Nusselt 62
   Prandtl 54, 62
   Reynolds 56, 58, 68
   Schmidt 55, 68

Opposing reactions 15, 28
Order of a reaction 10, 14, 25, 138, 289
Overventilated flames 262

Partial pressure 101, 159, 334, 365, 370
Particles, combustion of 200
Potential energy 340
Prandtl number 54, 62
Preexponential factor 28, 138
Preheat zone 298

Premixed flames 270
Propagation of flames 115, 122, 131, 272, 283, 285, 296, 301
Propulsion 2, 187, 195, 266, 305
Pulverized coal 195
Pyrolysis 196

Quenching 122, 125, 131, 135

Radicals in flames 16, 21, 26, 112, 300
Rankine–Hugoniot equations 273
Reaction
   chain 21, 26, 132
   coordinate 38, 40
   kinetics 8, 81, 125, 131, 135, 138, 322
   mechanisms 10, 17, 18, 19, 20, 23
   molecularity 10
   order 10, 14, 25, 138, 289
   progress 87
   rate 9, 12, 27
   thermal 12, 24
   time 13, 14
Reactor, well-stirred 135
Recirculation 310
Recombination 379
Reduced mass 52
Reservoir conditions 149
Reversible process 15, 331
Reynolds number 56, 58, 62
Rosin–Rammler distribution 148, 180

Scale of turbulence 69
Schlieren technique 284
Schmidt number 56, 68
Screaming 312
Screeching 313
Second law of thermodynamics 322, 329, 357
Second order reactions 10, 14, 25, 138, 289
Semenov theory of ignition 95, 99
Shadowgraph techniques 284
Shear stress 44, 56
Shock wave 272, 275
Shwab-Zeldovich transformation 171, 193
Solid fuel combustion 195
Space heating rate 42, 135, 138
Spalding's B-number 155, 162, 171, 173, 206, 208
Spark ignition 121
Species, active 17, 19, 22, 280, 293, 301
Specific reaction rate constant 13

Subject Index

Speed, molecular 30, 34, 276
Spherical Symmetry 135, 152, 197, 205
Spontaneous ignition 91, 92, 104, 110, 131
Spray combustion 179, 213, 265
Stability of flames 279, 282, 305, 312
Stagnation point 63, 310
Standard heat of formation 341
State, equation of 59, 322
Steric factor 36
Stoichiometry 9, 130, 169, 203, 208, 343
Streamline pattern 310
Surface
  combustion 205
  conditions 22, 35, 132, 148
  equilibrium 164
  shear stress 44, 56
Suspensions 192

Temperature, adiabatic flame 131, 175, 211, 247, 279, 291, 297, 373, 376
Termination of chain carriers 17, 18, 19, 24
Thermal
  conductivity 45, 50, 197, 297, 384, 385, 386, 389
  diffusivity 45
  explosions 41
  theory 92, 95, 104, 116, 297
Thermochemistry 340
Thermodynamic functions 321, 332
Thermodynamics
  first law of 323, 340
  second law of 329, 357
  zeroth law of 324
Third order reactions 11
Time
  burning 178, 207
  ignition 93, 109
  vaporization 164, 266
Transport
  properties 253, 382
  phenomena 43, 48, 51
Turbulence 62, 68, 230, 233, 243, 301
  large scale 304
  small scale 302

Turbulent
  boundary layer 68
  burning velocity 303
  eddies 304
  flames 241, 248, 301

Unburned
  gases 115, 272, 278, 304
  fuel 5
Underventilated flame 262
Unimolecular reactions 10
Units of gas constant 323
Universal gas constant 322

Vapor pressure 158, 159
Velocity
  blow-off 279, 282, 283, 305, 307, 312
  burning 301
  critical 249, 279, 282, 305
  distribution 30, 44, 55, 57, 220, 228, 277, 283
  interfacial 71, 72, 149
  gradient 306
  molecular 30, 276
  profile 44, 69, 220, 224, 228, 235, 258, 260
  propagation 116, 274, 277, 296
Viscosity
  dynamic 44, 50, 383, 385, 388
  kinematic 44, 54
Vaporization
  rate 152, 156, 187
  droplet 155, 164, 180, 265
Volumetric source 42, 92, 135, 138, 297

Wall conditions 148
Well-stirred
  reactor 135
  blow-off limits 136, 311
Wrinkles 248, 304

Zeldovich–Frank-Kamanetsky theory 297
Zeroth law of thermodynamics 324
Zone
  flame 86, 115, 135, 172, 240, 247, 277, 365
  preheat 298
  reaction 124, 177, 197, 247, 299